HIGHER GCSE MATHEMATICS

Revision and Practice

David Rayner

OXFORD

OXFORD

UNIVERSITY PRESS

Great Clarendon Street, Oxford OX2 6DP

Oxford University Press is a department of the University of Oxford.
It furthers the University's objective of excellence in research, scholarship,
and education by publishing worldwide in

Oxford   New York

Auckland   Bangkok   Buenos Aires   Cape Town   Chennai
Dar es Salaam   Delhi   Hong Kong   Istanbul   Karachi   Kolkata
Kuala Lumpar   Madrid   Melbourne   Mexico City   Mumbai   Nairobi
São Paulo   Shanghai   Taipei   Tokyo   Toronto

Oxford is a registered trade mark of Oxford University Press
in the UK and in certain other countries

10 9 8 7

The moral rights of the author have been asserted

Database right Oxford University Press (maker)

First published 2001

British Library Cataloguing in Publication Data

Data available

ISBN 0 19 914791 4 (School edition)
     0 19 914792 2 (Bookshop edition)

Additonal artwork by Stephen Hill, Nick Bawden, Oxford
Illustrators and Julian Page
Typeset by Tech-Set, Gateshead, Tyne and Wear
Printed and bound in Great Britain by Butler & Tanner Ltd, Frome
and London

Preface

This book is for candidates working through Key Stage 4 towards a GCSE in Mathematics. It can be used both in the classroom and by students working on their own. There are explanations, worked examples and numerous exercises which, it is hoped, will help students to build up confidence. The author believes that people learn mathematics by doing mathematics. The questions are graded in difficulty throughout the exercises.

† This symbol is used to highlight the most difficult questions.

This revised edition has been written to cater for the new non-calculator papers. Future developments in the National Curriculum have also been anticipated by reference to the latest documents from the Department of Education. In response to the requests of many teachers a section of past examination questions has also been included. Teachers will note that the specifications and questions set by all the main GCSE boards are very similar. This book can be used with confidence when preparing for all GCSEs.

The book can be used either as a course book over the last two years before the GCSE examinations or as a revision text in the final year. The contents list shows where all the topics appear and an index at the back of the book provides further reference.

The work is collected into sections on Number, Algebra, Shape, space and measures, Handling Data, Probability, and Using and Applying Mathematics. This is done for ease of reference but the material can be taught in any convenient order. Most teachers will prefer to alternate freely between topics from each of the sections.

At the end of the book, in addition to past examination questions, there are several revision exercises which provide mixed questions across the curriculum. There are also multiple choice questions for variety.

The section on Using and Applying Mathematics contains a selection of starters for coursework projects which have been tried and tested. They can be used to provide practice in the strategies involved in attempting 'open-ended' problems.

The author is indebted to the many students and colleagues who have assisted him in this work. He is particularly grateful to Christine Godfrey for her help and many suggestions.

Thanks are also due to the following examination boards for kindly allowing the use of questions from their past mathematics papers and sample questions:

University of Cambridge Local Examinations Syndicate International   [IGCSE]
London Examinations, a division of Edexcel Foundation   [L]
Midland Examinations Group   [M]
Northern Examinations and Assessment Board   [N]
Northern Ireland Council for the Curriculum Examinations and Assessment   [NI]
Southern Examining Group   [S]
Welsh Joint Education Committee   [WJEC]

The sample questions from these boards represent one stage in the development of nationally approved material and do not necessarily reflect the style currently accepted. The answers are solely the work of the author and not ratified by the examining groups.

D. Rayner 2000

Contents

1 NUMBER 1

1.1 Properties of numbers

- **Factors**　　Any number which divides exactly into 8 is a *factor* of 8.
The factors of 8 are 1, 2, 4, 8.
- **Multiples**　　Any number in the 8-times table is a *multiple* of 8.
The first five multiples of 8 are 8, 16, 24, 32, 40.
- **Prime**　　A *prime* number has just two different factors: 1 and itself.
The number 1 is *not* prime. [It does not have two different factors.]
The first five prime numbers are 2, 3, 5, 7, 11.
- **Prime factor**　　The factors of 8 are 1, 2, 4, 8. The only *prime factor* of 8 is 2. It is the only prime number which is a factor of 8.
- **Square numbers**　　A square number can be written in the form $n \times n$, where n is an integer. [Examples: 4, 9, 100]
- **Cube numbers**　　A cube number can be written in the form $n \times n \times n$, where n is an integer. [Examples: 1, 8, 27, 64]

(a) Express 6930 as a product of primes.

$$2\,)\,69^13^10 \quad \text{(divide by 2)}$$
$$3\,)\,34^16^15 \quad \text{(divide by 3)}$$
$$3\,)\,11^25^15 \quad \text{(divide by 3)}$$
$$5\,)\quad 38^35 \quad \text{(divide by 5)}$$
$$7\,)\quad\;\;\; 77 \quad \text{(divide by 7)}$$
$$\qquad\quad 11 \quad \text{(stop because 11 is prime)}$$

$$\therefore \quad 6930 = 2 \times 3 \times 3 \times 5 \times 7 \times 11$$

(b) In prime factors,

$$90 = 2 \times 3 \times 3 \times 5$$
$$42 = 2 \times 3 \times 7$$

By inspection we see that the *Highest Common Factor* (H.C.F.) of 90 and 42 is 6　[i.e. 2×3]

Exercise 1

1. Write down all the factors of the following numbers.
　(a) 6　　　(b) 15　　　(c) 18　　　(d) 21　　　(e) 40

2. Write down all the prime numbers less than 20.

3. Write down two prime numbers which add up to another prime number. Do this in *two* ways.

4. Which of the following are prime numbers?

3, 11, 15, 19, 21, 23, 27, 29, 31, 37, 39, 47, 51, 59, 61, 67, 72, 73, 87, 99

5. Express each of the following numbers as a product of primes. See the example above.
　(a) 600　　　　　(b) 693　　　　　(c) 2464
　(d) 3510　　　　(e) 4000　　　　(f) 22 540

6. Copy the grid and then write the numbers 1 to 9, one in each box, so that all the numbers satisfy the conditions for both the row and the column.

	Prime number	Multiple of 3	Factor of 16
Number greater than 5			
Odd number			
Even number			

7. Find each of the mystery numbers below.
   (a) I am an odd number and a prime number. I am a factor of 14.
   (b) I am a two-digit multiple of 50.
   (c) I am one less than a prime number which is even.
   (d) I am odd, greater than one and a factor of both 20 and 30.

8. The first few multiples of 4 are   4, 8, 12, 16, (20), 24, 28 ...

   The first few multiples of 5 are   5, 10, 15, (20), 25, 30, 35 ...

   The *Lowest Common Multiple* (L.C.M.) of 4 and 5 is 20.
   It is the lowest number which is in both lists.
   (a) Write down the first four multiples of 6.
   (b) Write down the first four multiples of 9.
   (c) Write down the L.C.M. of 6 and 9.

9. Find the L.C.M. of:
   (a) 6 and 15   (b) 20 and 30   (c) 6 and 10.

10. Write down the first 15 square numbers. *Learn* these numbers.

11. Write down the first 5 cube numbers.

12. Work out which of the following are prime numbers.
    (a) 91   (b) 101   (c) 143   (d) 151   (e) 293
    [Hint: Divide by the prime numbers 2, 3, 5, 7, 11 etc.]

13. Put three different numbers in the circles so that when you add the numbers at the end of each line you always get a square number.

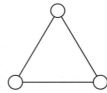

14. (a) Express 1008 and 840 as products of their prime factors.
    (b) Find the H.C.F. (highest common factor) of 1008 and 840.
    (c) Find the smallest number which can be multiplied by 1008 to give a square number.

15. (a) Express 19 800 and 12 870 as products of their prime factors.
    (b) Find the H.C.F. of 19 800 and 12 870.
    (c) Find the smallest number which can be multiplied by 19 800 to give a square number.

16. The first of three consecutive numbers is n. Find the smallest value of n if the three numbers are all non-prime.

17. Below the odd numbers are written in a special way.

Row 1   1
Row 2   3   5
Row 3   7   9   11
Row 4   13   15   17   19

(a) (i) Write down the sum of each of the first four rows.
   (ii) What is the special name of the numbers that you have written down?
   (iii) What is the sum of the numbers in the hundredth row?
(b) (i) Write down the sum of all the numbers in the first two rows,
             in the first three rows,
             in the first four rows.
   (ii) What is the special name of the numbers that you have written down?
   (iii) What is the sum of all the numbers in the first ten rows?

18. Make six prime numbers using the digits

   1, 2, 3, 4, 5, 6, 7, 8, 9

once each.

19. Consider the expression $5^n + 3$.
Substitute six different values for n and find out if the result is a multiple of 4 each time.

20. Substitute six pairs of values of a and n in the expression $a^n - a$.
Is the result always a multiple of n?

Operations and inverses

Operation	Inverse	Example	
Add n	Subtract n	$9 + 7 = 16$,	$16 - 7 = 9$
Subtract n	Add n	$12 - 5 = 7$,	$7 + 5 = 12$
Multiply by n	Divide by n	$20 \times 4 = 80$,	$80 \div 4 = 20$
Divide by n	Multiply by n	$12 \div 3 = 4$,	$4 \times 3 = 12$
Square	Square root	$7^2 = 49$,	$\sqrt{49} = 7$
Raise to power n	Find nth root $\left[\text{raise to the power } \dfrac{1}{n}\right]$	$10^4 = 10000$,	$\sqrt[4]{10000} = 10$
[This operation is discussed more fully in the section on indices.]			

Also: The 'reciprocal' of n is $\dfrac{1}{n}$. $(n \neq 0)$

The reciprocal of $\dfrac{1}{n}$ is n.

Note that zero has no reciprocal because division by zero is not defined.
The reciprocal is sometimes called the 'multiplicative inverse'.

Exercise 2

In the flow charts, the boxes A, B, C and D each contain a single mathematical operation (like $+5$, $\times 4$, -15, $\div 2$).

Look at flow charts (i) and (ii) together and work out what is the same operation which will replace A. Complete the flow chart by replacing B, C and D.

Now copy and complete each flow chart on the right, using the same operations.

1. (i) $\xrightarrow{2}$ A $\xrightarrow{17}$ B $\xrightarrow{34}$ C $\xrightarrow{12}$ D $\xrightarrow{3}$

  (ii) $\xrightarrow{4}$ A $\xrightarrow{19}$ B $\xrightarrow{38}$ C $\xrightarrow{16}$ D $\xrightarrow{4}$

  (a) $\xrightarrow{7}$ A $\xrightarrow{?}$ B $\xrightarrow{?}$ C $\xrightarrow{?}$ D $\xrightarrow{?}$

  (b) $\xrightarrow{10}$ A $\xrightarrow{?}$ B $\xrightarrow{?}$ C $\xrightarrow{?}$ D $\xrightarrow{?}$

  (c) $\xrightarrow{?}$ A $\xrightarrow{?}$ B $\xrightarrow{62}$ C $\xrightarrow{?}$ D $\xrightarrow{?}$

  (d) $\xrightarrow{?}$ A $\xrightarrow{15\frac{1}{2}}$ B $\xrightarrow{?}$ C $\xrightarrow{?}$ D $\xrightarrow{?}$

2. (i) $\xrightarrow{2}$ A $\xrightarrow{4}$ B $\xrightarrow{12}$ C $\xrightarrow{2}$ D $\xrightarrow{1}$

  (ii) $\xrightarrow{3}$ A $\xrightarrow{9}$ B $\xrightarrow{27}$ C $\xrightarrow{17}$ D $\xrightarrow{8\frac{1}{2}}$

  (a) $\xrightarrow{4}$ A $\xrightarrow{16}$ B $\xrightarrow{?}$ C $\xrightarrow{?}$ D $\xrightarrow{?}$

  (b) $\xrightarrow{5}$ A $\xrightarrow{?}$ B $\xrightarrow{?}$ C $\xrightarrow{?}$ D $\xrightarrow{?}$

  (c) $\xrightarrow{?}$ A $\xrightarrow{?}$ B $\xrightarrow{108}$ C $\xrightarrow{?}$ D $\xrightarrow{?}$

  (d) $\xrightarrow{?}$ A $\xrightarrow{?}$ B $\xrightarrow{?}$ C $\xrightarrow{182}$ D $\xrightarrow{?}$

In questions **3**, **4** and **5** find the operations A, B, C, D.

3. (i) $\xrightarrow{4}$ A $\xrightarrow{16}$ B $\xrightarrow{4}$ C $\xrightarrow{-6}$ D $\xrightarrow{12}$

  (ii) $\xrightarrow{9}$ A $\xrightarrow{36}$ B $\xrightarrow{6}$ C $\xrightarrow{-4}$ D $\xrightarrow{8}$

4. (i) $\xrightarrow{5}$ A $\xrightarrow{\frac{1}{5}}$ B $\xrightarrow{1\cdot2}$ C $\xrightarrow{1\cdot44}$ D $\xrightarrow{0\cdot48}$

  (ii) $\xrightarrow{\frac{1}{2}}$ A $\xrightarrow{2}$ B $\xrightarrow{3}$ C $\xrightarrow{9}$ D $\xrightarrow{3}$

5. (i)

  $\xrightarrow{-1}$ A $\xrightarrow{2}$ B $\xrightarrow{8}$ C $\xrightarrow{-4}$ D $\xrightarrow{96}$

  (ii)

  $\xrightarrow{7}$ A $\xrightarrow{10}$ B $\xrightarrow{1000}$ C $\xrightarrow{-500}$ D $\xrightarrow{-400}$

1.2 Arithmetic without a calculator

Decimals

(a) 0.032×100
$= 3.2$

(b) $51 \div 1000$
$= 0.051$

(c) $26.1 \div 5$

$$5\overline{)26.\overset{1}{1}\overset{1}{0}} \quad \leftarrow \text{Add a zero}$$
$$\phantom{5\overline{)}}5.22$$

$26.1 \div 5 = 5.22$

(d) $\begin{array}{r} 2.5 \\ \times\ 0.3 \\ \hline 0.75 \\ \hline \end{array}$ ↖ two figures after
← the decimal points

↖ two figures after
the decimal point

(e) $8.64 \div 0.2$
[Multiply both numbers by 10.]
$86.4 \div 2$
$= 43.2$

(f) $6.9 \div 0.003$
[Multiply both numbers by 1000.]
$6900 \div 3$
$= 2300$

Exercise 3

Evaluate the following.

1. $7.6 + 0.31$
2. $15 + 7.22$
3. $7.004 + 0.368$
4. $0.06 + 0.006$
5. $4.2 + 42 + 420$
6. $3.84 - 2.62$
7. $11.4 - 9.73$
8. $4.61 - 3$
9. $17 - 0.37$
10. $8.7 + 19.2 - 3.8$
11. $25 - 7.8 + 9.5$
12. $3.6 - 8.74 + 9$
13. $20.4 - 20.399$
14. 2.6×0.6
15. 0.72×0.04
16. 27.2×0.08
17. 0.1×0.2
18. $(0.01)^2$

19. Sharon is paid a basic weekly wage of £85 and then a further 30p for each item completed. How many items must be completed in a week when she earns a total of £191.50?

20. James has the same number of 10p and 50p coins. The total value is £12. How many of each coin does he have?

21. A pile of 10p coins is 25.6 cm high. Calculate the total *value* of the coins if each coin is 0.2 cm thick.

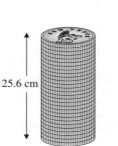

25.6 cm

Work out:

22. $0.34 \times 100\,000$
23. $3.6 \div 0.2$
24. $56.9 \div 100$
25. $0.1401 \div 0.06$
26. $3.24 \div 0.002$
27. $0.968 \div 0.11$
28. $600 \div 0.5$
29. $0.007 \div 4$
30. $2640 \div 200$
31. $1100 \div 5.5$
32. $(11 + 2.4) \times 0.06$
33. $(0.4)^2 \div 0.2$
34. $77 \div 1000$
35. $(0.3)^2 \div 100$
36. $(0.1)^4 \div 0.01$

37. (a) The calculation $\dfrac{480 \times 55}{22}$ is easier to perform after cancelling.

$$\frac{480 \times \overset{5}{\cancel{55}}}{\underset{2}{\cancel{22}}} = \frac{\overset{240}{\cancel{480}} \times 5}{\underset{1}{\cancel{2}}} = 1200$$

(b) Use cancelling to help work out the following.

(i) $\dfrac{180 \times 4}{36}$

(ii) $\dfrac{92 \times 4.6}{2.3}$

(iii) $\dfrac{35 \times 0.81}{45}$

(iv) $\dfrac{72 \times 3000 \times 24}{150 \times 64}$

(v) $\dfrac{63 \times 600 \times 0.2}{360 \times 7}$

38. Fill in the missing numbers so that the answer is always 0·7.

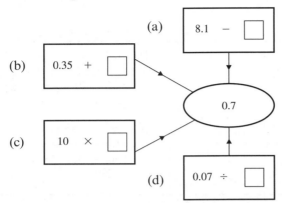

(a) 8.1 − ☐

(b) 0.35 + ☐

0.7

(c) 10 × ☐

(d) 0.07 ÷ ☐

39. ● When you multiply by a number greater than 1 you make it bigger.

so $5·3 \times 1·03 > 5·3$ and $6·75 \times 0·89 < 6·75$

● When you divide by a number greater than 1 you make it smaller.

so $8·92 \div 1·13 < 8·92$ and $11·2 \div 0·73 > 11·2$

State whether true or false:
(a) $3·72 \times 1·3 > 3·72$ (b) $253 \times 0·91 < 253$
(c) $0·92 \times 1·04 > 0·92$ (d) $8·5 \div 1·4 > 8·5$
(e) $113 \div 0·73 < 113$ (f) $17·4 \div 2·2 < 17·4$
(g) $0·73 \times 0·73 < 0·73$ (h) $2511 \div 0·042 < 2511$
(i) $614 \times 0·993 < 614$

40. Write down each statement with either $>$ or $<$ in lace of the box.

(a) $17·83 \div 0·07$ ☐ $17·83$ (b) $9·04 \times 1·13$ ☐ $9·04$

(c) $207 \times 0·96$ ☐ 207 (d) $308 \div 2·8$ ☐ 308

(e) $1·19 \div 0·06$ ☐ $1·19$ (f) $0·74 \div 0·95$ ☐ $0·74$

Long multiplication and division

To work out 327×53 we will use the fact that $327 \times 53 = (327 \times 50) + (327 \times 3)$.
Set out the working like this.

$$
\begin{array}{r}
327 \\
53 \times \\
\hline
16\,350 \rightarrow \text{This is } 327 \times 50 \\
981 \rightarrow \text{This is } 327 \times 3 \\
\hline
17331 \rightarrow \text{This is } 327 \times 53
\end{array}
$$

Here is another example.

$$
\begin{array}{r}
541 \\
84 \times \\
\hline
43280 \rightarrow \text{This is } 541 \times 80 \\
2164 \rightarrow \text{This is } 541 \times 4 \\
\hline
45444 \rightarrow \text{This is } 541 \times 84
\end{array}
$$

Exercise 4

Work out:

1. 35×23	**2.** 27×17	**3.** 26×25
4. 31×43	**5.** 45×61	**6.** 52×24
7. 323×14	**8.** 416×73	**9.** 504×56
10. 306×28	**11.** 624×75	**12.** 839×79
13. 694×83	**14.** 973×92	**15.** 415×235

With ordinary 'short' division, we divide and find remainders. The method for 'long' division is really the same but we set it out so that the remainders are easier to find.

Work out $736 \div 32$	
$\begin{array}{r} 23 \\ 32\overline{)736} \\ 64\downarrow \\ \hline 96 \\ 96 \\ \hline 0 \end{array}$	(a) 32 into 73 goes 2 times (b) $2 \times 32 = 64$ (c) $73 - 64 = 9$ (d) 'bring down' 6 (e) 32 into 96 goes 3 times

Exercise 5

Work out:

1. $672 \div 21$	**2.** $425 \div 17$	**3.** $576 \div 32$
4. $247 \div 19$	**5.** $875 \div 25$	**6.** $574 \div 26$
7. $806 \div 34$	**8.** $748 \div 41$	**9.** $666 \div 24$
10. $707 \div 52$	**11.** $951 \div 27$	**12.** $803 \div 31$
13. $2917 \div 45$	**14.** $2735 \div 18$	**15.** $56\,274 \div 19$

Exercise 6

Solve each problem.

1. A shop owner buys 56 tins of paint at 84p each. How much does he spend altogether?

2. Eggs are packed eighteen to a box.

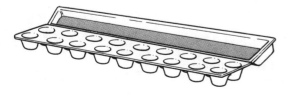 

   How many boxes are needed for 828 eggs?

3. On average a man smokes 146 cigarettes a week. How many does he smoke in a year?

4. Sally wants to buy as many 23p stamps as possible. She has £5 to buy them. How many can she buy and how much change is left?

5. How many 49-seater coaches will be needed for a school trip for a party of 366?

6. An office building has 24 windows on each of eight floors. A window cleaner charges 42p for each window. How much is he paid for the whole building?

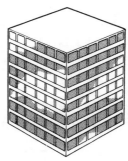

7. A lottery prize of £238 million was won by a syndicate of 17 people who shared the prize equally between them. How much did each person receive?

8. It costs £7905 to hire a plane for a day. A trip is organised for 93 people. How much does each person pay?

9. The headmaster of a school discovers an oil well in the school playground. As is the custom in such cases, he receives all the money from the oil. The oil comes out of the well at a rate of £15 for every minute of the day and night. How much does the headmaster receive in a 24-hour day?

10. Tins of peaches are packed 24 to a box. How many boxes are needed for 1285 tins?

11. Work out:
(a) $2 \cdot 1 \times 3 \cdot 2$ (b) $0 \cdot 65 \times 4 \cdot 3$ (c) $0 \cdot 71 \times 5 \cdot 6$

12. Work out the following, giving your answer correct to 1 decimal place.
(a) $8 \cdot 32 \div 1 \cdot 7$ (b) $0 \cdot 7974 \div 0 \cdot 23$ (c) $1 \cdot 4 \div 0 \cdot 71$

13. Susie breeds gerbils for pet shops. The number of gerbils trebles each year. Susie has 40 gerbils at the end of year 1.
(a) Copy and complete the table below.

End of year	1	2	3	4	5
Number of gerbils	40				

(b) A gerbil cage can hold 16 gerbils. Work out the minimum number of cages needed by the end of year 6.

1.3 Fractions

Fractions arithmetic

- Common fractions are added or subtracted from one another directly only when they have a common denominator.

(a) $\frac{3}{4} + \frac{2}{5} = \frac{15}{20} + \frac{8}{20}$
$= \frac{23}{20}$
$= 1\frac{3}{20}$

(b) $\frac{3}{7} - \frac{1}{3} = \frac{9}{21} - \frac{7}{21}$
$= \frac{2}{21}$

so $2\frac{3}{8} - 1\frac{5}{12} = \frac{19}{8} - \frac{17}{12}$
$= \frac{57}{24} - \frac{34}{24}$
$= \frac{23}{24}$

- Multiplying and dividing fractions is made easier by 'cancelling down'.

(d) $\frac{2}{5} \times \frac{6}{7} = \frac{12}{35}$

(no cancelling here)

(e) $\frac{\cancel{12}^{3}}{13} \times \frac{5}{\cancel{8}_{2}} = \frac{15}{26}$

(cancel down)

(f) $2\frac{2}{5} \div 6 = \frac{12}{5} \div 6$

$= \frac{\cancel{12}^{2}}{5} \times \frac{1}{\cancel{6}_{1}} = \frac{2}{5}$

- Notice in part (f): $2\frac{2}{5}$ is written as $\frac{12}{5}$;

  'Dividing by 6' is the same as 'multiplying by $\frac{1}{6}$'.

Exercise 7

1. Copy and complete.

(a) $\frac{1}{4} + \frac{1}{8}$

$= \frac{\square}{8} + \frac{1}{8} =$

(b) $\frac{1}{2} + \frac{2}{5}$

$= \frac{\square}{10} + \frac{\square}{10} =$

(c) $\frac{2}{5} + \frac{1}{3}$

$= \frac{\square}{15} + \frac{\square}{15} =$

2. Work out:

(a) $\frac{1}{2} + \frac{1}{3}$ (b) $\frac{1}{4} + \frac{1}{6}$ (c) $\frac{1}{3} + \frac{1}{4}$ (d) $\frac{1}{4} + \frac{1}{5}$

(e) $\frac{3}{4} + \frac{4}{5}$ (f) $\frac{5}{6} + \frac{2}{3}$ (g) $1\frac{1}{4} + \frac{1}{3}$ (h) $2\frac{1}{3} + 1\frac{1}{4}$

3. Work out:

(a) $\frac{2}{3} - \frac{1}{2}$ (b) $\frac{3}{4} - \frac{1}{3}$ (c) $\frac{4}{5} - \frac{1}{3}$ (d) $\frac{1}{2} - \frac{2}{5}$ (e) $1\frac{3}{4} - \frac{2}{3}$ (f) $1\frac{2}{3} - 1\frac{1}{2}$

4. Copy and complete the addition square:

$+$			$\frac{1}{3}$
$\frac{1}{2}$	$\frac{3}{4}$		$\frac{5}{6}$
$\frac{1}{8}$	$\frac{3}{8}$	$\frac{1}{2}$	
$\frac{1}{5}$			

5. Work out these multiplications.

(a) $\frac{2}{3} \times \frac{4}{5}$ (b) $\frac{1}{7} \times \frac{5}{6}$ (c) $\frac{5}{8} \times \frac{12}{13}$ (d) $1\frac{3}{4} \times \frac{2}{3}$

(e) $2\frac{1}{2} \times \frac{1}{4}$ (f) $3\frac{2}{3} \times 2\frac{1}{2}$ (g) $\frac{2}{3}$ of $\frac{2}{5}$ (h) $\frac{3}{7}$ of $\frac{1}{2}$

6. Work out these divisions.

(a) $\frac{1}{3} \div \frac{4}{5}$ (b) $\frac{3}{4} \div \frac{1}{6}$ (c) $\frac{5}{6} \div \frac{1}{2}$ (d) $\frac{5}{8} \div 2$

(e) $\frac{3}{8} \div \frac{1}{5}$ (f) $1\frac{3}{4} \div \frac{2}{3}$ (g) $3\frac{1}{2} \div 2\frac{3}{5}$ (h) $6 \div \frac{3}{4}$

7. Fill in the missing numbers so that the answer is always $\frac{3}{8}$

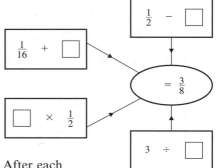

8. A rubber ball is dropped from a height of 300 cm. After each bounce, the ball rises to $\frac{4}{5}$ of its previous height.
How high, to the nearest cm, will it rise after the fourth bounce?

9. Steve Braindead spends his income as follows:

(a) $\frac{2}{5}$ of his income goes in tax

(b) $\frac{2}{3}$ of what is left goes on food, rent and transport

(c) he spends the rest on cigarettes, beer and betting.
What fraction of his income is spent on cigarettes, beer and betting?

10. Given $a = \frac{3}{4}$, $b = \frac{2}{5}$, $c = \frac{1}{3}$, work out:

(a) a^2 (b) $a - b$ (c) $\dfrac{b}{c}$ (d) $\dfrac{1}{b} - \dfrac{1}{a}$

(e) abc (f) $2b - c$ (g) $\dfrac{1}{a + b}$ (h) $\dfrac{b}{ac}$

11. In the equation below all the asterisks stand for the same number. What is the number?

$$\left[\frac{*}{*} - \frac{*}{6} = \frac{*}{30} \right]$$

12. Find the value of n if
$$\left(1\frac{1}{3}\right)^n - \left(1\frac{1}{3}\right) = \frac{28}{27}$$

13. A formula used by opticians is
$$\frac{1}{f} = \frac{1}{u} + \frac{1}{v}$$

Given that $u = 3$ and $v = 5\frac{1}{2}$ find the exact value of f.

14. When it hatches from its egg, the shell of a certain crab is 1 cm across. When fully grown the shell is approximately 10 cm across. Each new shell is one-third bigger than the previous one. How many shells does a fully grown crab have during its life?

15. In each equation find two *positive* integers a and b.

(a) $\dfrac{a}{4} + \dfrac{b}{3} = \dfrac{11}{12}$ (b) $\dfrac{a}{6} + \dfrac{b}{21} = \dfrac{17}{42}$ (c) $\dfrac{a}{12} + \dfrac{b}{18} = \dfrac{29}{36}$

16. Copy and complete the multiplication square.

$\times$	$\frac{2}{5}$		
$\frac{1}{3}$		$\frac{1}{4}$	
		$\frac{3}{8}$	
$\frac{1}{4}$			$\frac{1}{6}$

17. A cylinder is $\frac{1}{4}$ full of water. After 60 ml of water is added the cylinder is $\frac{2}{3}$ full. Calculate the total volume of the cylinder.

†18. (a) Here is a sequence:

$$\frac{1}{2} + \frac{1}{4} = \frac{3}{4}$$

$$\frac{1}{2} + \frac{1}{4} + \frac{1}{8} = \frac{7}{8}$$

$$\frac{1}{2} + \frac{1}{4} + \frac{1}{8} + \frac{1}{16} = \boxed{}$$

$$\frac{1}{2} + \frac{1}{4} + \frac{1}{8} + \frac{1}{16} + \frac{1}{32} = \boxed{}$$

Fill in the missing answers.

(b) Write down the answer to the nth line of the sequence.

1.4 Estimating

It is always sensible to check that the answer to a calculation is 'about the right size'.

Estimate the value of $\dfrac{57 \cdot 2 \times 110}{2 \cdot 146 \times 46 \cdot 9}$, correct to one significant figure.

We have approximately, $\dfrac{60 \times 100}{2 \times 50} = 60$

On a calculator the value if 62·52 (to 4 significant figures).

Exercise 8

In this exercise there are 25 questions, each followed by three possible
answers. In each case only one answer is correct.
Write down each question and decide (by estimating) which answer is correct.

	Question	Answer A	Answer B	Answer C
1.	$7 \cdot 2 \times 9 \cdot 8$	52·16	98·36	70·56
2.	$2 \cdot 03 \times 58 \cdot 6$	118·958	87·848	141·116
3.	$23 \cdot 4 \times 19 \cdot 3$	213·32	301·52	451·62
4.	$313 \times 107 \cdot 6$	3642·8	4281·8	33678·8
5.	$6 \cdot 3 \times 0 \cdot 098$	0·6174	0·0622	5·98
6.	$1200 \times 0 \cdot 89$	722	1068	131
7.	$0 \cdot 21 \times 93$	41·23	9·03	19·53
8.	$88 \cdot 8 \times 213$	18914·4	1693·4	1965·4
9.	$0 \cdot 04 \times 968$	38·72	18·52	95·12
10.	$0 \cdot 11 \times 0 \cdot 089$	0·1069	0·0959	0·00979
11.	$13 \cdot 92 \div 5 \cdot 8$	0·52	4·2	2·4
12.	$105 \cdot 6 \div 9 \cdot 6$	8·9	11	15
13.	$8405 \div 205$	4·6	402	41
14.	$881 \cdot 1 \div 99$	4·5	8·9	88
15.	$4 \cdot 183 \div 0 \cdot 89$	4·7	48	51
16.	$6 \cdot 72 \div 0 \cdot 12$	6·32	21·2	56
17.	$20 \cdot 301 \div 1010$	0·0201	0·211	0·0021
18.	$0 \cdot 28896 \div 0 \cdot 0096$	312	102·1	30·1
19.	$0 \cdot 143 \div 0 \cdot 11$	2·3	1·3	11·4
20	$159 \cdot 65 \div 515$	0·11	3·61	0·31
21.	$(5 \cdot 6 - 0 \cdot 21) \times 39$	389·21	210·21	20·51
22.	$\dfrac{17 \cdot 5 \times 42}{2 \cdot 5}$	294	504	86
23.	$(906 + 4 \cdot 1) \times 0 \cdot 31$	473·21	282·131	29·561
24.	$\dfrac{543 + 472}{18 \cdot 1 + 10 \cdot 9}$	65	35	85
25.	$\dfrac{112 \cdot 2 \times 75 \cdot 9}{6 \cdot 9 \times 5 \cdot 1}$	242	20·4	25·2

Exercise 9

1. For a wedding the caterers provided food at £39·75 per head. There
   were 207 guests at the wedding. Estimate the total cost of the food.

2. 985 people share the cost of hiring an ice rink. About how much
   does each person pay if the total cost is £6017?

3. On a charity walk, Susie walked 31 miles in 11 hours 7 minutes.
   Estimate the number of minutes it took to walk one mile.

In Questions **4** to **11**, estimate which answer is closest to the actual answer.

4. The height of a double-decker bus:

A	B	C
3 m	6 m	10m

5. The height of the tallest player in the Olympic basketball competition:

A	B	C
1·8 m	3·0 m	2·2 m

6. The mass of a £1 coin:

A	B	C
1 g	10 g	100 g

7. The volume of your classroom:

A	B	C
20 m^3	200 m^3	2000 m^3

8. The top speed of a Grand Prix racing car:

A	B	C
600 km/h	80 km/h	300 km/h

9. The number of times your heart beats in one day (24 h):

A	B	C
10 000	100 000	1 000 000

10. The thickness of one page in this book:

A	B	C
0·01 cm	0·001 cm	0·0001 cm

11. The number of cars in a traffic jam 10 km long on a 3-lane motorway:

A	B	C
3000	30 000	300 000

[Assume each car takes up 10 m of road.]

In Questions **12** and **13** there are six calculations and six answers. Write down each calculation and insert the correct answer from the list given. Use estimation.

12. (a) $8·9 \times 10·1$ (b) $7·98 \div 1·9$ (c) $112 \times 3·2$
  (d) $11·6 + 47·2$ (e) $2·82 \div 9·4$ (f) $262 \div 100$

Answers: 2·62, 58·8, 0·3, 89·89, 358·4, 4·2

13. (a) $49·5 \div 11$ (b) 21×22 (c) $9·1 \times 104$
  (d) $86 - 8·2$ (e) $2·4 \div 12$ (f) $651 \div 31$

Answers: 21, 946·4, 0·2, 4·5, 462, 77·8

14. Mr Gibson, the famous maths teacher, has won the pools.
He decides to give a rather unusual prize for the person
who comes top in his next maths test.
The prize winner receives his or her own weight in coins
and they can choose to have either 1p, 2p, 5p, 10p, 20p,
50p or £1 coins. All the coins must be the same.

Approximate masses	
1 p	3·6 g
2 p	7·2 g
5 p	3·2 g
10 p	6·5 g
20 p	5·0 g
50 p	7·5 g
£1	9·0 g

Shabeza is the winner and she weighs 47 kg.
Estimate the highest value of her prize.

15. The largest tree in the world has a
diameter of 11 m.
Estimate the number of 'average' 15-year-olds
required to circle the tree so that they form an
unbroken chain.

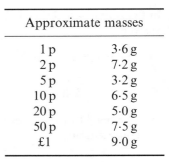

16. There are about 7000 cinemas in the U.K. and
every day about 300 people visit each one.
The population of the U.K. is about 60 million.

A film magazine report said:

'Over 3% of British people
go to the cinema everyday'.

Is the magazine report fair?
Show the working you did to decide.

17. The petrol consumption of a large car is 4 miles per litre and petrol
costs 79p per litre.
Jasper estimates that the petrol costs of a round trip of about
1200 miles will be £240. Is this a reasonable estimate?

18. The 44 teachers in a rather difficult school decide to buy 190 canes
at £2·42 each. They share the cost equally between them.
The headmaster used a calculator to work out the cost per teacher
and got an answer of £1·05 to the nearest penny.
Without using a calculator, work out an estimate for the answer to
check whether or not he got it right. Show your working.

19. The surface area, A, of an object is given by the formula
$$A = \pi r\left[10r + \sqrt{(r^2 + d^2)}\right].$$
Estimate the value of A correct to one significant figure if
$r = 1·08$ and $d = 5·87$.

Approximations

(a) Reminders: 35·2 | 6 = 35·3 to 3 sig. fig.

                     ↑

          '5 or more'

          0·041 | 2 = 0·041 to 2 sig. fig.

          ↑↑

Do not count zeros
at the beginning.

          15·26 | 66 = 15·27 to 2 decimal places

            ↑

          0·349 | 7 = 0·350 to 3 decimal places

            ↑

(b) Suppose you timed a race with a stopwatch and got 13·2 seconds
while your friend with an electronic watch got 13·20 seconds.
Is there any difference? Yes!
When you write 13·20, the figure is accurate to 2 decimal places
even though the last figure is a zero.
The figure 13·2 is accurate to only one decimal place.

(c) Similarly a weight, given as 12·00 kg, is accurate to 4 significant
figures while '12 kg' is accurate to only 2 significant figures.

(d) Suppose you measure the length of a line in cm and you want to
show that it is accurate to one decimal place. The measured length
using a ruler might be 7 cm. To show that it is accurate to one
decimal place, you must write 'length = 7·0 cm'.

Exercise 10

1. Round off to the number of decimal places indicated.

  (a) 0·672 (1 d.p.)   (b) 8·814 (2 d.p.)   (c) 0·7255 (3 d.p.)
  (d) 1·1793 (2 d.p.)   (e) 0·863 (1 d.p.)   (f) 8·2222 (2 d.p.)
  (g) 0·07518 (3 d.p.) (h) 11·7258 (3 d.p.) (i) 20·154 (1 d.p.)
  (j) 6·6666 (2 d.p.)   (k) 0·342 (1 d.p.)   (l) 0·07248 (4 d.p.)

2. Round off to the number of significant figures indicated

  (a) 2·658 (2 s.f.)   (b) 188·79 (3 s.f.)   (c) 2·87 (1 s.f.)
  (d) 0·3569 (2 s.f.)   (e) 1·7231 (2 s.f.) (f) 0·041551 (3 s.f.)
  (g) 0·0371 (1 s.f.)   (h) 811·1 (1 s.f.)   (i) 9320 (2 s.f.)
  (j) 8·051 (2 s.f.)   (k) 6·0955 (3 s.f.)   (l) 8·205 (1 s.f.)

Common sense

When you find the answer to a problem, you should choose the degree
of accuracy appropriate for that situation.

(a) Suppose you were calculating how much tax someone should pay in
a year and your actual answer was £2153·6752. It would be sensible
to give the answer as £2154 to the nearest pound.

(b) In calculating the average speed of a car journey from Bristol to
Cardiff, it would not be realistic to give an answer of 38·241 km/h.
A more sensible answer would be 38 km/h.

A mad idea

In order to celebrate his mother's 100th birthday, Signor Gibsoni, the head of the largest mafia family in Sicily, decides to invite the entire population of the earth, which is about 5 000 000 000 people to a surprise party on the island.

The land area of Sicily is approximately 26 000 km^2.
Will there be room for everyone to meet on her birthday?
You may need to conduct an experiment in your class to find what area is required for say 20 or 30 pupils. What would be the effect of including babies and adults?
Make a list of some of the practical difficulties which the organisers would have to overcome.

1.5 Ratio and proportion

Ratio

● The word 'ratio' is used to describe a fraction. If the *ratio* of a boy's height to his father's height is $4:5$, then he is $\frac{4}{5}$ as tall as his father.

(a) Change the ratio $2:5$ into the form

(i) $1:n$ (ii) $m:1$

(i) $2:5 = 1:\frac{5}{2}$ (ii) $2:5 = \frac{2}{5}:1$

       $= 1:2\cdot5$ $= 0\cdot4:1$

(b) Divide £60 between two people A and B in the ratio $5:7$.

    Consider £60 as 12 equal parts (i.e. $5+7$). Then A receives 5 parts and B receives 7 parts.

    ∴   A receives $\frac{5}{12}$ of £60 = £25

          B receives $\frac{7}{12}$ of £60 = £35

Exercise 11

Express the following ratios in the form $1 : n$.

1. $2 : 6$ **2.** $5 : 30$ **3.** $2 : 100$

4. $5 : 8$ **5.** $4 : 3$ **6.** $8 : 3$

Express the following ratios in the form $n : 1$.

7. $12 : 5$ **8.** $5 : 2$ **9.** $4 : 5$

In Questions **10** to **13**, divide the quantity in the ratio given.

10. £40; $(3 : 5)$ **11.** £120; $(3 : 7)$

12. 180 kg; $(1 : 5 : 6)$ **13.** 184 minutes; $(2 : 3 : 3)$

14. When £143 is divided in the ratio $2 : 4 : 5$, what is the difference between the largest share and the smallest share?

15. If $\frac{5}{8}$ of the children in a school are boys, what is the ratio of boys to girls?

16. Find the ratio (shaded area) : (unshaded area) for each diagram.

(a) 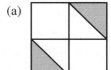   (b) 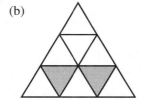   (c)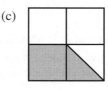

17. A man and a woman share a bingo prize of £1000 between them in the ratio $1 : 4$. The woman shares her part between herself, her mother and her daughter in the ratio $2 : 1 : 1$.
How much does her daughter receive?

18. A man and his wife share a sum of money in the ratio $3 : 2$. If the sum of money is doubled, in what ratio should they divide it so that the man still receives the same amount?

19. In a herd of x cattle, the ratio of the number of bulls to cows is $1 : 6$. Find the number of bulls in the herd in terms of x.

20. If $x : 3 = 12 : x$, calculate the positive value of x.

21. £400 is divided between Ann, Brian and Carol so that Ann has twice as much as Brian, and Brian has three times as much as Carol. How much does Brian receive?

22. A cake weighing 550 g has three ingredients: flour, sugar and raisins. There is twice as much flour as sugar and one and a half times as much sugar as raisins. How much flour is here?

23. A brother and sister share out their collection of 5000 stamps in the ratio $5 : 3$. The brother then shares his stamps with two friends in the ratio $3 : 1 : 1$, keeping most for himself. How many stamps do each of his friends receive?

Map scales

A map is drawn to a scale of 1 to 50 000.
Calculate the length of a road which appears as
3 cm long on the map.

1 cm on the map is equivalent to 50 000 cm on the earth.

$$\therefore \quad 3\,\text{cm} \equiv 3 \times 50\,000 = 150\,000\,\text{cm}$$
$$= 1500\,\text{m}$$
$$= 1\cdot5\,\text{km}$$

The road is 1·5 km long.

Exercise 12

1. On a map of scale 1 : 100 000, the distance between Tower Bridge
   and Hammersmith Bridge is 12·3 cm.
   What is the actual distance in km?

2. On a map of scale 1 : 15 000, the distance between Buckingham
   Palace and Brixton Underground Station is 31·4 cm.
   What is the actual distance in km?

3. If the scale of a map is 1 : 10 000, what will be the length on this
   map of a road which is 5 km long?

4. The distance from Hertford to St Albans is 32 km.
   How far apart will they be on a map of scale 1 : 50 000?

5. The 17th hole at the famous St Andrews golf course is 420 m in length.
   How long will it appear on a plan of the course of scale 1 : 8000?

6. The scale of a map is 1 : 1000. What are the actual dimensions of a
   rectangle which appears as 4 cm by 3 cm on the map? What is the
   area on the map in cm^2? What is the actual area in m^2?

7. The scale of a map is 1 : 100. What area does 1 cm^2 on the map
   represent? What area does 6 cm^2 represent?

8. The scale of a map is 1 : 20 000. What area does 8 cm^2 represent?

Proportion

(a) If 9 litres of petrol costs £5·76,
   find the cost of 20 litres.

   The cost of petrol is *directly* proportional to
   the quantity bought.

   9 litres costs £5·76
∴ 1 litre costs £5·76 ÷ 9 = £0·64
∴ 20 litres costs £0·64 × 20 = £12·80

(b) If five men can paint a bridge in 12 days,
   how long would it take three men?

   The length of time required to paint the
   bridge is *inversely* proportional to the
   number of men who paint.

   5 men take 12 days
∴ 1 man takes 60 days.
∴ 3 men take 20 days.

In the first example we found the cost of *one* litre of petrol.

In the second example we found the time needed for *one* man.

Exercise 13

This exercise contains a mixture of questions involving both *direct* and *inverse* proportion.

1. If 7 discs cost £1·54, find the cost of 5 discs.

2. Twelve people are needed to harvest a crop in 2 hours. How long would it take 4 people?

3. Three men build a wall in 10 days. How long would it take five men?

4. Nine milk bottles contain $4\frac{1}{2}$ litres of milk between them. How much do five bottles hold?

5. A car uses 10 litres of petrol in 75 km. How far will it go on 8 litres?

6. A wire 11 cm long has a mass of 187 g. What is the mass of 7 cm of this wire?

7. A train travels 30 km in 120 minutes. How long will it take to travel 65 km at the same speed?

8. A ship has sufficient food to supply 600 passengers for 3 weeks. How long would the food last for 800 people?

9. Usually it takes 12 hours for 5 men to tarmac a road. How many men are needed to tarmac the same road in 10 hours?

10. 80 machines can produce 4800 identical pens in 5 hours. At this rate:
    (a) how many pens would one machine produce in one hour?
    (b) how many pens would 25 machines produce in 7 hours?

11. Three men can build a wall in 10 hours. How many men would be needed to build the wall in $7\frac{1}{2}$ hours?

12. If it takes 6 men 4 days to dig a hole 3 feet deep, how long will it take 10 men to dig a hole 7 feet deep?

13. A car travels 279 km on 31 litres of petrol. How much petrol is needed for a journey of 396 km?

14. 5 machines produce 10 000 bottles in 10 hours. How many bottles would 8 machines produce in 6 hours?

15. It takes x ants n hours to build a nest. How long will it take $x + 1000$ ants to build the same size nest?

16. Usually it takes 20 hours for 4 men to build a wall. How many men are needed to build the same wall in 16 hours?

1.6 Negative numbers

● For adding and subtracting use the number line.

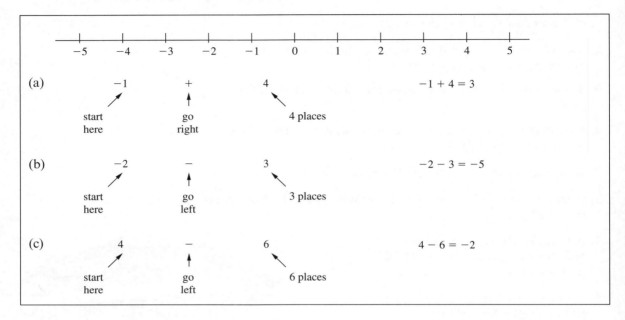

Exercise 14

Work out:

1. $-6+2$	**2.** $-7-5$	**3.** $-3-8$	**4.** $-5+2$
5. $-6+1$	**6.** $8-4$	**7.** $4-9$	**8.** $11-19$
9. $4+15$	**10.** $-7-10$	**11.** $16-20$	**12.** $-7+2$
13. $-6-5$	**14.** $10-4$	**15.** $-4+0$	**16.** $-6+12$
17. $-7+7$	**18.** $2-20$	**19.** $8-11$	**20.** $-6-5$
21. $-8-4$	**22.** $-3+7$	**23.** $-6+10$	**24.** $-5+5$
25. $-11+3$	**26.** $7-10$	**27.** $-5+8$	**28.** $-12+0$
29. $-1+19$	**30.** $20-25$	**31.** $-6-60$	**32.** $-2+100$

● When you have two $(+)$ or $(-)$ signs together use this rule:

$$++ = + \qquad +- = -$$
$$-- = + \qquad -+ = -$$

(a) $3-(-6) = 3+6 = 9$

(b) $-4+(-5) = -4-5 = -9$

(c) $-5-(+7) = -5-7 = -12$

Exercise 15

Work out.

1. $-3 + (-5)$	**2.** $-5 - (+2)$	**3.** $4 - (+3)$	**4.** $-3 - (-4)$
5. $6 - (-3)$	**6.** $16 + (-5)$	**7.** $-4 + (-4)$	**8.** $20 - (-22)$
9. $-6 - (-10)$	**10.** $95 + (-80)$	**11.** $-3 - (+4)$	**12.** $-5 - (+4)$
13. $6 + (-7)$	**14.** $-4 + (-3)$	**15.** $-7 - (-7)$	**16.** $3 - (-8)$
17. $-8 + (-6)$	**18.** $7 - (+7)$	**19.** $12 - (-5)$	**20.** $9 - (+6)$
21. $-3 - (-2)$	**22.** $8 + (-11)$	**23.** $10 - (-2)$	**24.** $-7 + (-2)$
25. $9 - (+6)$	**26.** $7 + (-7)$	**27.** $0 - (-8)$	**28.** $-6 - (-8)$

29. Copy and complete these addition squares.

(a)

+	−5	1	6	−2
3	−2	4		
−2				
6				
−10				

(b)

+			−4	
5			1	
		0		5
	7		6	17
		−4		

Multiplying and dividing

A directed number is a positive or negative number.

- When two directed numbers with the same sign are multiplied together, the answer is positive.

  $+7 \times (+3) = +21$
  $-6 \times (-4) = +24$

- When two directed numbers with different signs are multiplied together, the answer is negative.

  $-8 \times (+4) = -32$
  $+7 \times (-5) = -35$

- When dividing directed numbers, the rules are the same as in multiplication.

  $-70 \div (-2) = +35$
  $+12 \div (-3) = -4$
  $-20 \div (+4) = -5$

Exercise 16

1. $-3 \times (+2)$	**2.** $-4 \times (+1)$	**3.** $+5 \times (-3)$	**4.** $-3 \times (-3)$
5. $-4 \times (+2)$	**6.** $-5 \times (+3)$	**7.** $6 \times (-4)$	**8.** $3 \times (+2)$
9. $-3 \times (-4)$	**10.** $6 \times (-3)$	**11.** $-7 \times (+3)$	**12.** $-5 \times (-5)$
13. $6 \times (-10)$	**14.** $-3 \times (-7)$	**15.** $8 \times (+6)$	**16.** $-8 \times (+2)$
17. $-7 \times (+6)$	**18.** $-5 \times (-4)$	**19.** $-6 \times (+7)$	**20.** $11 \times (-6)$

21. $8 \div (-2)$ **22.** $-9 \div (+3)$ **23.** $-6 \div (-2)$ **24.** $10 \div (-2)$

25. $-12 \div (-3)$ **26.** $-16 \div (+4)$ **27.** $4 \div (-1)$ **28.** $8 \div (-8)$

29. $16 \div (-8)$ **30.** $-20 \div (-5)$ **31.** $-16 \div (+1)$ **32.** $18 \div (-9)$

33. $36 \div (-9)$ **34.** $-45 \div (-9)$ **35.** $-70 \div (+7)$ **36.** $-11 \div (-1)$

37. $-16 \div (-1)$ **38.** $1 \div (-\frac{1}{2})$ **39.** $-2 \div (+\frac{1}{2})$ **40.** $50 \div (-10)$

41. $-8 \times (-8)$ **42.** $-9 \times (+3)$ **43.** $10 \times (-60)$ **44.** $-8 \times (-5)$

45. $-12 \div (-6)$ **46.** $-18 \times (-2)$ **47.** $-8 \div (+4)$ **48.** $-80 \div (+10)$

49. Copy and complete these multiplication squares.

(a)

$\times$	4	-3	0	-2
-5				
2				
10				
-1				

(b)

$\times$			-1	
3			-3	
		-15		18
		-14	-7	-42
		10		

Questions on negative numbers are more difficult when the different sorts are mixed together. The remaining questions are given in the form of four short tests.

Test 1

1. $-8 - 8$ **2.** $-8 \times (-8)$ **3.** -5×3 **4.** $-5 + 3$

5. $8 - (-7)$ **6.** $20 - 2$ **7.** $-18 \div (-6)$ **8.** $4 + (-10)$

9. $-2 + 13$ **10.** $+8 \times (-6)$ **11.** $-9 + (+2)$ **12.** $-2 - (-11)$

13. $-6 \times (-1)$ **14.** $2 - 20$ **15.** $-14 - (-4)$ **16.** $-40 \div (-5)$

17. $5 - 11$ **18.** -3×10 **19.** $9 + (-5)$ **20.** $7 \div (-7)$

Test 2

1. $-2 \times (+8)$ **2.** $-2 + 8$ **3.** $-7 - 6$ **4.** $-7 \times (-6)$

5. $+36 \div (-9)$ **6.** $-8 - (-4)$ **7.** $-14 + 2$ **8.** $5 \times (-4)$

9. $11 + (-5)$ **10.** $11 - 11$ **11.** $-9 \times (-4)$ **12.** $-6 + (-4)$

13. $3 - 10$ **14.** $-20 \div (-2)$ **15.** $16 + (-10)$ **16.** $-4 - (+14)$

17. $-45 \div 5$ **18.** $18 - 3$ **19.** $-1 \times (-1)$ **20.** $-3 - (-3)$

Test 3

1. $-10 \times (-10)$ **2.** $-10 - 10$ **3.** $-8 \times (+1)$ **4.** $-8 + 1$

5. $5 + (-9)$ **6.** $15 - 5$ **7.** $-72 \div (-8)$ **8.** $-12 - (-2)$

9. $-1 + 8$ **10.** $-5 \times (-7)$ **11.** $-10 + (-10)$ **12.** $-6 \times (+4)$

13. $6 - 16$ **14.** $-42 \div (+6)$ **15.** $-13 + (-6)$ **16.** $-8 - (-7)$

17. $5 \times (-1)$ **18.** $2 - 15$ **19.** $21 + (-21)$ **20.** $-16 \div (-2)$

Test 4

Write down each statement and find the missing number.

1. $(-6) \times \boxed{} = 30$ 　　　 **2.** $\boxed{} + (-2) = 0$ 　　　 **3.** $\boxed{} \div (-2) = 10$

4. $\boxed{} - (-3) = 7$ 　　　 **5.** $(-1) \times \boxed{} = \frac{1}{2}$ 　　　 **6.** $(-2) - \boxed{} = 3$

7. $(-2) \div \boxed{} = -2$ 　　　 **8.** $\boxed{} - 8 = -6$ 　　　 **9.** $(-12) \times (-\frac{1}{2}) = \boxed{}$

10. $\boxed{} + (-10) = 2$ 　　 **11.** $6 \times \boxed{} = 0$ 　　　 **12.** $\boxed{} - (-8) = 0$

13. $(-1)^4 = \boxed{}$ 　　　 **14.** $0{\cdot}2 \times \boxed{} = -200$ 　　 **15.** $(-1)^{13} = \boxed{}$

1.7 Mixed numerical problems

Exercise 17

1. I have lots of 1p, 2p, 3p and 4p stamps. How many different combinations of stamps can I make which total 5p?

2. Find n if:
$$8 + 9 + 10 + \ldots + n = 5^3$$

3. Copy and complete.
$$3^2 + 4^2 + 12^2 = 13^2$$
$$5^2 + 6^2 + 30^2 = 31^2$$
$$6^2 + 7^2 + \quad = $$
$$x^2 + \quad + \quad = $$

4. You are told that 8 cakes and 6 biscuits cost 174 pence and 2 cakes and 4 biscuits cost 66 pence. Without using simultaneous equations, work out the cost of each of the following.
  (a) 4 cakes and 3 biscuits   (b) 10 cakes and 10 biscuits
  (c) 3 cakes and 3 biscuits   (d) 1 cake
  (e) 1 biscuit

5. The total mass of a jar one-quarter full of jam is 250 g. The total mass of the same jar three-quarters full of jam is 350 g.

$\frac{1}{4}$
250g

$\frac{3}{4}$
350g

What is the mass of the empty jar?

6. On an aircraft a certain weight of luggage is carried free and a charge of £10 per kilogram is made for any excess luggage.
  (a) If the charge for 60 kg of luggage is £400, find the charge for 35 kg.
  (b) If the charge for 30 kg of luggage is £100, find the charge for 15 kg.
  (c) If a fifth of the luggage is carried free, what is the average cost per kilogram of the luggage?

7. Evaluate

(a) $\frac{1}{3} \times \frac{2}{4} \times \frac{3}{5} \times \ldots \times \frac{9}{11} \times \frac{10}{12}$.

8. Find the least positive integer n for which $(829\,642 + n)$ is exactly divisible by 7.

9. Find all the missing digits in these multiplications.

(a)
$$5\,\square$$
$$9\,\times$$
$$\overline{\square\,\square\,6}$$

(b)
$$\square\,7$$
$$\square\,\times$$
$$\overline{4\,\square\,6}$$

(c)
$$5\,\square$$
$$\square\,\times$$
$$\overline{1\,\square\,4}$$

10. (a) Work out $\frac{1}{4} + \frac{1}{12}$ as a single fraction in its lowest terms.

(b) Find integers a and b such that $\dfrac{1}{a} + \dfrac{1}{b} = \dfrac{5}{8}$.

Exercise 18

1. Copy and complete this multiplication square.

$\times$	$\frac{2}{3}$		
$\frac{1}{2}$		$\frac{3}{8}$	
		$\frac{3}{16}$	
$\frac{2}{5}$			$\frac{2}{25}$

2. Look at this number pattern.

$$7^2 = 49$$
$$67^2 = 4489$$
$$667^2 = 444\,889$$
$$6667^2 = 44\,448\,889$$

This pattern continues.
(a) Write down the next line of the pattern.
(b) Use the pattern to work out $6\,666\,667^2$.
(c) What is the square root of $4\,444\,444\,488\,888\,889$?

3. Find a pair of positive integers a and b for which
(a) $18a + 65b = 1865$
(b) $23a + 7b = 2314$

4. If the number in the space is prime, write PRIME next to it. If it is not prime, write it as the product of its prime factors. The first two have been done for you.

47 ...PRIME...   40 ....   63 ....

26 ...2×13...   25 ....   71 ....

5. Express $419\,965$ in terms of its prime factors.

6. How many prime numbers are there between 120 and 130?

7. Find three consecutive square numbers whose sum is 149.

8. The diagrams show magic squares in which the sum of the numbers in any row, column or diagonal is the same. Find the value of x in each square.

(a)

	x	6
3		7
		2

(b)

4		5	16
x		10	
	7	11	2
1			13

9. Work out $100 - 99 + 98 - 97 + 96 - \ldots + 4 - 3 + 2 - 1$.

10. The smallest three-digit product of a one-digit prime and a two-digit prime is:

    (a) 102   (b) 103   (c) 104   (d) 105   (e) 106

Exercise 19

1. A group of 17 people share some money equally and each person gets £415. What is the total amount of money they have shared?

2. The 'reciprocal' of 2 is $\frac{1}{2}$. The reciprocal of 7 is $\frac{1}{7}$. The reciprocal of x is $\dfrac{1}{x}$.

Find the square root of the reciprocal of the square root of the reciprocal of ten thousand.

3. The purpose of the set of steps given below is to work out some numbers in a number sequence.

    Step 1   Write down the number 1.
    Step 2   Add 2 to the number you have just written.
    Step 3   Write down the answer obtained in step 2.
    Step 4   If your answer in step 2 is more than 10 then stop.
    Step 5   Go to step 2.

    (a) Write down the numbers produced from the set of steps.
    (b) Change one line of the set of steps so that it could be used to produce the first six even numbers.

4. Put four different numbers in the circles so that when you add the numbers at the end of each line you always get a square number.

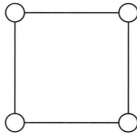

5. You are given that $41 \times 271 = 11111$.
Work out the following *in your head*.
    (a) 246×271
    (b) $22\,222 \div 271$
    (c) This time you can write down a (little!) working.
        Work out $41^2 \times 271$.

6. Copy and complete the crossnumber puzzle.

Across
1. Square number
4. (1 across) × (3 down) − 7000
6. Next in the sequence
1, 3, 7, 15, 31, 63, 127, __
7. Sum of the prime numbers
between 30 and 40

Down
2. South-East as a bearing
3. Number of days in 13 weeks
5. One-eighth of 6048
6. Cube number

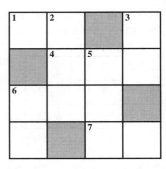

7. (a) Write down the value of sin A as a fraction
   (b) Work out, as a fraction

$$\frac{5\sin A}{(1 - \sin A)(1 + \sin A)}$$

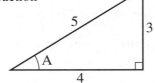

8. S_1 is the sum of all the even numbers from 2 to 1000 inclusive.
S_2 is the sum of all the odd numbers from 1 to 999 inclusive.
Work out $S_1 - S_2$.

9. Pages 6 and 27 are on the same (double) sheet of a newspaper.
What are the page numbers on the opposite side of the sheet? How
many pages are there in the newspaper altogether?

10. Use the numbers 1, 2, 3, 4, 5, 6, 7, 8, 9 once each and in their
natural order to obtain an answer of 100. You may use only the
operations +, −, ×, ÷.

Exercise 20

1. (a) The sum of the factors of n (including 1 and n) is 7. Find n.
   (b) The sum of the factors of p (including 1 and p) is 6. Find p.

2. Each packet of washing powder carries a token and four tokens can
be exchanged for a free packet. How many free packets will I
receive if I buy 64 packets?

3. At birth, the mass of a kitten was 0·3 kg.
A few weeks later its mass has increased so that the ratio

mass now : mass at birth = 7 : 4

Calculate the kitten's mass now.

4. Apart from 1, 3 and 5, all odd numbers less than 100 can be written
in the form $p + 2^n$ where p is a prime number and n is greater than
or equal to 2.

e.g. $43 = 11 + 2^5$
       $27 = 23 + 2^2$

For the odd numbers 7, 9, 11, ... 99, write as many as you can in
the form $p + 2^n$.

5. The digits 1, 9, 9, 4 are used to form fractions less than 1. Each of the four digits must be used,

e.g. $\dfrac{9}{419}$ or $\dfrac{4}{919}$

(a) Write down, smallest first, the three smallest fractions that can be made.
(b) Write down the largest fraction less than one that can be made.

6. (a) 411 fans went to a Capital Radio pop concert. The tickets cost £16 each.
Calculate the total amount paid by the fans.
(b) (i) All of these fans travelled on 52-seater coaches.
What is the least number of coaches that could be used?
(ii) On the return journey two coaches failed to turn up.
How many fans had to be left behind?

7. Find the least positive whole number n for which $582\,416\,035 + n$ is exactly divisible by 11.

Operator squares

Each empty square contains either a number or a mathematical symbol $(+, -, \times, \div)$. Copy each square and fill in the details.

1.

0.5	−	0.01	→	
		×		
	×		→	35
↓		↓		
4	÷	0.1	→	

2.

	−	1.8	→	3.4
	−		÷	
	×		→	
↓		↓		
	+	0.36	→	1

3.

		×	30	→	21
×			−		
			−		→ 35
↓			↓		
			−	49	→

4.

	×	−6	→	72
÷		+		
4	+		→	
↓		↓		
	+	1	→	−2

5.

	÷	4	→	
×		÷		
$\frac{1}{5}$	+		→	−3
↓		↓		
$-\frac{1}{25}$	−		→	1.21

6.

	+	0.21	→	1.07
		÷		
$\frac{1}{3}$			→	$-\frac{7}{24}$
↓		↓		
2.58	+		→	2.916

2 NUMBER 2

2.1 Percentages

Percentages, fractions and decimals

Percentages are simply a convenient way of expressing fractions or decimals.

$$25\% = \frac{25}{100} = \frac{1}{4} = 0.25$$

(a) Change $\frac{3}{8}$ to a percentage.

$$\frac{3}{8} = \left(\frac{3}{8} \times \frac{100}{1}\right)\% = 37\frac{1}{2}\%$$

(b) Change $\frac{7}{8}$ to a decimal.

$\frac{7}{8}$ means 'divide 8 into 7'

$$8\overline{\smash{)}7 \cdot 000}^{\,0.875}, \quad \frac{7}{8} = 0.875$$

(c) Change 0·35 to a fraction.

$$0.35 = \frac{\overset{7}{\cancel{35}}}{\underset{20}{\cancel{100}}} = \frac{7}{20}$$

(d) Change 6% to a decimal.

$$6\% = \frac{6}{100} = 0.06$$

Exercise 1

1. Change the fractions to decimals.
   (a) $\frac{1}{4}$ (b) $\frac{2}{5}$ (c) $\frac{3}{8}$ (d) $\frac{5}{12}$ (e) $\frac{1}{6}$ (f) $\frac{2}{7}$

2. Change the decimals to fractions and simplify.
   (a) 0·2 (b) 0·45 (c) 0·36 (d) 0·125 (e) 1·05 (f) 0·007

3. Change to percentages.
   (a) $\frac{1}{4}$ (b) $\frac{1}{10}$ (c) 0·72 (d) 0·075 (e) 0·02 (f) $\frac{1}{3}$

4. Arrange in order of size (smallest first).
   (a) $\frac{1}{2}$; 45%; 0·6   (b) 0·38; $\frac{6}{16}$; 4%
   (c) 0·111; 11%; $\frac{1}{9}$ (d) 32%; 0·3; $\frac{1}{3}$

Change the fractions to decimals and then evaluate the following, giving the answer to 2 decimal places:

5. $\frac{1}{4} + \frac{1}{3}$ 6. $\frac{2}{3} + 0.75$ 7. $\frac{8}{9} - 0.24$

8. $\frac{7}{8} - \frac{5}{9}$ 9. $\frac{1}{3} \times 0.2$ 10. $\frac{5}{8} \times 1.1$

11. What percentage of the integers from 1 to 10 inclusive are prime numbers? [1 is *not* prime.]

12. What percentage of this shape is shaded?

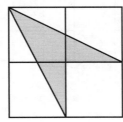

Arrange the numbers in order of size (smallest first).

13. $\frac{1}{3}$, 0·33, 30%

14. $\frac{2}{7}$, 0·3, 29%

15. 0·71, $\frac{7}{11}$, 0·705

16. $\frac{1}{1000}$, 0·2%, 0·0005

17. The pass mark in a physics test was 60%.
Sam got 23 out of 40. Did she pass?

18. In 2000 the prison population was 48 700 men and 1600 women.
What percentage of the total prison population were men?

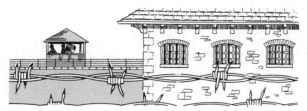

19. The table shows the results of a test for
pulse rate conducted on 274 children.

	Boys	Girls	Total
High pulse	58	71	129
Low pulse	83	62	145
Total	141	133	274

(a) What percentage of the boys had a low pulse rate?

(b) What percentage of the children with a high pulse rate were girls?

20. Jasper reads a newspaper and then says, 'Great! Now prices will fall for a change'. Explain whether or not you agree with Jasper's statement.

> **Inflation falls from 5% to 3.5%**

Working out percentages

(a) Work out 7% of £3200.

Either: 7% of £3200

$$= \frac{7}{100} \times \frac{3200}{1} = £224$$

Or: 7% of £3200

$$= 0·07 \times 3200 = £224$$

(b) The price of a car costing £8400 is increased by 5%. Find the new price.

Either: 5% of £8400 $= \frac{5}{100} \times 8400$

$$= £420$$

New price $= £8400 + £420$
$$= £8820$$

Or: New price $= 105\%$ of £8400

$$= 1·05 \times 8400$$
$$= £8820$$

Exercise 2

1. Calculate:
   (a) 30% of £50
   (b) 45% of 2000 kg
   (c) 4% of $70
   (d) 2·5% of 5000 people

2. In a sale, a jacket costing £40 is reduced by 20%.
   What is the sale price?

3. The charge for a telephone call costing 12p is increased by 10%.
   What is the new charge?

4. In peeling potatoes 4% of the mass of the potatoes is lost as 'peel'.
   How much is *left* for use from a bag containing 55 kg?

5. Write down the missing number, as a *decimal*.

   (a) 53% of 650 = $\boxed{}$ × 650
   (b) 3% of 2600 = $\boxed{}$ × 2600

   (c) 8·5% of 700 = $\boxed{}$ × 700
   (d) 122% of 285 = $\boxed{}$ × 285

6. Work out, to the nearest penny:
   (a) 6·4% of £15·95
   (b) 11·2% of £192·66
   (c) 8·6% of £25·84
   (d) 2·9% of £18·18

7. Find the total bill:
       5 golf clubs at £18·65 each
       60 golf balls at £16·50 per dozen
       1 bag at £35·80
   V.A.T. at $17\frac{1}{2}$% is added to the total cost.

8. In 1994 a club has 250 members who each pay £95 annual
   subscription. In 2000 the membership increases by 4% and the
   annual subscription is increased by 6%. What is the total income
   from subscriptions in 2000?

9. The cash price for a car was £7640. Mr Elder bought the car on the
   following hire purchase terms: 'A deposit of 20% of the cash price
   and 36 monthly payments of £191·60'. Calculate the total amount
   Mr Elder paid.

10. A quarterly telephone bill consists of £19·15 rental plus 4·7p for
    each dialled unit. V.A.T. is added at $17\frac{1}{2}$%. What is the total bill for
    Mrs Jones who used 915 dialled units?

11. A motor bike costs £820. After a 12% increase the new price is
    112% of £820. The 'quick' way to work this out is as follows:

    $$\begin{aligned} \text{New price} &= 112\% \text{ of } £820 \\ &= 1·12 \times 820 \\ &= £918·40 \end{aligned}$$

    Use this quick method to find the new price of a lorry costing £6500
    when the price is increased by 5%.

12. Find the new price of the following items

Item	Old price	Price change
Video	£190	6% increase
House	£150 000	11% increase
Boat	£2500	8% increase
Tree	£210	5% decrease
Phone	£65	15% decrease

13. Over a period of 6 months, a colony of rabbits increases in number by 25% and then by a further 30%. If there were originally 200 rabbits in the colony how many were there at the end?

14. In the last two weeks of a sale, prices are reduced first by 30% and then by a *further* 40% of the new price. What is the final sale price of a shirt which originally cost £15?

15. Work out the following:
  (a) 8% of 3·2 kg   (answer in grams)
  (b) 16% of £4·50   (answer in pence)
  (c) 28% of 5 cm   (answer in mm)
  (d) 5% of 10 hours   (answer in minutes)
  (e) 12% of 2 minutes   (answer in seconds)

16. A 12% increase in the value of N, followed by a 12% decrease in value can be calculated as $1.12 \times N \times 0.88$. Copy and complete:
  (a) An 8% increase in the value of P, followed by a 10% decrease in value can be calculated as ⬚.

  (b) A ⬚ increase in the value of X, followed by a further ⬚ increase in value can be calculated as $1.15 \times X \times 1.06$.

  (c) A 25% increase in the value of Q, followed by a 5% decrease in value can be calculated as ⬚.

Percentage profit/loss

In the next exercise use the formulae:

- Percentage profit $= \dfrac{\text{actual profit}}{\text{original price}} \times \dfrac{100}{1}$

- Percentage loss $= \dfrac{\text{actual loss}}{\text{original price}} \times \dfrac{100}{1}$

A radio is bought for £16 and sold for £20.
What is the percentage profit?

  Actual profit $= £4$

∴ Percentage profit $= \frac{4}{16} \times \frac{100}{1} = 25\%$

The radio is sold at a 25% profit.

Exercise 3

1. The first figure is the cost price and the second figure is the selling price. Calculate the percentage profit or loss in each case.

   (a) £20, £25
   (b) £400, £500
   (c) £60, £54
   (d) £9000, £10 800
   (e) £460, £598
   (f) £512, £550·40
   (g) £45, £39·60
   (h) 50p, 23p

2. A car dealer buys a car for £500, gives it a clean, and then sells it for £640. What is the percentage profit?

3. A damaged carpet which cost £180 when new, is sold for £100. What is the percentage loss?

4. During the first four weeks of her life, a baby girl increases her weight from 3·2 kg to 4·7 kg. What percentage increase does this represent? (Give your answer to 3 sig. fig.)

5. When V.A.T. is added to the cost of a car tyre, its price increases from £16·50 to £18·48. What is the rate at which V.A.T. is charged?

6. In order to increase sales, the price of a Concorde airliner is reduced from £30 000 000 to £28 400 000. What percentage reduction is this?

7. A picture has the dimensions shown.

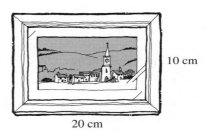

10 cm

20 cm

   Calculate the percentage increase in the area of the picture after both the length and width are increased by 20%.

8. A box has a square base of side 30 cm and height 20 cm. Calculate the percentage increase in the volume of the box after the length and width of the base are both increased by 10% and the height is increased by 5%.

9. The rental for a television set changed from £80 per year to £8 per month. What is the percentage increase in the yearly rental?

10. If an employer reduces the working week from 40 hours to 35 hours, with no loss of weekly pay, calculate the percentage increase in the hourly rate of pay.

11. Given that $G = ab$, find the percentage increase in G when both a and b increase by 10%.

12. The formula connecting P, a and n is $P = an^3$.
    (a) Calculate the value of P when $a = 200$ and $n = 5$.
    (b) Calculate the percentage increase in P when a is increased by 2% and n is increased by 20%.

Reverse percentages

(a) After an increase of 6%, the price of a motor bike is £9010.
What was the price before the increase?

A common mistake here is to work out 6% of £9010. This is wrong because the increase is 6% of the *old* price, not 6% of the new price.

$$106\% \text{ of old price} = £9010$$

$$\therefore \quad 1\% \text{ of old price} = \frac{9010}{106}$$

$$\therefore \quad 100\% \text{ of old price} = \frac{9010}{106} \times 100$$

The old price = £8500

(b) To increase sales, the price of a magazine is reduced by 5%.
Find the original price if the new price is 190p.

This is a reduction, so the new price is 95% of the old price.

$$95\% \text{ of old price} = 190p$$

$$1\% \text{ of old price} = \frac{190}{95}$$

$$100\% \text{ of old price} = \frac{190}{95} \times 100$$

Original price = 200p

Exercise 4

1. After an increase of 8%, the price of a car is £6696.
   Find the price of the car before the increase.

2. After a 12% pay rise, the salary of Mr Brown was £28 560.
   What was his salary before the increase?

3. Find the missing prices.

Item	Old price	New price	% change
(a) Jacket	?	£55	10% increase
(b) Dress	?	£212	6% increase
(c) CD player	?	£56·16	4% increase
(d) T.V.	?	£195	30% increase
(e) Car	?	£3960	65% increase

4. Between 1990 and 1997 the population of an island fell by 4%. The population in 1997 was 201 600. Find the population in 1990.

5. After being ill for 3 months, a man's weight went down by 12%.
   Find his original weight if he weighed 74·8 kg after the illness.

6. The diagram shows two rectangles.
   The width and height of rectangle B are both 20% greater than the width and height of rectangle A.

   Use the figures given to find the width and height of rectangle A.

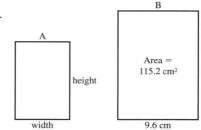

7. The label on a carton of yoghurt is shown.
   The figure on the right is smudged out.
   Work out what the figure should be.

8. A television costs £376 including $17\frac{1}{2}\%$ V.A.T.
   How much of the cost is tax?

9. During a Grand Prix car race, the tyres on a car are reduced in
   weight by 3%. If they weigh 388 kg at the end of the race, how
   much did they weigh at the start?

10. A heron has discovered Sergio's goldfish pond. Sergio lost
    70% of his collection of goldfish. If he has 60 survivors,
    how many did he have originally?

11. The average attendance at Everton football club fell by
    7% in 2000. If 2030 fewer people went to matches in 2000,
    how many went in 1999?

12. When heated an iron bar expands by 0·2%. If the increase in length
    is 1 cm, what is the original length of the bar?

13. An oven is sold for £600, thereby making a profit of 20% on the
    cost price. What was the cost price?

14. In 1999 Corus plc had sales of £516 000 000 which was $7\frac{1}{2}\%$ higher
    than the figure for 1998. Calculate the value of sales in 1998.

Compound interest

Suppose a bank pays a fixed interest of 10% on money in deposit
accounts. A man puts £500 in the bank.

After one year he has
   500 + 10% of 500 = £550

After two years he has
   550 + 10% of 550 = £605

   [Check that this is $1 \cdot 10^2 \times 500$]

After three years he has
   605 + 10% of 605 = £665·50

   [Check that this is $1 \cdot 10^3 \times 500$]

In general after n years the money in the bank will be £($1 \cdot 10^n \times 500$).

Exercise 5

1. A bank pays interest of 10% on money in deposit accounts. Mrs
   Wells puts £2000 in the bank. How much has she after:
   (a) one year   (b) two years   (c) three years?

2. A bank pays interest of 12%. Mr Olsen puts £5000 in the bank.
How much has he after:
(a) one year (b) three years?

3. A computer operator is paid £10 000 a year. Assuming her pay is
increased by 7% each year, what will her salary be in four years time?

4. Mrs Bergkamp's salary in 2000 is £30 000 per year. Every year her
salary is increased by 5%.
In 2001 her salary will be $30\,000 \times 1{\cdot}05$ $= £31\,500$
In 2002 her salary will be $30\,000 \times 1{\cdot}05 \times 1{\cdot}05$ $= £33\,075$
In 2003 her salary will be $30\,000 \times 1{\cdot}05 \times 1{\cdot}05 \times 1{\cdot}05 = £34\,728{\cdot}75$
And so on.
(a) What will her salary be in 2004?
(b) What will her salary be in 2006?

5. The price of a house was £90 000 in 1998. At the end of each
year the price has increased by 6%.
(a) Find the price of the house after 1 year.
(b) Find the price of the house after 3 years.
(c) Find the price of the house after 10 years.

6. Assuming an average inflation rate of 8%, work out the
probable cost of the following items in 10 years:
(a) car £6500 (b) T.V. £340 (c) house £50 000

7. A new car is valued at £15 000. At the end of each year its value
is reduced by 15% of its value at the start of the year.
What will it be worth after 3 years?

8. Twenty years ago a bus driver was paid £50 a week. He is now paid
£185 a week. Assuming an average rate of inflation of 7%, has his
pay kept up with inflation?

9. The population of an island increases by 10% each year. After how
many years will the original population be doubled?

10. A bank pays interest of 11% on £6000 in a deposit account.
After how many years will the money have trebled?

*11. (a) Draw the graph of $y = 1{\cdot}08^x$ for values of x from 0 to 10.
(b) Solve approximately the equation $1{\cdot}08^x = 2$.
(c) Money is invested at 8% interest.
   After how many years will the money have doubled?

12. A tree grows in height by 21% per year. It is 2 m tall after one year.
After how many more years will the tree be 20 m tall?

13. Which is the better investment over ten years:
   £20 000 at 12% compound interest
or £30 000 at 8% compound interest?

2.2 Using a calculator

To use a calculator efficiently you sometimes have to think ahead and make use of the memory, inverse $\boxed{1/x}$ and $\boxed{+/-}$ buttons.

In the example below the buttons are:

$\boxed{\sqrt{}}$ square root $\boxed{y^x}$ raises number y to the power x

$\boxed{x^2}$ square $\boxed{\text{M in}}$ puts number in memory [It automatically clears any number already in the memory when it puts in the new number.]

$\boxed{1/x}$ reciprocal $\boxed{\text{MR}}$ recalls number from memory

---

Evaluate the following to 4 significant figures:

(a) $\dfrac{2 \cdot 3}{4 \cdot 7 + 3 \cdot 61}$ Find the denominator first.

$\boxed{2 \cdot 3}$ $\boxed{\div}$ $\boxed{(}$ $\boxed{4 \cdot 7}$ $\boxed{+}$ $\boxed{3 \cdot 61}$ $\boxed{)}$ $\boxed{=}$ $\dfrac{2 \cdot 3}{4 \cdot 7 + 3 \cdot 61} = 0 \cdot 2768$ (to four sig. fig.)

(b) $\left(\dfrac{1}{0 \cdot 084}\right)^4$

$\boxed{0 \cdot 084}$ $\boxed{1/x}$ $\boxed{y^x}$ $\boxed{4}$ $\boxed{=}$ $\left(\dfrac{1}{0 \cdot 084}\right)^4 = 20\,090$ (to four sig. fig.)

(c) $\sqrt[3]{[3 \cdot 2 \times (1 \cdot 7 - 1 \cdot 64)]}$

$\boxed{1 \cdot 7}$ $\boxed{-}$ $\boxed{1 \cdot 64}$ $\boxed{=}$ $\boxed{\times}$ $\boxed{3 \cdot 2}$ $\boxed{=}$ $\boxed{y^x}$ $\boxed{0 \cdot 333\,333}$ $\boxed{=}$ $\sqrt[3]{[3 \cdot 2 \times (1 \cdot 7 - 1 \cdot 64)]} = 0 \cdot 5769$ (to four sig. fig.)

Note: To find a cube root, raise to the power $\frac{1}{3}$.
   or as a decimal $0 \cdot 333 \ldots$
   Of course, if your calculator has a $\boxed{\sqrt[3]{}}$ button, use that.

Exercise 6

Use a calculator to evaluate the following, giving the answers to 4 significant figures:

1. $\dfrac{7 \cdot 351 \times 0 \cdot 764}{1 \cdot 847}$

2. $\dfrac{0 \cdot 0741 \times 14700}{0 \cdot 746}$

3. $\dfrac{0 \cdot 0741 \times 9 \cdot 61}{23 \cdot 1}$

4. $\dfrac{417 \cdot 8 \times 0 \cdot 00841}{0 \cdot 07324}$

5. $\dfrac{8 \cdot 41}{7 \cdot 601 \times 0 \cdot 00847}$

6. $\dfrac{4 \cdot 22}{1 \cdot 701 \times 5 \cdot 2}$

7. $\dfrac{9 \cdot 61}{17 \cdot 4 \times 1 \cdot 51}$

8. $\dfrac{8 \cdot 71 \times 3 \cdot 62}{0 \cdot 84}$

9. $\dfrac{0 \cdot 76}{0 \cdot 412 - 0 \cdot 317}$

10. $\dfrac{81 \cdot 4}{72 \cdot 6 + 51 \cdot 92}$

11. $\dfrac{111}{27 \cdot 4 + 2960}$

12. $\dfrac{27 \cdot 4 + 11 \cdot 61}{5 \cdot 9 - 4 \cdot 763}$

13. $\dfrac{6 \cdot 51 - 0 \cdot 1114}{7 \cdot 24 + 1 \cdot 653}$

14. $\dfrac{5 \cdot 71 + 6 \cdot 093}{9 \cdot 05 - 5 \cdot 77}$

15. $\dfrac{0 \cdot 943 - 0 \cdot 788}{1 \cdot 4 - 0 \cdot 766}$

16. $\dfrac{2 \cdot 6}{1 \cdot 7} + \dfrac{1 \cdot 9}{3 \cdot 7}$

17. $\dfrac{8 \cdot 06}{5 \cdot 91} - \dfrac{1 \cdot 594}{1 \cdot 62}$

18. $\dfrac{4 \cdot 7}{11 \cdot 4 - 3 \cdot 61} + \dfrac{1 \cdot 6}{9 \cdot 7}$

19. $\dfrac{3 \cdot 74}{1 \cdot 6 \times 2 \cdot 89} - \dfrac{1}{0 \cdot 741}$

20. $\dfrac{1}{7 \cdot 2} - \dfrac{1}{14 \cdot 6}$

21. $\dfrac{1}{0 \cdot 961} \times \dfrac{1}{0 \cdot 412}$

22. $\dfrac{1}{7} + \dfrac{1}{13} - \dfrac{1}{8}$

23. $4 \cdot 2 \left(\dfrac{1}{5 \cdot 5} - \dfrac{1}{7 \cdot 6} \right)$

24. $\sqrt{(9 \cdot 61 + 0 \cdot 1412)}$

25. $\sqrt{\left(\dfrac{8 \cdot 007}{1 \cdot 61} \right)}$

26. $(1 \cdot 74 + 9 \cdot 611)^2$

27. $\left(\dfrac{1 \cdot 63}{1 \cdot 7 - 0 \cdot 911} \right)^2$

28. $\left(\dfrac{9 \cdot 6}{2 \cdot 4} - \dfrac{1 \cdot 5}{0 \cdot 74} \right)^2$

29. $\sqrt{\left(\dfrac{4 \cdot 2 \times 1 \cdot 611}{9 \cdot 83 \times 1 \cdot 74} \right)}$

30. $(0 \cdot 741)^3$

31. $(1 \cdot 562)^5$

32. $(0 \cdot 32)^3 + (0 \cdot 511)^4$

33. $(1 \cdot 71 - 0 \cdot 863)^6$

34. $\left(\dfrac{1}{0 \cdot 971} \right)^4$

35. $\sqrt[3]{(4 \cdot 714)}$

36. $\sqrt[3]{(0 \cdot 9316)}$

37. $\sqrt[3]{\left(\dfrac{4 \cdot 114}{7 \cdot 93} \right)}$

38. $\sqrt[4]{(0 \cdot 8145 - 0 \cdot 799)}$

39. $\sqrt[5]{(8 \cdot 6 \times 9 \cdot 71)}$

40. $\sqrt[3]{\left(\dfrac{1 \cdot 91}{4 \cdot 2 - 3 \cdot 766} \right)}$

41. $\left(\dfrac{1}{7 \cdot 6} - \dfrac{1}{18 \cdot 5} \right)^3$

42. $\dfrac{\sqrt{(4 \cdot 79)} + 1 \cdot 6}{9 \cdot 63}$

43. $\dfrac{(0 \cdot 761)^2 - \sqrt{(4 \cdot 22)}}{1 \cdot 96}$

44. $\sqrt[3]{\left(\dfrac{1 \cdot 74 \times 0 \cdot 761}{0 \cdot 0896} \right)}$

45. $\left(\dfrac{8 \cdot 6 \times 1 \cdot 71}{0 \cdot 43} \right)^3$

46. $\dfrac{9 \cdot 61 - \sqrt{(9 \cdot 61)}}{9 \cdot 61^2}$

47. $\dfrac{9 \cdot 6 \times 10^4 \times 3 \cdot 75 \times 10^7}{8 \cdot 88 \times 10^6}$

48. $\dfrac{8 \cdot 06 \times 10^{-4}}{1 \cdot 71 \times 10^{-6}}$

49. $\dfrac{3 \cdot 92 \times 10^{-7}}{1 \cdot 884 \times 10^{-11}}$

50. $\left(\dfrac{1 \cdot 31 \times 2 \cdot 71 \times 10^5}{1 \cdot 91 \times 10^4} \right)^5$

51. $\left(\dfrac{1}{9 \cdot 6} - \dfrac{1}{9 \cdot 99} \right)^{10}$

52. $\dfrac{\sqrt[3]{(86 \cdot 6)}}{\sqrt[4]{(4 \cdot 71)}}$

53. $\dfrac{23 \cdot 7 \times 0 \cdot 0042}{12 \cdot 48 - 9 \cdot 7}$

54. $\dfrac{0 \cdot 482 + 1 \cdot 6}{0 \cdot 024 \times 1 \cdot 83}$

55. $\dfrac{8 \cdot 52 - 1 \cdot 004}{0 \cdot 004 - 0 \cdot 0083}$

56. $\dfrac{1 \cdot 6 - 0 \cdot 476}{2 \cdot 398 \times 41 \cdot 2}$

57. $\left(\dfrac{2 \cdot 3}{0 \cdot 791} \right)^7$

58. $\left(\dfrac{8 \cdot 4}{28 \cdot 7 - 0 \cdot 47} \right)^3$

59. $\left(\dfrac{5 \cdot 114}{7 \cdot 332} \right)^5$

60. $\left(\dfrac{4 \cdot 2}{2 \cdot 3} + \dfrac{8 \cdot 2}{0 \cdot 52} \right)^3$

61. $\dfrac{1}{8 \cdot 2^2} - \dfrac{3}{19^2}$

62. $\dfrac{100}{11^3} + \dfrac{100}{12^3}$

63. $\dfrac{7 \cdot 3 - 4 \cdot 291}{2 \cdot 6^2}$

64. $\dfrac{9 \cdot 001 - 8 \cdot 97}{0 \cdot 95^3}$

65. $\dfrac{10 \cdot 1^2 + 9 \cdot 4^2}{9 \cdot 8}$

66. $(3 \cdot 6 \times 10^{-8})^2$

Rounding errors

- A very common error occurs when numbers are rounded in the intermediate steps of a calculation.

Suppose we need to find the area of a circle of circumference 25 cm, correct to 3 significant figures.

$$\text{radius} = \frac{25}{2\pi} = 3.9788736\ldots \qquad \textcircled{1}$$

Suppose we round this number to 3 significant figures.

i.e.   Take radius = 3·98 cm

Then area = $\pi \times 3.98^2$

$$= 49.764084$$

$$= 49.8 \text{ cm}^2 \text{ to 3 s.f.}$$

The *correct* method is to use the radius shown in line $\textcircled{1}$.

Then area = 49·7 cm² to 3 s.f.

Area ?

circumference = 25 cm

- Remember: Avoid *early approximation* in calculations with several steps.

Exercise 7

1. (a) Find the area of a circle of circumference 80 cm, giving your answer correct to 3 significant figures.
   (b) Repeat the calculation but this time round off the radius to 3 significant figures before you find the area. Compare your two answers.

2. Different makes of calculator use different logic chips which may produce slightly different answers.
   Work out $\left(\frac{1}{7}\right)^2 \times 321\,489$ and compare your answer with other people in the class (assuming there are different makes of calculator in the room!).

3. Many calculators have a $\boxed{\text{FIX}}$ mode in which the calculator will round off a number to 1 or 2 or 3 ... decimal places.
   Experiment with your calculator in $\boxed{\text{FIX}}$ mode, working out:

   (a) $7 \div 9$ (b) 11.73×6.947 (c) 0.001×0.007

   Write down what you observe.

Exercise 8

On a calculator work out $9508^2 + 192^2 + 10^2 + 6$.
If you turn the calculator upside down and use a little imagination, you
can see the word 'HEDGEHOG'.
Find the words given by the clues below.

1. $19 \times 20 \times 14 - 2 \cdot 66$ (Not an upstanding man)
2. $(84 + 17) \times 5$ (Dotty message)
3. $904^2 + 89\,621\,818$ (Prickly customer)
4. $(559 \times 6) + (21 \times 55)$ (What a surprise!)
5. $566 \times 711 - 23\,617$ (Bolt it down)

6. $\dfrac{9999 + 319}{8 \cdot 47 + 2 \cdot 53}$ (Sit up and plead)

7. $\dfrac{2601 \times 6}{4^2 + 1^2}$; $(401 - 78) \times 5^2$ (two words) (Not a great man)

8. $0 \cdot 4^2 - 0 \cdot 1^2$ (Little Sidney)

9. $\dfrac{(27 \times 2000 - 2)}{(0 \cdot 63 \div 0 \cdot 09)}$ (Not quite a mountain)

10. $(5^2 - 1^2)^4 - 14239$ (Just a name)

11. $48^4 + 102^2 - 4^2$ (Pursuits)
12. $615^2 + (7 \times 242)$ (Almost a goggle)
13. $(130 \times 135) + (23 \times 3 \times 11 \times 23)$ (Wobbly)
14. $164 \times 166^2 + 734$ (Almost big)
15. $8794^2 + 25 \times 342 \cdot 28 + 120 \times 25$ (Thin skin)
16. $0 \cdot 08 - (3^2 \div 10^4)$ (Ice house)
17. $235^2 - (4 \times 36 \cdot 5)$ (Shiny surface)
18. $(80^2 + 60^2) \times 3 + 81^2 + 12^2 + 3013$ (Ship gunge)
19. $3 \times 17 \times (329^2 + 2 \times 173)$ (Unlimited)
20. $230 \times 230\frac{1}{2} + 30$ (Fit feet)

21. $33 \times 34 \times 35 + 15 \times 3$ (Beleaguer)
22. $0 \cdot 32^2 + \frac{1}{1000}$ (Did he or didn't he?)
23. $(23 \times 24 \times 25 \times 26) + (3 \times 11 \times 10^3) - 20$ (Help)
24. $(16^2 + 16)^2 - (13^2 - 2)$ (Slander)
25. $(3 \times 661)^2 - (3^6 + 22)$ (Pester)
26. $(22^2 + 29 \cdot 4) \times 10$; $(3 \cdot 03^2 - 0 \cdot 02^2) \times 100^2$ (Four words) (Goliath)
27. $1 \cdot 25 \times 0 \cdot 2^6 + 0 \cdot 2^2$ (Tissue time)
28. $(710 + (1823 \times 4)) \times 4$ (Liquor)
29. $(3^3)^2 + 2^2$ (Wriggler)
30. $14 + (5 \times (83^2 + 110))$ (Bigger than a duck)

31. $2 \times 3 \times 53 \times 10^4 + 9$ (Opposite to hello, almost!)
32. $(177 \times 179 \times 182) + (85 \times 86) - 82$ (Good salesman)
33. $14^4 - 627 + 29$ (Good book, by God!)
34. $6 \cdot 2 \times 0 \cdot 987 \times 1\,000\,000 - 860^2 + 118$ (Flying ace)
35. $(426 \times 474) + (318 \times 487) + 22\,018$ (Close to a bubble)

36. $\dfrac{36^3}{4} - 1530$ (Foreign-sounding girl's name)

37. $(7^2 \times 100) + (7 \times 2)$ (Lofty)

38. $240^2 + 134;\ 241^2 - 7^3$ (two words)   (Devil of a chime)

39. $(2 \times 2 \times 2 \times 2 \times 3)^4 + 1929$ (Unhappy ending)

40. $141\,918 + 83^3$ (Hot stuff in France)

2.3 Standard form

When dealing with either very large or very small numbers, it is not convenient to write them out in full in the normal way. It is better to use standard form. Most calculators represent large and small numbers in this way:

The number $a \times 10^n$ is in standard form when $1 \leqslant a < 10$ and n is a positive or negative integer.

e.g.   This calculator shows $2 \cdot 3 \times 10^8$.

$$2.3 \quad {}^{08}$$

Write the following numbers in standard form:

(a) $2000 = 2 \times 1000 = 2 \times 10^3$

(b) $150 = 1 \cdot 5 \times 100 = 1 \cdot 5 \times 10^2$

(c) $0 \cdot 0004 = 4 \times \dfrac{1}{10\,000} = 4 \times 10^{-4}$

(d) $450 \times 10^6 = (4 \cdot 5 \times 10^2) \times 10^6$
$$= 4 \cdot 5 \times 10^8$$

(e) $0 \cdot 07 \times 10^{-4} = (7 \times 10^{-2}) \times 10^{-4}$
$$= 7 \times 10^{-6}$$

Exercise 9

Write the following numbers in standard form:

1. 4000	**2.** 500	**3.** 70 000
4. 60	**5.** 2400	**6.** 380
7. 46 000	**8.** 46	**9.** 900 000
10. 2560	**11.** 0·007	**12.** 0·0004
13. 0·0035	**14.** 0·421	**15.** 0·000 055
16. 0·01	**17.** 564 000	**18.** 19 million

19. The population of China is estimated at $1100\,000\,000$. Write this in standard form.

20. A hydrogen atom has a mass of $0 \cdot 000\,000\,000\,000\,000\,000\,000\,001\,67$ grams. Write this mass in standard form.

21. The area of the surface of the earth is about $510\,000\,000\,\text{km}^2$. Express this in standard form.

22. A certain virus is $0\cdot000\,000\,000\,25\,\text{cm}$ in diameter. Write this in standard form.

23. Avogadro's number is $602\,300\,000\,000\,000\,000\,000\,000$. Express this in standard form.

24. The speed of light is $300\,000\,\text{km/s}$. Express this speed in cm/s in standard form.

25. A very rich oil sheikh leaves his fortune of $£3\cdot6 \times 10^8$ to be divided between 100 relatives. How much does each relative receive? Give the answer in standard form.

26. If $a = 512 \times 10^2$
  $b = 0\cdot478 \times 10^6$
  $c = 0\cdot0049 \times 10^7$
arrange a, b and c in order of size (smallest first).

27. If the number $2\cdot74 \times 10^{15}$ is written out in full, how many zeros follow the 4?

28. If the number $7\cdot31 \times 10^{-17}$ is written out in full, how many zeros would there be between the decimal point and the first significant figure?

29. To work out $(3 \times 10^2) \times (2 \times 10^5)$, *without* a calculator:

A Multiply: $3 \times 2 = 6$
B Add the powers of 10: $10^2 \times 10^5 = 10^7$
   So $(3 \times 10^2) \times (2 \times 10^5) = 6 \times 10^7$

Without a calculator, work out:
(a) $(2 \times 10^8) \times (4 \times 10^3)$ (b) $(2\cdot5 \times 10^3) \times (2 \times 10^{-10})$
(c) $(1\cdot1 \times 10^{-6}) \times (5 \times 10^2)$ (d) $(8 \times 10^{12}) \div (2 \times 10^3)$
(e) $(9 \times 10^6) \div (3 \times 10^{-2})$ (f) $(2 \times 10^6)^2$

30. Write out these numbers in full.
(a) $2\cdot3 \times 10^4$ (b) 3×10^{-2} (c) $5\cdot6 \times 10^2$
(d) 8×10^5 (e) $2\cdot2 \times 10^{-3}$ (f) 9×10^8
(g) 6×10^{-1} (h) 7×10^3 (i) $3\cdot14 \times 10^6$

31. Without a calculator, work out the following and give your answers in standard form.
(a) $(6 \times 10^5) + (5 \times 10^4)$ (b) $(3 \times 10^3) - (3 \times 10^2)$
(c) $(3\cdot5 \times 10^6) + (4 \times 10^4)$ (d) $(8 \times 10^{-3}) + (2 \times 10^{-2})$
(e) $(6\cdot2 \times 10^{-3}) + (8 \times 10^{-4})$ (f) $(7 \times 10^{22}) + (7 \times 10^{20})$

32. Rewrite the following numbers in standard form.
(a) $0\cdot6 \times 10^4$ (b) 11×10^{-2}
(c) 450×10^6 (d) $0\cdot085 \times 10^{-4}$
(e) 6 billion   (f) $\frac{1}{2}$

Using a calculator

(a) Work out $(5 \times 10^7) \times (3 \times 10^{12})$

Use the $\boxed{\text{EXP}}$ button as follows:

$\boxed{5}$ $\boxed{\text{EXP}}$ $\boxed{7}$ $\boxed{\times}$ $\boxed{3}$ $\boxed{\text{EXP}}$ $\boxed{12}$ $\boxed{=}$ Answer $= 1 \cdot 5 \times 10^{20}$

Notice that you do NOT press the $\boxed{\times}$ button after the $\boxed{\text{EXP}}$ button!

(b) Work out $(3 \cdot 2 \times 10^3) \div (8 \times 10^{-7})$

$\boxed{3 \cdot 2}$ $\boxed{\text{EXP}}$ $\boxed{3}$ $\boxed{\div}$ $\boxed{8}$ $\boxed{\text{EXP}}$ $\boxed{7}$ $\boxed{+/-}$ $\boxed{=}$ Answer $= 4 \times 10^9$

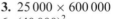

[Press this key for negative.]

(c) $(5 \times 10^{-8})^2$

$\boxed{5}$ $\boxed{\text{EXP}}$ $\boxed{8}$ $\boxed{+/-}$ $\boxed{x^2}$ Answer $= 2 \cdot 5 \times 10^{-15}$

Exercise 10

Use a calculator to work out the following and write the answer in standard form.

1. $2000 \times 30\,000$

2. $40\,000 \times 500$

3. $25\,000 \times 600\,000$

4. $3500 \times 2\,\text{million}$

5. $600\,000 \times 1500$

6. $(40\,000)^2$

7. $18\,000 \div 400$

8. $(4 \times 10^5) \times (3 \times 10^8)$

9. $(6 \cdot 2 \times 10^4) \times (3 \times 10^6)$

10. $(5 \times 10^{-4}) \times (4 \times 10^{-7})$

11. $(3 \times 10^8) \times (2 \cdot 5 \times 10^{-20})$

12. $(5 \times 10^7) \div (2 \times 10^{-2})$

13. $(3 \times 10^5)^3$

14. $(7 \times 10^{-2}) \div (2 \times 10^4)$

15. $(9 \times 10^{-11})^2$

16. $(4 \cdot 2 \times 10^8) \times (1 \cdot 5 \times 10^5)$

17. $(3 \times 10^{-4}) \div (2 \times 10^{-20})$

18. $(4 \times 10^5) \times (2 \times 10^{-2})$

19. $(1 \cdot 4 \times 10^{-1}) \times (2 \times 10^{17})$

20. $(2 \times 10^5) \times (4 \times 10^{-8})$

21. $(8 \times 10^{-5}) \div (2 \times 10^{-3})^2$

22. $(2 \times 10^{-4})^3 \div (1 \cdot 6 \times 10^8)$

23. $10^5 \div (2 \times 10^8)$

24. $(3 \times 10^{-7}) \times 10^{-4}$

25. A micrometre is 10^{-6} metres.
Write 200 micrometres in metres. [Answer in standard form]

26. A patient in hospital is very ill. Between noon and midnight one day the number of viruses in her body increased from 3×10^7 to $5 \cdot 6 \times 10^9$.
Work out the increase in the number of viruses, giving your answer in standard form.

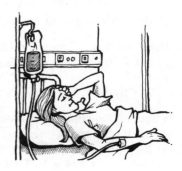

27.

Continent	Population	Area (m^2)
Europe	$6 \cdot 82 \times 10^8$	$1 \cdot 05 \times 10^{10}$
Asia	$2 \cdot 96 \times 10^9$	$4 \cdot 35 \times 10^{10}$

$$\text{Population density} = \frac{\text{Population}}{\text{Area}}$$

Which of these two continents has the larger population density?

28. Here is a statement about roads:

> *At a constant 80 kilometres per hour it would take 35 years*
> *to travel over the entire road network of the world.*

Taking a year as 365 days, calculate the length of the road network of the world, giving your answer in standard form.

29. Oil flows through a pipe at a rate of $40 \, m^3/s$. How long will it take to fill a tank of volume $1 \cdot 2 \times 10^5 \, m^3$?

30. Given that $L = 2\sqrt{\dfrac{a}{k}}$, find the value of L in standard form when

$a = 4 \cdot 5 \times 10^{12}$ and $k = 5 \times 10^7$.

31. The mean weight of all the men, women, children and babies in the UK is 42·1 kg. The population of the UK is 56 million. Work out the total weight of the entire population giving your answer in kg in standard form. Give your answer to a sensible degree of accuracy.

32. A light year is the distance travelled by a beam of light in a year. Light travels at a speed of approximately $3 \times 10^5 \, km/s$.
(a) Work out the length of a light year in km.
(b) Light takes about 8 minutes to reach the Earth from the Sun. How far is the Earth from the Sun in km?

†**33.** (a) The number 10 to the power 100 (10 000 sexdecillion) is called a 'Googol'! Write 60 googols in standard form.
If it takes $\frac{1}{5}$ second to write a zero and $\frac{1}{10}$ second to write a 'one', how long would it take to write the number 100 'Googols' in full?
(b) The number 10 to the power of a 'Googol' is called a 'Googolplex'. Using the same speed of writing, how long in years would it take to write 1 'Googolplex' in full? You may assume that your pen has enough ink.

2.4 Rational and irrational numbers

- A rational number can always be written exactly in the form $\dfrac{a}{b}$ where a and b are whole numbers.

$\frac{3}{7}$	$1\frac{1}{2} = \frac{3}{2}$	$5\cdot14 = \frac{257}{50}$	$0\cdot\dot{6} = \frac{2}{3}$

All these are rational numbers.

- An irrational number cannot be written in the form $\dfrac{a}{b}$.
  $\sqrt{2}, \sqrt{5}, \pi, \sqrt[3]{2}$ are all irrational numbers.
- In general $\sqrt{n}$ is irrational unless n is a square number.

In this triangle the length of the hypotenuse is *exactly* $\sqrt{5}$. On a calculator, $\sqrt{5} = 2 \cdot 236068$. This value of $\sqrt{5}$ is *not* exact and is correct only to 6 decimal places.

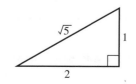

The recurring decimal $0 \cdot 631\,631\,631 \ldots$ can be written in the form $\dfrac{a}{b}$.

Let $r = 0 \cdot 631\,631\,631 \ldots$

Multiply by 1000, $1000\,r = 631 \cdot 631\,631\,631 \ldots$

Subtract, $999\,r = 631$

$\therefore \qquad\qquad\qquad r = \dfrac{631}{999}$ so the number is rational.

Exercise 11

1. Which of the following numbers are rational?

$\dfrac{\pi}{2}$ $\sqrt{5}$ $(\sqrt{17})^2$ $\sqrt{3}$

$3 \cdot 14$ $\dfrac{\sqrt{12}}{\sqrt{3}}$ π^2 $3^{-1} + 3^{-2}$

$7^{-\frac{1}{2}}$ $\dfrac{22}{7}$ $\sqrt{2} + 1$ $\sqrt{2 \cdot 25}$

2. (a) Write down any rational number between 4 and 6.
(b) Write down any irrational number between 4 and 6.
(c) Find a rational number between $\sqrt{2}$ and $\sqrt{3}$.
(d) Write down any rational number between π and $\sqrt{10}$.

3. (a) For each triangle use Pythagoras' theorem to calculate the length x.
(b) For each triangle state whether the *perimeter* is rational or irrational.
(c) For each triangle state whether the *area* is rational or irrational.
(d) In which triangle is $\sin \theta$ an irrational number?

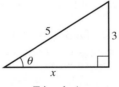

Triangle A

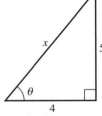

Triangle B

4. The diagram shows a circle of radius 3 cm drawn inside a square. Write down the exact value of the following and state whether the answer is rational or not:
(a) the circumference of the circle   (b) the diameter of the circle
(c) the area of the square   (d) the area of the circle
(e) the shaded area.

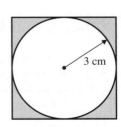

3 cm

5. Write in the form $\dfrac{a}{b}$.

(a) $0 \cdot \dot{2}$ (b) $0 \cdot \dot{2}\dot{9}$ (c) $0 \cdot 5\dot{4}\dot{1}$

6. Think of two *irrational* numbers x and y such that $\dfrac{x}{y}$ is a *rational* number.

7. Explain the difference between a rational number and an irrational number.

8. (a) Is it possible to multiply a rational number by an irrational number to give an answer which is rational?

(b) Is it possible to multiply two irrational numbers together to give a rational answer?

(c) If either or both are possible, give an example.

9. (a) Write down two rational numbers, both between 0 and 1, one of which is equal to a non-recurring decimal and other is equal to a recurring decimal.

(b) Add together the two numbers you found above and show that the answer is rational.

10. n is a positive integer such that $\sqrt{n} = 8.6$, correct to 1 decimal place.

(a) Find the value of n.

(b) State whether $\sqrt{n}$ is rational or irrational, giving your reason.

11. The number N is rational and is not zero. Decide whether $\dfrac{1}{N}$ is rational or irrational.

12. Write down a number between 5 and 6 that has a rational square root.

13. Adding or subtracting two irrational numbers usually gives an irrational result. The result can, however, be rational, for example:

$$(\sqrt{3} + 2) + (5 - \sqrt{3}) = 7$$

or $(3 - \sqrt{2}) + \sqrt{2} = 3$

Find two examples of adding two irrational numbers so that the result is a rational number.

14. Find an irrational number which, when multiplied by the number below, gives a rational number.

(a) $\sqrt{7}$ (b) $\dfrac{1}{\sqrt{3}}$ (c) $5\sqrt{2}$ (d) $\dfrac{4}{\pi}$

Surds

- Numbers like $\sqrt{4}$, $\sqrt{3}$, $\sqrt{7}$ are called *surds*. The following rules apply.

- $\sqrt{a} \times \sqrt{b} = \sqrt{ab}$

- $\dfrac{\sqrt{a}}{\sqrt{b}} = \sqrt{\dfrac{a}{b}}$

(a) $\sqrt{2} \times \sqrt{3} = \sqrt{6}$

(b) $\sqrt{80} = \sqrt{16 \times 5}$

$\qquad = \sqrt{16} \times \sqrt{5}$

$\qquad = 4\sqrt{5}$

$\dfrac{\sqrt{27}}{\sqrt{3}} = \sqrt{\dfrac{27}{3}}$

$\qquad = \sqrt{9} = 3$

- The fraction $\dfrac{6}{\sqrt{3}}$ can be written with a rational denominator by multiplying numerator and denominator by $\sqrt{3}$:

$$\frac{6}{\sqrt{3}} = \frac{6 \times \sqrt{3}}{\sqrt{3} \times \sqrt{3}} = \frac{6\sqrt{3}}{3} = 2\sqrt{3}$$

- A *common mistake* occurs with surds.
  $\sqrt{4}$ means 'the positive square root of 4'
  So $\sqrt{4} = 2$ only and *not* ± 2.

  Notice that the solutions of the equation $x^2 = 4$ are $x = 2, -2$
  You *can* write $x = \sqrt[+]{4}$ or $\sqrt[-]{4}$.

Exercise 12

1. Sort these into four pairs of equal value.

   A $\sqrt{20}$ B $\sqrt{12}$ C $\sqrt{3}$ D $2\sqrt{3}$

   E 2   F $\sqrt{10} \times \sqrt{2}$ G $\frac{\sqrt{20}}{\sqrt{5}}$ H $\frac{\sqrt{15}}{\sqrt{5}}$

2. Answer 'True' or 'False'.

   (a) $\sqrt{28} = 2\sqrt{7}$ (b) $(\sqrt{8})^2 = 8$ (c) $\sqrt{16} = \pm 4$

   (d) $\sqrt{\frac{39}{3}} = \sqrt{13}$ (e) $\sqrt{4} + \sqrt{4} = \sqrt{8}$ (f) If $x^2 = 9$, $x = \pm 3$

3. Remove the brackets and simplify.

   (a) $(1 + \sqrt{2})^2$ (b) $(2 - \sqrt{3})^2$ (c) $(\sqrt{2} + \sqrt{8})^2$

   (d) $(3 - \sqrt{5})^2$ (e) $(\sqrt{2} + 2)^2$ (f) $(\sqrt{18} - \sqrt{2})^2$

4. Answer 'True' or 'False'.

   (a) $\sqrt{2} + \sqrt{2} = 2\sqrt{2}$ (b) $\frac{\sqrt{8}}{2} = \sqrt{2}$

   (c) $\sqrt{100} = \pm 10$ (d) $(\sqrt{2})^4 = 4$

   (e) $\sqrt{9 + 16} = \sqrt{9} + \sqrt{16}$ (f) $(1 + \sqrt{3})^2 = 4 + 2\sqrt{3}$

5. Without using a calculator, simplify the following. Write your
   answers using surds where necessary.

   (a) $\sqrt{8} \times \sqrt{2}$ (b) $\sqrt{32}$ (c) $\sqrt{300}$

   (d) $\sqrt{18} \times 3$ (e) $\sqrt{5} + 4\sqrt{5}$ (f) $7\sqrt{3} - 2\sqrt{3}$

   (g) $\sqrt{20} + \sqrt{45}$ (h) $\sqrt{75} - \sqrt{48}$ (i) $\frac{\sqrt{8}}{\sqrt{2}}$

   (j) $\frac{\sqrt{27}}{\sqrt{12}}$ (k) $\frac{\sqrt{125}}{\sqrt{20}}$ (l) $\frac{\sqrt{80}}{\sqrt{45}}$

6. (a) Work out $\frac{6}{\sqrt{3}} \times \frac{\sqrt{3}}{\sqrt{3}}$.

   (b) Rationalise the denominators of these fractions.

      (i) $\frac{8}{\sqrt{2}}$ (ii) $\frac{12}{\sqrt{3}}$ (iii) $\frac{12}{\sqrt{8}}$ (iv) $\frac{2}{\sqrt{3}}$

†7. The denominator of the fraction $\dfrac{4}{(\sqrt{2} + 1)}$ can be rationalised by
   multiplying numerator and denominator by $(\sqrt{2} - 1)$.

   (a) Work out $\dfrac{4}{(\sqrt{2} + 1)} \times \dfrac{(\sqrt{2} - 1)}{(\sqrt{2} - 1)}$.

   (b) Simplify:

      (i) $\dfrac{4}{(\sqrt{3} + 1)}$ (ii) $\dfrac{1}{(\sqrt{5} + 2)}$ (iii) $\dfrac{10}{(\sqrt{7} - 2)}$

2.5 Substitution

Expressions

In general an expression contains algebraic terms and numbers but there is *no equals sign*.

Here are some expressions: $3x^2$, $2x(x-1)$, $\dfrac{x+1}{x}$, $y^3 - 3y^2$

When $a = 3$, $b = -2$, $c = 5$, find the value of each expression.

(a) $3a + b = (3 \times 3) + (-2)$ (b) $ac + b^2 = (3 \times 5) + (-2)^2$ (c) $a(c - b)$
$\qquad\quad = 9 - 2$ $\qquad\qquad = 15 + 4$ $\qquad\quad = 3[5 - (-2)]$
$\qquad\quad = 7$ $\qquad\qquad = 19$ $\qquad\quad = 3[7]$
$\qquad\qquad\qquad\qquad\qquad\qquad\qquad\qquad\qquad\qquad\qquad\qquad = 21$

Notice that working *down* the page is often easier to follow.

Exercise 13

Evaluate the following expressions.
For Questions **1** to **12** $a = 3$, $c = 2$, $e = 5$.

1. $3a - 2$ **2.** $4c + e$ **3.** $2c + 3a$ **4.** $5e - a$

5. $e - 2c$ **6.** $e - 2a$ **7.** $4c + 2e$ **8.** $7a - 5e$

9. $c - e$ **10.** $10a + c + e$ **11.** $a + c - e$ **12.** $a - c - e$

For Questions **13** to **24** $h = 3$, $m = -2$, $t = -3$.

13. $2m - 3$ **14.** $4t + 10$ **15.** $3h - 12$ **16.** $6m + 4$

17. $9t - 3$ **18.** $4h + 4$ **19.** $2m - 6$ **20.** $m + 2$

21. $3h + m$ **22.** $t - h$ **23.** $4m + 2h$ **24.** $3t - m$

For Questions **25** to **36** $x = -2$, $y = -1$, $k = 0$.

25. $3x + 1$ **26.** $2y + 5$ **27.** $6k + 4$ **28.** $3x + 2y$

29. $2k + x$ **30.** xy **31.** xk **32.** $2xy$

33. $2(x + k)$ **34.** $3(k + y)$ **35.** $5x - y$ **36.** $3k - 2x$

$2x^2$ means $2(x^2)$
$(2x)^2$ means 'work out $2x$ and *then* square it'
$-7x$ means $-7(x)$
$-x^2$ means $-1(x^2)$

When $x = -2$, find the value of:

(a) $2x^2 - 5x = 2(-2)^2 - 5(-2)$ (b) $(3x)^2 - x^2 = (3 \times -2)^2 - 1(-2)^2$
$\qquad\qquad\quad = 2(4) + 10$ $\qquad\qquad\qquad = (-6)^2 - 1(4)$
$\qquad\qquad\quad = 18$ $\qquad\qquad\qquad = 36 - 4$
$\qquad\qquad\qquad\qquad\qquad\qquad\qquad\qquad\qquad = 32$

Exercise 14

If $x = -3$ and $y = 2$, evaluate the following expressions.

1. x^2 2. $3x^2$ 3. y^2 4. $4y^2$ 5. $(2x)^2$
6. $2x^2$ 7. $10 - x^2$ 8. $10 - y^2$ 9. $20 - 2x^2$ 10. $20 - 3y^2$
11. $5 + 4x$ 12. $x^2 - 2x$ 13. $y^2 - 3x^2$ 14. $x^2 - 3y$ 15. $(2x)^2 - y^2$
16. $4x^2$ 17. $(4x)^2$ 18. $1 - x^2$ 19. $y - x^2$ 20. $x^2 + y^2$
21. $x^2 - y^2$ 22. $2 - 2x^2$ 23. $(3x)^2 + 3$ 24. $11 - xy$ 25. $12 + xy$
26. $(2x)^2 - (3y)^2$ 27. $2 - 3x^2$ 28. $y^2 - x^2$

When $a = -2$, $b = 3$, $c = -3$, evaluate: (a) $\dfrac{2a(b^2 - a)}{c}$ (b) $\sqrt{(a^2 + b^2)}$

(a) $(b^2 - a) = 9 - (-2)$
$\quad\quad\quad\quad = 11$

$\therefore \quad \dfrac{2a(b^2 - a)}{c} = \dfrac{2 \times (-2) \times (11)}{-3}$

$\quad\quad\quad\quad\quad = 14\frac{2}{3}$

(b) $a^2 + b^2 = (-2)^2 + (3)^2$
$\quad\quad\quad\quad = 4 + 9$
$\quad\quad\quad\quad = 13$

$\therefore \quad \sqrt{(a^2 + b^2)} = \sqrt{13}$

Notice that $\sqrt{13}$ means 'the positive square root of 13' *not* $\pm\sqrt{13}$

Exercise 15

Evaluate the following:
In Questions **1** to **16**, $a = 4$, $b = -2$, $c = -3$.

1. $a(b + c)$ 2. $a^2(b - c)$ 3. $2c(a - c)$ 4. $b^2(2a + 3c)$
5. $c^2(b - 2a)$ 6. $2a^2(b + c)$ 7. $2(a + b + c)$ 8. $3c(a - b - c)$
9. $b^2 + 2b + a$ 10. $c^2 - 3c + a$ 11. $2b^2 - 3b$ 12. $\sqrt{(a^2 + c^2)}$

13. $\sqrt{(ab + c^2)}$ 14. $\sqrt{(c^2 - b^2)}$ 15. $\dfrac{b^2}{a} + \dfrac{2c}{b}$ 16. $\dfrac{c^2}{b} + \dfrac{4b}{a}$

In Questions **17** to **32**, $k = -3$, $m = 1$, $n = -4$.

17. $k^2(2m - n)$ 18. $5m\sqrt{(k^2 + n^2)}$ 19. $\sqrt{(kn + 4m)}$ 20. $kmn(k^2 + m^2 + n^2)$
21. $k^2m^2(m - n)$ 22. $k^2 - 3k + 4$ 23. $m^3 + m^2 + n^2 + n$ 24. $k^3 + 3k$
25. $m(k^2 - n^2)$ 26. $m\sqrt{(k - n)}$ 27. $100k^2 + m$ 28. $m^2(2k^2 - 3n^2)$
29. $\dfrac{2k + m}{k - n}$ 30. $\dfrac{kn - k}{2m}$ 31. $\dfrac{3k + 2m}{2n - 3k}$ 32. $\dfrac{k + m + n}{k^2 + m^2 + n^2}$

33. Find $K = \sqrt{\left(\dfrac{a^2 + b^2 + c^2 - 2c}{a^2 + b^2 + 4c}\right)}$

    if $a = 3$, $b = -2$, $c = -1$.

34. Find $W = \dfrac{kmn(k + m + n)}{(k + m)(k + n)}$

    if $k = \frac{1}{2}$, $m = -\frac{1}{3}$, $n = \frac{1}{4}$.

Formulas

When a calculation is repeated many times it is often helpful to use a formula.

When a building society offers a mortgage it may use a formula like '$2\frac{1}{2}$ times the main salary plus the second salary'.

Publishers use a formula to work out the selling price of a book based on the production costs and the expected sales of the book.

In a formula letters stand for defined quantities or variables.

Exercise 16

1. The final speed v of a car is given by the formula $v = u + at$. [u = initial speed, a = acceleration, t = time taken]. Find v when $u = 15$, $a = 0.2$, $t = 30$.

2. The time period T of a simple pendulum is given by the formula

   $T = 2\pi\sqrt{\left(\dfrac{l}{g}\right)}$, where l is the length of the pendulum and g is

   the gravitational acceleration.
   Find T when $l = 0.65$, $g = 9.81$ and $\pi = 3.142$.

3. The total surface area A of a cone is related to the radius r and the slant height l by the formula $A = \pi r (r + l)$.
   Find A when $r = 7$ and $l = 11$.

4. The sum S of the squares of the integers from 1 to n is given by $S = \frac{1}{6}n(n+1)(2n+1)$. Find S when $n = 12$.

5. The acceleration a of a train is found using the formula

   $a = \dfrac{v^2 - u^2}{2s}$. Find a when $v = 20$, $u = 9$ and $s = 2.5$.

6. Einstein's famous equation relating energy, mass and the speed of light is $E = mc^2$. Find E when $m = 0.0001$ and $c = 3 \times 10^8$.

7. The area A of a parallelogram with sides a and b is given by $A = ab \sin \theta$, where θ is the angle between the sides. Find A when $a = 7$, $b = 3$ and $\theta = 30°$.

8. The distance s travelled by an accelerating rocket is given by $s = ut + \frac{1}{2}at^2$. Find s when $u = 3$, $t = 100$ and $a = 0.1$.

9. The formula for the velocity of sound in air is $V = 72\sqrt{T + 273}$ where V is the velocity in km/h and T is the temperature of the air in °C.
   (a) Find the velocity of sound where the temperature is 26°C.
   (b) Find the temperature if the velocity of sound is 1200 km/h.
   (c) Find the velocity of sound where the temperature is −77°C.

10. (a) Find a formula for the area, A, of the shape opposite, in terms of a, b and c.
    (b) Find the value of A when $a = 2$, $b = 7$ and $c = 10$.

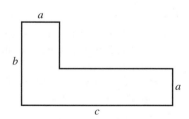

11. (a) Find a formula for the shaded part, S, below, in terms of p, q and r.

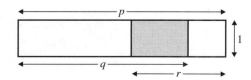

    (b) Find the value of S when $r = 8$, $p = 12.1$ and $q = 8.2$.

2.6 Compound measures

Speed, distance and time

Calculations involving these three quantities are simpler when the speed is *constant*. The formulae connecting the quantities are as follows:

(a) distance = speed × time

(b) speed = $\dfrac{\text{distance}}{\text{time}}$

(c) time = $\dfrac{\text{distance}}{\text{speed}}$

A helpful way of remembering these formulae is to write the letters D, S and T in a triangle,

thus: to find D, cover D and we have ST.

to find S, cover S and we have $\dfrac{D}{T}$

to find T, cover T and we have $\dfrac{D}{S}$

Great care must be taken with the units in these questions.

(a) A man is running at a speed of 8 km/h for a distance of 5200 metres. Find the time taken in minutes.

$$5200 \text{ metres} = 5\cdot2 \text{ km}$$

$$\text{time taken in hours} = \left(\frac{D}{S}\right) = \frac{5\cdot2}{8}$$

$$= 0\cdot65 \text{ hours}$$

$$\text{time taken in minutes} = 0\cdot65 \times 60$$

$$= 39 \text{ minutes}$$

(b) Change the units of a speed of 54 km/h into metres per second.

$$54 \text{ km/hour} = 54\,000 \text{ metres/hour}$$

$$= \frac{54\,000}{60} \text{ metres/minute}$$

$$= \frac{54\,000}{60 \times 60} \text{ metres/second}$$

$$= 15 \text{ m/s}$$

Exercise 17

1. Find the time taken for the following journeys:
   (a) 100 km at a speed of 40 km/h
   (b) 250 miles at a speed of 80 miles per hour
   (c) 15 metres at a speed of 20 cm/s (answer in seconds)
   (d) 10^4 metres at a speed of 2·5 km/h

2. A grand prix driver completed a lap of 5·2 km in exactly 2 minutes.
   What was his average speed in km/h?
   [Hint: change 2 minutes into hours.]

3. Find the speeds of the bodies which move as follows:
   (a) a distance of 600 km in 8 hours
   (b) a distance of 4×10^4 m in 10^{-2} seconds
   (c) a distance of 500 m in 10 minutes (in km/h)

4. Find the distance travelled (in metres) in the following:
   (a) at a speed of 40 km/h for $\frac{1}{4}$ hour
   (b) at a speed of 338·4 km/h for 10 minutes
   (c) at a speed of 15 m/s for 5 minutes
   (d) at a speed of 14 m/s for 1 hour

5. A car travels 60 km at 30 km/h and then a further 180 km at
   160 km/h. Find:
   (a) the total time taken
   (b) the average speed for the whole journey

6. A cyclist travels 25 kilometres at 20 km/h and then a further 80
   kilometres at 25 km/h.

Find:
   (a) the total time taken
   (b) the average speed for the whole journey.

7. Sebastian Coe ran two laps around a 400 m track. He completed the
   first lap in 50 seconds and then decreased his speed by 5% for the
   second lap. Find:
   (a) his speed on the first lap
   (b) his speed on the second lap
   (c) his total time for the two laps.

8. The airliner Concorde flies 2000 km at a speed of 1600 km/h and
   then returns due to bad weather at a speed of 1000 km/h. Find the
   average speed for the whole trip.

9. A train travels from A to B, a distance of 100 km, at a speed of
   20 km/h. If it had gone two and a half times as fast, how much
   earlier would it have arrived at B?

10. Two men running towards each other at 4 m/s and 6 m/s
   respectively are one kilometre apart. How long will it take before
   they meet?

11. A car travelling at 90 km/h is 500 m behind another car travelling at
   70 km/h in the same direction. How long will it take the first car to
   catch the second?

12. How long is a train which passes a signal in twenty seconds at a
   speed of 108 km/h?

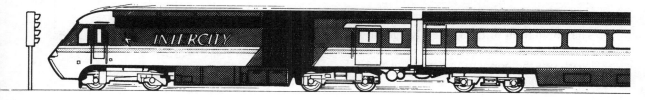

13. A train of length 180 m approaches a tunnel of length 620 m.
    How long will it take the train to pass completely through the
    tunnel at a speed of 54 km/h?

14. An earthworm of length 15 cm is crawling along at 2 cm/s. An ant
    overtakes the worm in 5 seconds. How fast is the ant walking?

15. A train of length 100 m is moving at a speed of 50 km/h.
    A horse is running alongside the train at a speed of 56 km/h.
    How long will it take the horse to overtake the train?

16. A car completes a journey at an average speed of 40 m.p.h.
    At what speed must it travel on the return journey if the average
    speed for the complete journey (out and back) is 60 m.p.h.?

Density and other compound measures

● The density of a material is its mass per unit volume.

It is helpful to use an 'MDV' triangle similar to the one used for speed.

$$M = D \times V \qquad D = \frac{M}{V} \qquad V = \frac{M}{D}$$

Exercise 18

1. The volume of a solid object is 5 cm^3 and its mass is 50 g.
   Calculate the density of the material.

2. A statue is made of metal
   of density 8 g/cm^3.
   Find the mass of the statue
   if its volume is 30 cm^3.

3. Water of density 1000 kg/m^3 fills a pool.
   The mass of the water is 35 tonnes.
   Calculate the volume of the pool. [1 tonne = 1000 kg]

4. The solid cuboid shown is made of metal
   with density 10 g/cm^3. Calculate the mass of
   the cuboid, giving your answer in kg.

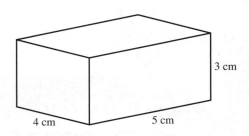

3 cm

4 cm   5 cm

5. The population of the United Kingdom is about
   60 million and the land area is about 240 000 km^2.
   Work out the population density of the U.K.
   [number of people/km^2]

6. Hong Kong has a population of about 6 million and land area $1000\,km^2$. The U.S.A. has a land area of about $9\,300\,000\,km^2$.
   (a) Work out the population density of Hong Kong.
   (b) Work out what the population of the United States would be if it was as densely populated as Hong Kong.

7. The picture shows the rate at which grass seed is sown. How much seed is needed for a field of area 2 hectares? [1 hectare $= 10\,000\,m^2$]

8. A certain plastic has a density of $3\,g/cm^3$. Convert the density into kg/m^3.

9. Oil has a density of $800\,kg/m^3$. Convert this density into g/cm^3.

10. Change the units of the following speeds as indicated:
   (a) $72\,km/h$ into m/s
   (b) $30\,m/s$ into km/h
   (c) $0\cdot012\,m/s$ into cm/s
   (d) $9000\,cm/s$ into m/s

2.7 Measurements and bounds

Measurement is approximate

(a) A length of some cloth is measured for a dress. You might say the length is 145 cm to the nearest cm.

   The actual length could be anything from 144·5 cm to 145·49999 . . . cm using the normal convention which is to round up a figure of 5 or more. Clearly 145·4999 . . . is effectively 145·5 and we say the *upper bound* is 145·5.

   The *lower bound* is 144·5.

   As an inequality we can write $144\cdot5 \leqslant \text{length} < 145\cdot5$

The upper limit often causes confusion. We use 145·5 as the upper bound simply because it is *inconvenient* to work with 145·49999...

(b) When measuring the length of a page in a book, you might say the length is 437 mm to the nearest mm.

   In this case the actual length could be anywhere from 436·5 mm to 437·5 mm. We write 'length is between 436·5 mm and 437·5 mm'.

● In both cases (a) and (b), the measurement expressed to a given unit is in *possible error* of *half* a unit.

(c) (i) Similarly if you say your weight is 57 kg to the nearest kg, you could actually weigh anything from 56·5 kg to 57·5 kg.

> The 'unit' is 1 so 'half a unit' is 0·5.

(ii) If your brother was weighed on more sensitive scales and the result was 57·2 kg, his actual weight could be from 57·15 kg to 57·25 kg.

> The 'unit' is 0·1 so 'half a unit' is 0·05.

(iii) The weight of a butterfly might be given as 0·032 g. The actual weight could be from 0·0315 g to 0·0325 g.

> The 'unit' is 0·001 so 'half a unit' is 0·0005.

Here are some further examples:

Measurement	Lower bound	Upper bound
The diameter of a CD is 12 cm to the nearest cm.	11·5 cm	12·5 cm
The mass of a coin is 6·2 g to the nearest 0·1 g.	6·15 g	6·25 g
The length of a fence is 330 m to the nearest 10 m.	325 m	335 m

Exercise 19

1. In a DIY store the height of a door is given as 195 cm to the nearest cm. Write down the upper bound for the height of the door.

2. A vet weighs a sick goat at 37 kg to the nearest kg. What is the least possible weight of the goat?

3. A cook's weighing scales weigh to the nearest 0·1 kg. What is the upper bound for the weight of a chicken which she weighs at 3·2 kg?

4. A surveyor using a laser beam device can measure distances to the nearest 0·1 m. What is the least possible length of a warehouse which he measures at 95·6 m?

5. In the county sports Jill was timed at 28·6 s for the 200 m. What is the upper bound for the time she could have taken?

6. Copy and complete the table.

	Measurement	Lower bound	Upper bound
(a)	temperature in a fridge = 2°C to the nearest degree		
(b)	mass of an acorn = 2·3 g to 1 d.p.		
(c)	length of telephone cable = 64 m to nearest m		
(d)	time taken to run 100 m = 13·6 s to nearest 0·1 s		

7. The length of a telephone is measured as 193 mm, to the nearest mm. The length lies between:

A	B	C
192 and 194 mm	192·5 and 193·5 mm	188 and 198 mm

8. The weight of a labrador is 35 kg, to the nearest kg. The weight lies between:

A	B	C
30 and 40 kg	34 and 36 kg	34·5 and 35·5 kg

9. Liz and Julie each measure a different worm and they both say that their worm is 11 cm long to the nearest cm.
(a) Does this mean that both worms are the same length?
(b) If not, what is the maximum possible difference in the length of the two worms?

10. To the nearest cm, the length, l of a stapler is 12 cm. As an inequality we can write $11·5 \leqslant l < 12·5$.

For parts (a) to (j) you are given a measurement. Write the possible values using an inequality as above.

(a) mass = 17 kg   (2 s.f.)   (b) $d = 256$ km   (3 s.f.)
(c) length = 2·4 m   (1 d.p.)   (d) $m = 0·34$ grams   (2 s.f.)
(e) $v = 2·04$ m/s   (2 d.p.)   (f) $x = 12·0$ cm   (1 d.p.)
(g) $T = 81·4$°C   (1 d.p.)   (h) $M = 0·3$ kg   (1 s.f.)
(i) mass = 0·7 tonnes (1 s.f.)   (j) $n = 52\,000$ (nearest thousand)

11. A card measuring 11·5 cm long (to the nearest 0·1 cm) is to be posted in an envelope which is 12 cm long (to the nearest cm).
Can you guarantee that the card will fit inside the envelope?
Explain your answer.

11.5 cm

12 cm

Errors in calculations

This section is concerned with the inaccuracy of numbers used in a calculation. We are not trying to prevent students from making mistakes!

Here is a rectangle with sides measuring 37 cm by 19 cm to the nearest cm.

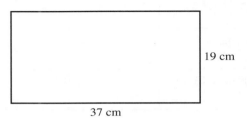

37 cm

19 cm

What are the largest and smallest possible areas of the rectangle consistent with this data?
We know the length is between 36·5 cm and 37·5 cm
and the width is between 18·5 cm and 19·5 cm.

largest possible area = 37·5 × 19·5   smallest possible area = 36·5 × 18·5
$= 731 \cdot 25 \, \text{cm}^2$ $= 675 \cdot 25 \, \text{cm}^2$

A common problem occurs when a measured quantity is multiplied by a large number.

A marble is weighed and found to be 12·3 grams. Find the weight of 100 identical marbles.

A simple answer is 100 × 12·3 = 1230 g. We must realise that each marble could weigh from 12·25 g to 12·35 g.

$$100 \times 12 \cdot 25 = 1225 \, \text{g}$$
$$100 \times 12 \cdot 35 = 1235 \, \text{g}$$

So there is a possible error of up to 5 grams above or below our initial answer of 1230 g.

Given that the numbers 8·6, 3·2 and 11·5 are accurate to 1 decimal place, calculate the upper and lower bounds for the calculation

$$\left(\frac{8 \cdot 6 - 3 \cdot 2}{11 \cdot 5} \right)$$

For the upper bound, make the top line of the fraction as large as possible and make the bottom line as small as possible.

upper bound $= \left(\dfrac{8 \cdot 65 - 3 \cdot 15}{11 \cdot 45} \right)$ lower bound $= \left(\dfrac{8 \cdot 55 - 3 \cdot 25}{11 \cdot 55} \right)$

$= 0 \cdot 480\,3493 \ldots$ $= 0 \cdot 458\,8744 \ldots$

Exercise 20

1. The sides of the triangle are measured correct to the nearest cm.
   (a) Write down the upper bounds for the lengths of the three sides.
   (b) Work out the maximum possible perimeter of the triangle.

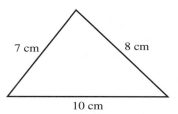

7 cm   8 cm

10 cm

2. The dimensions of a photo are measured correct to the nearest cm. Work out the minimum possible area of the photo.

9 cm

6 cm

3. In this question the value of a is either exactly 4 or 5, and the value of b is either exactly 1 or 2.
   Work out:
   (a) the maximum value of $a + b$ (b) the minimum value of $a + b$
   (c) the maximum value of ab (d) the maximum value of $a - b$
   (e) the minimum value of $a - b$ (f) the maximum value of $\dfrac{a}{b}$
   (g) the minimum value of $\dfrac{a}{b}$ (h) the maximum value of $a^2 - b^2$.

4. If $p = 7\,\text{cm}$ and $q = 5\,\text{cm}$, both to the nearest cm, find:
   (a) the largest possible value of $p + q$
   (b) the smallest possible value of $p + q$
   (c) the largest possible value of $p - q$
   (d) the largest possible value of $\dfrac{p^2}{q}$.

5. If $a = 3{\cdot}1$ and $b = 7{\cdot}3$, correct to 1 decimal place, find the largest possible value of:
   (i) $a + b$ (ii) $b - a$

6. If $x = 5$ and $y = 7$ to one significant figure, find the smallest possible values of
   (i) $x + y$ (ii) $y - x$ (iii) $\dfrac{x}{y}$

7. In the diagram, ABCD and EFGH are rectangles with $AB = 10\,\text{cm}$, $BC = 7\,\text{cm}$, $EF = 7\,\text{cm}$ and $FG = 4\,\text{cm}$, all figures accurate to the nearest cm.
Find the largest possible value of the shaded area.

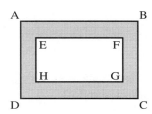

A   B
E   F
H   G
D   C

8. When a voltage V is applied to a resistance R the power consumed P is given by $P = \dfrac{V^2}{R}$.

If you measure V as $12{\cdot}2$ and R as $2{\cdot}6$, correct to 1 d.p., calculate the smallest possible value of P.

9. A cyclist was timed along a straight piece
of road. The time taken was 24·5 seconds,
to the nearest 0·1 second, and the distance
was 420 metres, to the nearest metre.
Calculate the maximum possible value
for the speed of the cyclist consistent
with this data.

10. The velocity v of a body is calculated from the
formula $v = \dfrac{2s}{t} - u$ where u, s and t are measured correct
to 1 decimal place. Find the largest possible value for v when
$u = 2 \cdot 1$, $s = 5 \cdot 7$ and $t = 2 \cdot 2$.
Find also the smallest possible value for v consistent with these
figures.

11. Use the formula $z = \dfrac{a - x^2}{2t}$ to find the largest value that z could
have when $a = 71 \cdot 4$, $x = 5 \cdot 3$ and $t = 5 \cdot 4$, all correct to one
decimal place.

12. A formula for velocity is $v = \sqrt{(u^2 + 2as)}$. Find the smallest possible
value for v if $u = 11 \cdot 5$, $a = -9 \cdot 8$ and $s = 4 \cdot 0$, all correct to one
decimal place.

13. The price of gold is $22·65 per gram, and the density of gold is
19·2 g/cm^3 (i.e. 1 cm^3 of gold weighs 19·2 g). These figures can be
used as exact. A solid gold bar in the shape of a cuboid has
dimensions $8 \cdot 6 \times 4 \cdot 1 \times 2 \cdot 4$, all measurements in cm correct to the
nearest 0·1 cm.

  (a) What is the range of possible prices for this gold bar?
      Write '... < price < ...' giving your answers to the nearest $10.
  (b) The gold bar was in fact weighed on some scales giving a weight
      of 1574 g to the nearest gram. Using this value, work out the
      range of possible prices for the gold bar.
      Again write '... < price < ...'.
  (c) Explain why the weighing method appears to give a smaller
      range of possible prices.

Percentage error

Suppose the answer to the calculation $4 \cdot 8 \times 33 \cdot 4$ is estimated to be 150.
The exact answer to the calculation is 160·32.
The percentage error is

$$\left(\frac{160 \cdot 32 - 150}{160 \cdot 32} \right) \times \frac{100}{1} = 6 \cdot 4\% \qquad \text{(2 s.f.)}$$

- In general, percentage error $= \left(\dfrac{\text{actual error}}{\text{exact value}} \right) \times 100$

Exercise 21

Give the answers correct to 3 significant figures, unless told otherwise.

1. A man's salary is estimated to be £20 000. His actual salary is £21 540. Find the percentage error.

2. The speedometer of a car showed a speed of 68 m.p.h. The actual speed was 71 m.p.h. Find the percentage error.

3. The rainfall during a storm was estimated at 3·5 cm. The exact value was 3·1 cm. Find the percentage error.

4. The dimensions of a rectangle were taken as 80 × 30. The exact dimensions were 81·5 × 29·5. Find the percentage error in the area.

5. The answer to the calculation 3·91 × 21·8 is estimated to be 80. Find the percentage error.

6. An approximate value for π is $\frac{22}{7}$. A more accurate value is given on most calculators. Find the percentage error.

7. A wheel has a diameter of 58·2 cm. Its circumference is estimated at 3 × 60. Find the percentage error.

8. A speed of 60 km/h is taken to be roughly 18 m/s. Find the percentage error.

9. The fraction $\frac{19}{11}$ gives an approximate value for $\sqrt{3}$.
Calculate the percentage error in using $\frac{19}{11}$ as the value of $\sqrt{3}$.
Give the answer to 2 s.f.

10. The most common approximate value for π is $\frac{22}{7}$.

Two better approximations are $\frac{355}{113}$ and $\sqrt{\left(\sqrt{\left(\frac{2143}{22}\right)}\right)}$

Which of these two values has the smaller percentage error?

2.8 Numerical problems

Exercise 22

Questions **1** to **5** are about foreign currency.
Use the rates of exchange given in the table.

1. Change the amount of British money into the foreign currency stated.
   (a) £20 [French francs]   (b) £70 [dollars]
   (c) £200 [pesetas]   (d) £1·50 [marks]
   (e) £2·30 [lire]   (f) 90p [euros]

Country	Rate of exchange
Europe (euro)	€1·54 = £1
France (franc)	Fr 10·9 = £1
Germany (mark)	DM 3·1 = £1
Italy (lire)	lire 2900 = £1
Spain (peseta)	Ptas 260 = £1
United States (dollar)	$1·64 = £1

2. Change the amount of foreign currency into British money.
   (a) 500 frs.   (b) $2500
   (c) DM 7·5   (d) €900
   (e) Lire 500 000   (f) 950 pta

3. An CD costs £12·99 in Britain and $15 in the United States. How much cheaper, in British money, is the CD when bought in the U.S.A?

4. A Renault car is sold in several countries at the prices given below.

Britain   £15 000
France   194 020 frs
USA $24 882

Write out in order a list of the prices converted into pounds.

5. A traveller in Switzerland exchanges 1300 Swiss francs for £400. What is the exchange rate?

6. A maths teacher bought 40 calculators at £8·20 each and a number of other calculators costing £2·95 each. In all she spent £387. How many of the cheaper calculators did she buy?

7. At a temperature of 20°C the common amoeba reproduces by splitting in half every 24 hours. If we start with a single amoeba, how many will there be after (a) 8 days (b) 16 days?

8. A humming bird moves its wings 130 times per second. How many times will it move its wings in one hour? Give your answer in standard form.

9. Here are four fractions: $\frac{3}{19}$, $\frac{37}{232}$, $\frac{1}{6}$, $\frac{83}{511}$
(a) Which of the fractions is the smallest?
(b) Which of the fractions is the largest?

10. A wicked witch stole a newborn baby from its parents. On the baby's first birthday the witch sent the grief-stricken parents 1 penny. On the second birthday she sent 2 pence. On the third birthday she sent 4 pence and so on, doubling the amount each time. How much did the witch send the parents on the twenty-first birthday?

Exercise 23

1. A videotape is 225 metres long. Its total playing time is three hours. Find the speed of the tape:
(a) in metres per minute   (b) in centimetres per second.

2. In 1999 there were 21 280 000 licensed vehicles on the road. Of these, 16 486 000 were private cars. What percentage of the licensed vehicles were private cars?

3. A map is 278 mm wide and 445 mm long. When reduced on a photocopier, the copy is 360 mm long. What is the width of the copy, to the nearest millimetre?

4. Taking 1 litre to be $1\frac{3}{4}$ pints, find:
(a) the number of pints in 20 litres
(b) the number of litres in 9 pints, giving your answer as a decimal.

5. Measured to the nearest millimetre, the sides of a triangle are 11 mm, 14 mm and 17 mm.
Work out the upper limit for the perimeter of the triangle.

6. (a) Fill in the empty boxes in the square opposite so that each row, each column and each diagonal adds up to 0.
(b) Multiply together the three numbers in the bottom row and write down your answer.

3	−4	1
	0	

7. In 2000 Anna picked 252 kg of plums.
This was 20% more than she picked in 1999.
Calculate how many kilograms of plums she picked in 1999.

8. Tiger arrived at the bus station at 11:15. His bus had left half an hour before.
(a) At what time had his bus left?
(b) Tiger's next bus is at 12.06. How long must he wait?

9. What is the smallest number greater than 1000 that is exactly divisible by 13 and 17?

10. In 2001 Mr Giggs drove 8360 miles in his car. His car does 37 miles per gallon. Petrol costs 68 pence per litre.
(a) By taking one gallon to be 4·55 litres, calculate, in £, how much Mr Giggs spent on petrol in 2001.
(b) Show how you can use approximation to check that your answer is of the right order of magnitude.
You **must** show all your working.

Exercise 24

1. A stick insect moves 3 centimetres in one minute.
(a) Find the speed of the stick insect in kilometres per hour, giving your answer as a decimal.
(b) Write your answer to part (a) in standard form.

2. Find the smallest value of n for which
$$1^2 + 2^2 + 3^2 + 4^2 + 5^2 + \ldots + n^2 > 800$$

3. Pythagoras, the Greek mathematician, was also a shrewd businessman. Suppose he deposited £1 in the Bank of Athens in the year 500 B.C. at 1% compound interest. What would the investment be worth to his descendants in the year 2000 A.D.?

4. Find the missing prices

Object	Old price	New price	% change
(a) Football	£9·50	?	3% increase
(b) Radio	?	£34·88	9% increase
(c) Roller blades	£52	?	15% decrease
(d) Golf club	?	£41·40	8% decrease

5. Booklets have a mass of 19 g each and they are posted in an envelope of mass 38 g. Postage charges are shown in the table below.

Mass (in grams) not more than	60	100	150	200	250	300	350	600
Postage (in pence)	24	30	37	44	51	59	67	110

   (a) A package consists of 15 booklets in an envelope.
       What is the total mass of the package?
   (b) The mass of a second package is 475 g.
       How many booklets does it contain?
   (c) What is the postage charge on a package of mass 320 g?
   (d) The postage on a third package was £1·10.
       What is the largest number of booklets it could contain?

6. A rabbit runs at $7 \, \mathrm{m \, s^{-1}}$ and a hedgehog at $\frac{1}{2} \mathrm{m \, s^{-1}}$.
They are 90 m apart and start to run towards each other.
How far does the hedgehog run before they meet?

7. Mark's recipe for a cake calls for 6 fluid ounces of milk. Mark knows that one pint is 20 fluid ounces, that one gallon is 8 pints and that 5 litres is roughly one gallon. He only has a measuring jug marked in millilitres. How many millilitres of milk does he need for the cake? Give your answer to a sensible degree of accuracy.

8. 8% of 2500 + 37% of P = 348. Find the value of P.

9. A train leaves Paris at 15:24 and arrives in Milan at 19:44.
A new train will cut 20% off the journey time.
At what time will the 15:24 train now arrive in Milan?

10. In France petrol for a car costs 5·54 francs per litre. The exchange rate is 9·60 francs to the pound and we can assume that one gallon equals 4·5 litres.
A family uses the car to travel a total distance of 1975 km and the owner of the car estimates the petrol consumption of the car to be 27 miles per gallon. Take 1 km equal to 0·62 miles.
Calculate the cost of the petrol in pounds to the nearest pound.

11. A group of friends share a bill for £13·69 equally between them.
How many were in the group?

12. Which is larger: 7^{77} or 77^7?

13. Write the recurring decimal 0·434 343 . . . as a fraction.

3 ALGEBRA 1

3.1 Basic algebra

Using letters for numbers

Exercise 1

1. Find what number I am left with if:
   (a) I start with x, double it and then subtract 6
   (b) I start with x, add 4 and then square the result
   (c) I start with x, take away 5, double the result and then divide by 3
   (d) I start with w, subtract x and then square the result
   (e) I start with n, add p, cube the result and then divide by a
   (f) I start with t, subtract 1, treble the result and then square the new result.

2. A brick weighs w kg. How much do n bricks weigh?

3. A man shares a sum of x pence equally between n children. How much does each child receive?

4. A cake weighing y kg is cut into n equal pieces. How much does each piece weigh?

5. A sum of £p is shared equally between you and four others. How much does each person receive?

6. Gary is n years older than Mark who is y years old. How old will Gary be in six years time?

7. Sangita used to earn £n per week. She then had a rise of £r per week. How much will she earn in w weeks?

8. The height of a balloon increases at a steady rate of x metres in t hours. How far will the balloon rise in n hours?

9. How many drinks costing x pence each can be bought for £n?

10. A prize of £n is shared equally between you and x other people. How much does each person receive in *pence*?

Collecting terms

- '$3x + 4$' is an *expression*. It is *not* an equation because there is no equals sign. $3x$ and 4 are two *terms* in the expression.
- The expression $7x + 2x$ consists of two *like terms*, $7x$ and $2x$. Like terms can be added or subtracted. So $7x + 2x = 9x$.
- The expression $5x - 3y$ consists of two *unlike* terms, $5x$ and $3y$. Unlike terms cannot be added or subtracted.

(a) $8a + 7a - 3a = 12a$ (b) $3x + y + 2x = 5x + y$
(c) $x^2 + 3x^2 + 2x = 4x^2 + 2x$ (d) $5 + 3y + 6 + y^2 = 11 + 3y + y^2$

Exercise 2

In Questions **1** to **20** collect like terms together.

1. $2x + 3 + 3x + 5$
2. $4x + 8 + 5x - 3$
3. $5x - 3 + 2x + 7$
4. $6x + 1 + x + 3$
5. $4x - 3 + 2x + 10 + x$
6. $5x + 8 + x + 4 + 2x$
7. $7x - 9 + 2x + 3 + 3x$
8. $5x + 7 - 3x - 2$
9. $4x - 6 - 2x + 1$
10. $10x + 5 - 9x - 10 + x$
11. $4a + 6b + 3 + 9a - 3b - 4$
12. $8m - 3n + 1 + 6n + 2m + 7$
13. $6p - 4 + 5q - 3p - 4 - 7q$
14. $12s - 3t + 2 - 10s - 4t + 12$
15. $a - 2b - 7 + a + 2b + 8$
16. $3x + 2y + 5z - 2x - y + 2z$
17. $6x - 5y + 3z - x + y + z$
18. $2k - 3m + n + 3k - m - n$
19. $12a - 3 + 2b - 6 - 8a + 3b$
20. $3a + x + e - 2a - 5x - 6e$

21. Simplify where possible.
   (a) $x^2 - 3x + 1 + 2x^2 + 3x$ (b) $5a + ab - 3a + 4ab$
   (c) $x^3 - 7x + 4x^2$ (d) $ab + 3a^2 - 7a - ab + a^2$

22. Which of these expressions is equivalent to $x - 2$?

A $x^2 - 7x - 1 - x^2 + 8x + 3$ B $x^2 + 7x + 2x - 8x - x^2 - 2$

C $5 + 7x - x - 4 - 6x + x^2$ D $5x - 7 + 4 - x + 1 - 3x$

Collect like terms together.

23. $x^2 + 5x + 2 - 2x + 1$
24. $x^2 + 2x + 2x^2 + 4x + 5$
25. $x^2 + 5x + x^2 + x - 7$
26. $2x^2 - 3x + 8 + x^2 + 4x + 4$
27. $3x^2 + 4x + 6 - x^2 - 3x - 3$
28. $5x^2 - 3x + 2 - 3x^2 + 2x - 2$
29. $2x^2 - 2x + 3 - x^2 - 2x - 5$
30. $6x^2 - 7x + 8 - 3x^2 + 5x - 10$
31. $3y^2 - 6x + y^2 + x^2 + 7x + 4x^2$
32. $8 - 5x - 2x^2 + 4 + 6x + 2x^2$
33. $5 + 2y + 3y^2 - 8y - 6 + 2y^2 + 3$
34. $ab + a^2 - 3b + 2ab - a^2$
35. $3c^2 - d^2 + 2cd - 3c^2 - d^2$
36. $ab + 2a^2 + 3ab - 4a^2 + 2a$
37. $x^3 + 2x^2 - x + 3x^2 + x^3 + x$
38. $5 - x^2 - 2x^3 + 6 + 2x^2 + 3x^3$
39. $xy + ab - cd + 2xy - ab + dc$
40. $pq - 3qp + p^2 + 2qp - q^2$

Simplifying terms and brackets

(a) $3 \times 4x = 12x$ $5x \times 4x = 20x^2$ $-3 \times 2x = -6x$
   $5(2a \times 3a) = 30a^2$ $3m \times 2n = 6mn$ $5b \times 2bc = 10b^2c$

(b) $3(2x - 1) = 6x - 3$ $x(3x + 4) = 3x^2 + 4x$
   $a(a + 2b) = a^2 + 2ab$ $t^2(3t - 2) = 3t^3 - 2t^2$

(c) $2(x + 3) + 5(x - 1)$ $5(a - b) - 2(2a - b)$
   $= 2x + 6 + 5x - 5$ $= 5a - 5b - 4a + 2b$
   $= 7x + 1$ $= a - 3b$

Exercise 3

1. Write in a more simple form.
(a) $2 \times 3x$ (b) $3a \times 5$ (c) $5t \times (-2)$ (d) $100 \times 10a$
(e) $x \times 2x$ (f) $x \times 4x$ (g) $x \times 2x^2$ (h) $4t \times 5t$

2. Remove the brackets.
(a) $2(2x+1)$ (b) $5(2x+3)$ (c) $a(a-3)$ (d) $2n(n+1)$
(e) $-2(2x+3)$ (f) $2x(x+y)$ (g) $-3(a-2)$ (h) $p(q-p)$
(i) $a(b+a)$ (j) $2b(b-1)$ (k) $2n(n+2)$ (l) $3x(2x+3)$

3. Find four pairs of equivalent expressions or terms.
A $2a \times 3a^2$ B $6 \times a \times a$ C $4a^2$ D $6a^2$
E $(2a)^2$ F $3a \times 3a$ G $6a^3$ H $9a^2$

4. Multiply these terms.
(a) $2a \times 3b$ (b) $x \times 3x^3$ (c) $y \times 2x$ (d) $2p \times 5q$
(e) $3x \times 5y$ (f) $6x \times 3x^2$ (g) $3a \times 8a^3$ (h) $ab \times 2a$
(i) $xy \times 3y$ (j) $cd \times 5c$ (k) $ab \times ab$ (l) $2xy \times xy$
(m) $3d \times d \times d$ (n) $5(2x \times xy)$ (o) $3a \times (-2a)$ (p) $-a \times (-2ab)$

5. A solid rectangular block measures x cm by x cm by $(x+3)$ cm. Find a simplified expression for its surface area in cm^2.

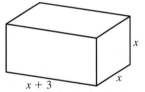

6. Remove the brackets and simplify.
(a) $3(x+2) + 4(x+1)$ (b) $5(x-2) + 3(x+4)$
(c) $2(a-3) + 3(a+1)$ (d) $5(a+1) + 6(a+2)$
(e) $7(a-2) + (a+4)$ (f) $3(t-2) + 5(2+t)$
(g) $3(x+2) - 2(x+1)$ (h) $4(x+3) - 3(x+2)$

7. Three rods A, B and C have lengths of x, $(x+1)$ and $(x-2)$ cm respectively, as shown.

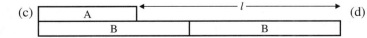

In the diagrams below express the length l in terms of x.
Give your answers in their simplest form.

(a)

(b)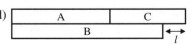

(c)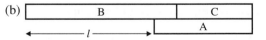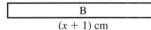

(d)

(e)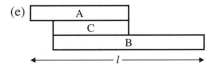

(f)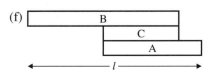

8. Simplify the following expressions.

(a) $x(2x + 1) + 3(x + 2)$ (b) $x(2x - 3) + 5(x + 1)$

(c) $a(3a + 2) + 2(2a - 2)$ (d) $y(5y + 1) + 3(y - 1)$

(e) $x(2x + 1) + x(3x + 1)$ (f) $a(2a + 3) - a(a + 1)$

Expanding brackets

To work out the expression $(x + 3)(x + 2)$, the area of a rectangle can be a help.

From the diagram, the total area of the rectangle

$$= (x + 3)(x + 2)$$
$$= x^2 + 2x + 3x + 6$$
$$= x^2 + 5x + 6$$

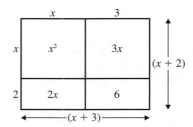

After a little practice, it is possible to do without the diagram.

(a) Remove the brackets and simplify $(3x - 2)(2x - 1)$.

$$(3x - 2)(2x - 1) = 3x(2x - 1) - 2(2x - 1)$$
$$= 6x^2 - 3x - 4x + 2$$
$$= 6x^2 - 7x + 2$$

(b) Remove the brackets and simplify $(x - 3)^2$

Be careful with this expression. It is not $x^2 - 9$, nor even $x^2 + 9$.

$$(x - 3)^2 = (x - 3)(x - 3)$$
$$= x(x - 3) - 3(x - 3)$$
$$= x^2 - 3x - 3x + 9$$
$$= x^2 - 6x + 9$$

Exercise 4

Remove the brackets and simplify.

1. $(x + 1)(x + 3)$ **2.** $(x + 3)(x + 2)$

3. $(y + 4)(y + 5)$ **4.** $(x - 3)(x + 4)$

5. $(x + 5)(x - 2)$ **6.** $(x - 3)(x - 2)$

7. $(a - 7)(a + 5)$ **8.** $(z + 9)(z - 2)$

9. $(x - 3)(x + 3)$ **10.** $(k - 11)(k + 11)$

11. $(2x + 1)(x - 3)$ **12.** $(3x + 4)(x - 2)$

13. $(2y - 3)(y + 1)$ **14.** $(7y - 1)(7y + 1)$

15. $(x + 4)^2$ **16.** $(x + 2)^2$

17. $(x - 2)^2$ **18.** $(2x + 1)^2$

19. $(x + 1)^2 + (x + 2)^2$ **20.** $(x - 2)^2 + (x + 3)^2$

21. $(x+2)^2 + (2x+1)^2$ **22.** $(y-3)^2 + (y-4)^2$
23. $(x+2)^2 - (x-3)^2$ **24.** $(x-3)^2 - (x+1)^2$
[Hint: $(x+2)^2 - [(x-3)^2]$]

25. Four identical tiles surround the shaded square.
   (a) Find an expression for the sum of the areas of the four tiles.
   (b) Find the area of square ABCD.
   (c) Hence find the area of the shaded square.

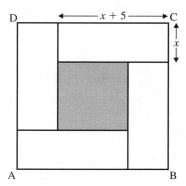

26. The expression $(x+1)(x+2)(x+3)$ is expanded in stages.
   Complete the working

$$(x+1)[(x+2)(x+3)]$$
$$= (x+1)(x^2 + 5x + 6)$$
$$= \ldots$$

27. Remove the brackets and simplify.
   (a) $(x+1)(x+3)(x+4)$
   (b) $(x+1)^3$
   (c) $(x-1)(x+2)^2$

28. (a) If n is an integer, explain why $(2n-1)$ is an odd number.
   (b) Write down the next odd number after $(2n-1)$ and show that
        the sum of these two numbers is a multiple of 4.
   (c) Write down the next odd number after these two and show that
        the sum of the three odd numbers is a multiple of 3.
   (d) Show that the sum of the squares of these three odd numbers is
        $12n^2 + 12n + 11$.
        Is it possible for this expression to be a multiple of 11? If so give
        two possible values of n.

Factorising expressions

Earlier we expanded expressions such as $x(3x-1)$ to give $3x^2 - x$.
The reverse of this process is called factorising.

Factorise: (a) $x^2 + 7x$ (b) $3y^2 - 12y$ (c) $6a^2b - 10ab^2$

(a) x is common to x^2 and $7x$. ∴ $x^2 + 7x = x(x+7)$.
   The factors are x and $(x+7)$.
(b) $3y$ is common ∴ $3y^2 - 12y = 3y(y-4)$
(c) $2ab$ is common ∴ $6a^2b - 10ab^2 = 2ab(3a - 5b)$

Exercise 5

Factorise the following expressions completely.

1. $x^2 + 5x$
2. $x^2 - 6x$
3. $7x - x^2$
4. $y^2 + 8y$
5. $2y^2 + 3y$
6. $6y^2 - 4y$
7. $3x^2 - 21x$
8. $16a - 2a^2$
9. $6c^2 - 21c$
10. $15x - 9x^2$
11. $56y - 21y^2$
12. $ax + bx + 2cx$
13. $x^2 + xy + 3xz$
14. $x^2y + y^3 + z^2y$
15. $3a^2b + 2ab^2$
16. $x^2y + xy^2$
17. $6a^2 + 4ab + 2ac$
18. $ma + 2bm + m^2$
19. $2kx + 6ky + 4kz$
20. $ax^2 + ay + 2ab$
21. $7x^2 + x$
22. $4y^2 - 4y$
23. $p^2 - 2p$
24. $6a^2 + 2a$
25. $4 - 8x^2$
26. $5x - 10x^3$
27. $4\pi r + \pi h$
28. $\pi r^2 + 2\pi r$
29. $3\pi r^2 + \pi rh$
30. $3xy + 2x$
31. $x^2k + xk^2$
32. $a^3b + 2ab^2$
33. $abc - 3b^2c$
34. $2a^2e - 5ae^2$
35. $a^3b + ab^3$
36. $x^3y + x^2y^2$
37. $6xy^2 - 4x^2y$
38. $3ab^3 - 3a^3b$
39. $2a^3b + 5a^2b^2$
40. $ax^2y - 2ax^2z$
41. $2abx + 2ab^2 + 2a^2b$
42. $ayx + yx^3 - 2y^2x^2$

43. In these two rectangles you are given one side and the area.
    Find the perimeter of each rectangle.

(a)

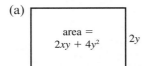

area =
$2xy + 4y^2$ $2y$

(b)

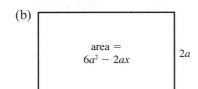

area =
$6a^2 - 2ax$ $2a$

Mathematical language

In algebra, letter symbols are used to represent unknowns in a variety of situations.

In *equations* the letters stand for particular numbers (the solutions of the equation), e.g. $3x - 1 = 2(x + 1)$.

In an *expression* there is no equals sign, e.g. $5x^2 - 3x + 1$.

In *formulae* letters stand for defined quantities or variables, e.g. $F = ma$.

In an *identity* there is an equals sign (sometimes written $\equiv$) but the equality holds for *all values* of the unknown, e.g.
$(x^2 - 1) = (x - 1)(x + 1)$

In the definition of *functions*, e.g. $y = x^2 + 7x$, $f(x) = 3x^2$.

Exercise 6

Copy and complete the table by putting the contents of each box in the correct column.

equation	expression	identity	formula

$$x(x+1) = x^2 + x$$

$$7y + 10$$

$$V = IR$$

$$7x + 11 = x - 9$$

$$x^2 - 3x + 10$$

$$(x+1)^2 = x^2 + 2x + 1$$

$$A = \pi r^2$$

$$x^2 - 7x = 0$$

3.2 Linear equations

Many questions in mathematics are easier to answer if algebra is used. It is often best to let the unknown quantity be x and to try to translate the question into the form of an equation.

We will start by concentrating on solving the equation. Later we will look at questions where we will have to make up the equation first and then solve it.

Basic rules

Rule 1. Treat both sides the same.

(a) We can add the same to both sides.
$$3x - 1 = 4$$
$$3x - 1 + 1 = 4 + 1 \qquad \text{[add 1]}$$
(b) We can subtract the same from both sides.
$$5x + 7 = 10$$
$$5x + 7 - 7 = 10 - 7 \qquad \text{[take away 7]}$$
(c) We can multiply both sides by the same factor.
$$\frac{x}{4} = 3$$
$$4\left(\frac{x}{4}\right) = 4 \times 3 \qquad \text{[multiply by 4]}$$
(d) We can divide both sides by the same factor.
$$3x = 8$$
$$\frac{3x}{3} = \frac{8}{3} \qquad \text{[divide by 3]}$$

Rule 2. If the x term is negative, take it to the other side where it becomes positive.

Rule 3. If there are x terms on both sides, collect them on one side.

(a) $4 - 3x = 2$

$\qquad 4 = 2 + 3x$

$\qquad 2 = 3x$

$\qquad \dfrac{2}{3} = x$

(b) $2x - 7 = 5 - 3x$

$\qquad 2x + 3x = 5 + 7$

$\qquad 5x = 12$

$\qquad x = \dfrac{12}{5} = 2\tfrac{2}{5}$

Exercise 7

Solve the following equations:

1. $2x - 5 = 11$ **2.** $3x - 7 = 20$ **3.** $2x + 6 = 20$ **4.** $5x + 10 = 60$

5. $8 = 7 + 3x$ **6.** $12 = 2x - 8$ **7.** $-7 = 2x - 10$ **8.** $3x - 7 = -10$

9. $12 = 15 + 2x$ **10.** $5 + 6x = 7$ **11.** $100x - 1 = 98$ **12.** $7 = 7 + 7x$

13. $\dfrac{x}{100} + 10 = 20$ **14.** $1000x - 5 = -6$ **15.** $-4 = -7 + 3x$ **16.** $2x + 4 = x - 3$

17 $x - 3 = 3x + 7$ **18.** $5x - 4 = 3 - x$ **19.** $4 - 3x = 1$ **20.** $5 - 4x = -3$

21. $7 = 2 - x$ **22.** $3 - 2x = x + 12$ **23.** $6 + 2a = 3$ **24.** $a - 3 = 3a - 7$

25. $2y - 1 = 4 - 3y$ **26.** $7 - 2x = 2x - 7$ **27.** $7 - 3x = 5 - 2x$ **28.** $8 - 2y = 5 - 5y$

29. $x - 16 = 16 - 2x$ **30.** $x + 2 = 3 \cdot 1$ **31.** $-x - 4 = -3$ **32.** $-3 - x = -5$

Rule 4. If there is a fraction in the x term, multiply out to simplify the equation.

(a) $\dfrac{2x}{3} = 10$

$\qquad 2x = 30$

$\qquad x = \dfrac{30}{2} = 15$

(b) $3 = \dfrac{x}{4} - 4$

$\qquad 7 = \dfrac{x}{4}$

$\qquad 28 = x$

(c) $\dfrac{a}{3} + 7 = 5$

$\qquad \dfrac{a}{3} = -2$

$\qquad a = -6$

Exercise 8

Solve the following equations:

1. $\dfrac{x}{5} = 7$ **2.** $\dfrac{x}{10} = 13$ **3.** $7 = \dfrac{x}{2}$ **4.** $\dfrac{x}{2} = \dfrac{1}{3}$

5. $\dfrac{3x}{2} = 5$ **6.** $\dfrac{4x}{5} = -2$ **7.** $7 = \dfrac{7x}{3}$ **8.** $\dfrac{3}{4} = \dfrac{2x}{3}$

9. $\dfrac{5x}{6} = \dfrac{1}{4}$ **10.** $-\dfrac{3}{4} = \dfrac{3x}{5}$ **11.** $\dfrac{x}{2} + 7 = 12$ **12.** $\dfrac{x}{3} - 7 = 2$

13. $\dfrac{x}{5} - 6 = -2$ **14.** $4 = \dfrac{x}{2} - 5$ **15.** $10 = 3 + \dfrac{x}{4}$ **16.** $\dfrac{a}{5} - 1 = -4$

17. $-\dfrac{x}{2} + 1 = -\dfrac{1}{4}$ **18.** $-\dfrac{3}{5} + \dfrac{x}{10} = -\dfrac{1}{5} - \dfrac{x}{5}$

Rule 5. When an equation has brackets, multiply out the brackets first.

(a) $x - 2(x - 1) = 1 - 4(x + 1)$ (b) $\dfrac{5}{x} = 2$

$\qquad x - 2x + 2 = 1 - 4x - 4 \qquad\qquad 5 = 2x$

$\qquad x - 2x + 4x = 1 - 4 - 2 \qquad\qquad \dfrac{5}{2} = x$

$\qquad\qquad\qquad 3x = -5$

$\qquad\qquad\qquad\ x = -\dfrac{5}{3}$

Exercise 9

Solve the following equations.

1. $x + 3(x + 1) = 2x$

2. $1 + 3(x - 1) = 4$

3. $2x - 2(x + 1) = 5x$

4. $2(3x - 1) = 3(x - 1)$

5. $4(x - 1) = 2(3 - x)$

6. $4(x - 1) - 2 = 3x$

7. $4(1 - 2x) = 3(2 - x)$

8. $3 - 2(2x + 1) = x + 17$

9. $4x = x - (x - 2)$

10. $7x = 3x - (x + 20)$

11. $5x - 3(x - 1) = 39$

12. $3x + 2(x - 5) = 15$

13. $7 - (x + 1) = 9 - (2x - 1)$

14. $10x - (2x + 3) = 21$

15. $3(2x + 1) + 2(x - 1) = 23$

16. $5(1 - 2x) - 3(4 + 4x) = 0$

17. $7x - (2 - x) = 0$

18. $3(x + 1) = 4 - (x - 3)$

19. $3y + 7 + 3(y - 1) = 2(2y + 6)$

20. $4(y - 1) + 3(y + 2) = 5(y - 4)$

Exercise 10

Solve the following equations.

1. $\dfrac{7}{x} = 21$

2. $30 = \dfrac{6}{x}$

3. $\dfrac{5}{x} = 3$

4. $\dfrac{9}{x} = -3$

5. $11 = \dfrac{5}{x}$

6. $-2 = \dfrac{4}{x}$

7. $\dfrac{x + 1}{3} = \dfrac{x - 1}{4}$

8. $\dfrac{x + 3}{2} = \dfrac{x - 4}{5}$

9. $\dfrac{2x - 1}{3} = \dfrac{x}{2}$

10. $\dfrac{3x + 1}{5} = \dfrac{2x}{3}$

11. $\dfrac{5}{x - 1} = \dfrac{10}{x}$

12. $\dfrac{12}{2x - 3} = 4$

13. $2 = \dfrac{18}{x + 4}$

14. $\dfrac{5}{x + 5} = \dfrac{15}{x + 7}$

15. $\dfrac{4}{x} + 2 = 3$

16. $\dfrac{6}{x} - 3 = 7$

17. $\dfrac{9}{x} - 7 = 1$

18. $-2 = 1 + \dfrac{3}{x}$

19. $4 - \dfrac{4}{x} = 0$

20. $5 - \dfrac{6}{x} = -1$

21. $\dfrac{x}{3} + \dfrac{x}{4} = 1$

Solving problems using linear equations

So far we have concentrated on solving given equations. Making up our own equations helps to solve problems which are difficult to solve. There are four steps.

(a) Let the unknown quantity be x (or any other suitable letter) and state the units where appropriate.
(b) Write the problem in the form of an equation.
(c) Solve the equation and give the answer in words.
(d) Check your solution using the problem and *not* your equation.

Find three consecutive even numbers which add up to 792.

(a) Let the smallest number be x.
  Then the other numbers are $(x + 2)$ and $(x + 4)$ because they are consecutive *even* numbers.

(b) Form an equation.
  $x + (x + 2) + (x + 4) = 792$

(c)   Solve: $3x + 6 = 792$
  $$3x = 786$$
  $$x = 262$$
  The three numbers are 262, 264 and 266.

(d) Check. $262 + 264 + 266 = 792 \ \checkmark$

Exercise 11

Solve each problem by forming an equation. The first questions are easy but should still be solved using an equation, in order to practise the method.

1. The length of a rectangle is twice the width.
   If the perimeter is 20 cm, find the width.

2. The width of a rectangle is one third of the length. If the perimeter is 96 cm, find the width.

3. The sum of three consecutive numbers is 276. Find the numbers. Let the first number be x.

4. The sum of four consecutive numbers is 90. Find the numbers.

5. The sum of three consecutive odd numbers is 177. Find the numbers.

6. Find three consecutive even numbers which add up to 1524.

7. When a number is doubled and then added to 13, the result is 38. Find the number.

8. If AB is a straight line, find x.

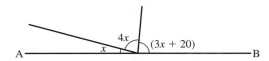

9. The difference between two numbers is 9. Find the numbers, if their sum is 46.

10. The three angles in a triangle are in the ratio $1 : 3 : 5$. Find the angles.

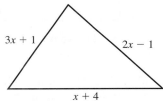

11. The sum of three numbers is 28. The second number is three times the first and the third is 7 less than the second. What are the numbers?

12. If the perimeter of the triangle is 22 cm, find the length of the shortest side.

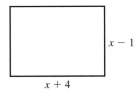

13. David weighs 5 kg less than John, who in turn is 8 kg lighter than Paul. If their total weight is 197 kg, how heavy is each person?

14. If the perimeter of the rectangle is 34 cm, find x.

15. The diagram shows a rectangular lawn surrounded by a footpath x m wide.
(a) Show that the area of the path is $4x^2 + 14x$.
(b) Find an expression, in terms of x for the distance around the outside edge of the path.
(c) Find the value of x when this perimeter is 20 m.

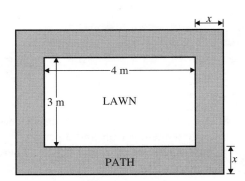

Exercise 12

1. Find the value of x so that the areas of the shaded rectangles are equal.

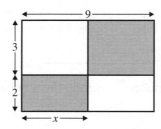

2. Two angles of an isosceles triangle are $a°$ and $(a + 10)°$. Find two possible values of a.

3. The perimeter of the rectangular picture is 37 cm. Find the value of x.

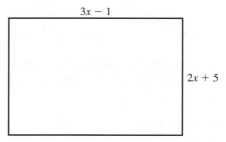

4. A man is 32 years older than his son. Ten years ago he was three times as old as his son was then. Find the present age of each.

5. A man runs to a telephone and back in 15 minutes. His speed on the way to the telephone is 5 m/s and his speed on the way back is 4 m/s. Find the distance to the telephone.

6. A car completes a journey in 10 minutes. For the first half of the distance the speed was 60 km/h and for the second half the speed was 40 km/h. How far is the journey?

7. A lemming runs from a point A to a cliff at 4 m/s, jumps over the edge at B and falls to C at an average speed of 25 m/s.
   If the total distance from A to C is 500 m and the time taken for the journey is 41 seconds, find the height BC of the cliff.

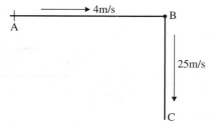

8. A bus is travelling with 48 passengers. When it arrives at a stop, x passengers get off and 3 get on. At the next stop half the passengers get off and 7 get on. There are now 22 passengers. Find x.

9. A bus is travelling with 52 passengers. When it arrives at a stop, y passengers get off and 4 get on. At the next stop one third of the passengers get off and 3 get on. There are now 25 passengers. Find y.

10. Mr Lee left his fortune to his 3 sons, 4 daughters and his wife. Each son received twice as much as each daughter and his wife received £6000, which was a quarter of the money. How much did each son receive?

11. In a regular polygon with n sides each interior angle is $180 - \dfrac{360}{n}$ degrees. How many sides does a polygon have if each angle is $156°$?

12. A sparrow flies to see a friend at a speed of $4\,\text{km/h}$. His friend is out, so the sparrow immediately returns home at a speed of $5\,\text{km/h}$. The complete journey took 54 minutes. How far away does his friend live?

13. Consider the equation $an^2 = 182$ where a is any number between 2 and 5 and n is a positive integer. What are the possible values of n?

14. Consider the equation $\dfrac{k}{x} = 12$ where k is any number between 20 and 65 and x is a positive integer. What are the possible values of x?

15.†

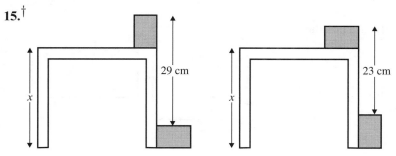

The diagrams show a table with two identical wooden blocks. Calculate the height of the table, x.

Solving equations involving brackets

Solve the equation $(x + 3)^2 = (x + 2)^2 + 3^2$

$$(x + 3)^2 = (x + 2)^2 + 3^2$$
$$(x + 3)(x + 3) = (x + 2)(x + 2) + 9$$
$$x^2 + 6x + 9 = x^2 + 4x + 4 + 9$$
$$6x + 9 = 4x + 13$$
$$2x = 4$$
$$x = 2$$

Exercise 13

Solve the following equations:

1. $x^2 + 4 = (x+1)(x+3)$

3. $(x+3)(x-1) = x^2 + 5$

5. $(x-2)(x+3) = (x-7)(x+7)$

7. $2x^2 + 3x = (2x-1)(x+1)$

9. $x^2 + (x+1)^2 = (2x-1)(x+4)$

2. $x^2 + 3x = (x+3)(x+1)$

4. $(x+1)(x+4) = (x-7)(x+6)$

6. $(x-5)(x+4) = (x+7)(x-6)$

8. $(2x-1)(x-3) = (2x-3)(x-1)$

10. $x(2x+6) = 2(x^2 - 5)$

In Questions **11** and **12**, form an equation in x by means of Pythagoras' Theorem, and hence find the length of each side of the triangle. (All the lengths are in cm.)

11.

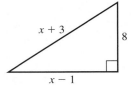

12.

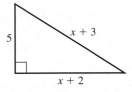

13. The area of the rectangle shown exceeds the area of the square by $2\,\text{cm}^2$. Find x.

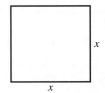

14. The area of the square exceeds the area of the rectangle by $13\,\text{m}^2$. Find y.

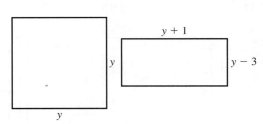

3.3 Trial and improvement

Some problems cannot be solved using linear equations. In such cases, the method of 'trial and improvement' is often a help.

Exercise 14

1. Think of a rectangle having an area of $72\,\text{cm}^2$ whose base is twice its height.

Write down the length of the base.

72 cm²   height

2. Find a rectangle of area $75\,\text{cm}^2$ so that its base is three times its height.

3. In each of the rectangles below, the base is twice the height. The area is shown inside the rectangle. Find the base and the height.

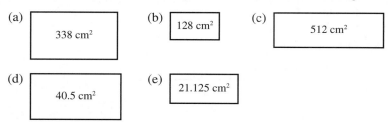

(a) 338 cm²   (b) 128 cm²   (c) 512 cm²

(d) 40.5 cm²   (e) 21.125 cm²

4. In each of the rectangles below, the base is 1 cm more than the height. Find each base and height.

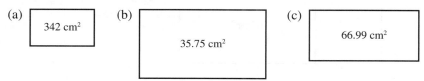

(a) 342 cm²   (b) 35.75 cm²   (c) 66.99 cm²

Inexact answers

In some questions it is not possible ever to find an answer which is precisely correct. However, we can find answers which are nearer and nearer to the exact one, perhaps to the nearest $0\cdot1\,\text{cm}$ or even to the nearest $0\cdot01\,\text{cm}$.

In this rectangle, the base is 1 cm more than the height h cm. The area is $80\,\text{cm}^2$. Find the height h, correct to 1 d.p.

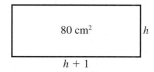

80 cm² h

$h + 1$

Here we are solving the equation $h(h + 1) = 80$.

(a) Try different values for h. BE SYSTEMATIC.

$h = 8\ \ : 8(8 + 1)\ \ \ \ = 72\ \ \ \ \ \text{Too small}$
$h = 9\ \ : 9(9 + 1)\ \ \ \ = 90\ \ \ \ \ \text{Too large}$
$h = 8\cdot5 : (8\cdot5 + 1)\ \ \ = 80\cdot75\ \ \text{Too large}$
$h = 8\cdot4 : 8\cdot4(8\cdot4 + 1) = 78\cdot96\ \ \text{Too small}$

(b) We now see that the answer is between $8\cdot4$ and $8\cdot5$
We also see that the value of $8\cdot5$ for h gave the value closest to 80.
We will take the answer as $h = 8\cdot5$, correct to 1 d.p.

Note: Strictly speaking, to ensure that our answer *is* correct to 1 decimal place, we should try $h = 8\cdot45$. This degree of complexity is not usually necessary in most situations.

Solve the equation $x^3 + 2x = 90$, correct to 1 d.p.

Try $x = 4$: $4^3 + (2 \times 4)$ $= 72$ Too small
     $x = 5$: $5^3 + (2 \times 5)$ $= 135$ Too large
     $x = 4 \cdot 5$: $4 \cdot 5^3 + (2 \times 4 \cdot 5)$ $= 100 \cdot 125$ Too large
     $x = 4 \cdot 4$: $4 \cdot 4^3 + (2 \times 4 \cdot 4)$ $= 93 \cdot 984$ Too large
     $x = 43$: $4 \cdot 3^3 + (2 \times 4 \cdot 3)$ $= 88 \cdot 107$ Too small
     $x = 4 \cdot 35$: $4 \cdot 35^3 + (2 \times 4 \cdot 35)$ $= 91 \cdot 013$ Too large

$\therefore$ The solution is $x = 4 \cdot 3$, correct to 1 decimal place

Exercise 15

1. In these two rectangles, the base is 1 cm more than the height.

(a)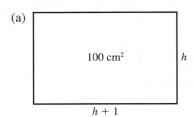

(b)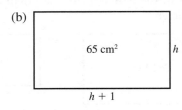

Find the value of h for each one, correct to 1 decimal place.

2. Find solutions to the following equations, giving the answer correct to 1 decimal place.
(a) $x(x - 3) = 11$ (b) $x(x - 2) = 7$
(c) $x^3 = 300$ (d) $x^2(x + 1) = 50$

3. Here you need a calculator with a $\boxed{x^y}$ button.

Find x, correct to 1 decimal place.
(a) $x^5 = 313$ (b) $5^x = 77$ (c) $x^x = 100$

4. The number n has two digits after the decimal point and n to the power 10 is 2110 to the nearest whole number. Find n.

5. An engineer wants to make a solid metal cube of volume $526 \, \text{cm}^3$. Call the edge of the cube x and write down an equation. Find x giving your answer correct to 1 d.p.

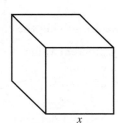

6. In this rectangle, the base is 2 cm more than the height. If the diagonal is 15 cm, find h correct to 2 d.p.

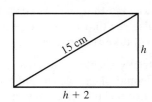

7. A designer for a supermarket chain wants to make a cardboard box of depth 6 cm. He has to make the length of the box 10 cm more than the width.

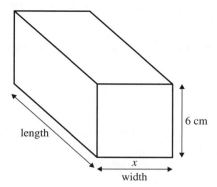

(a) What is the length of the box in terms of x? The box is designed so that its volume is to be 9000 cm³.

(b) Form an equation involving x.

(c) Solve the equation and hence give the dimensions of the box to be nearest cm.

8. The diagram represents a rectangular piece of paper ABCD which has been folded along EF so that C has moved to G.

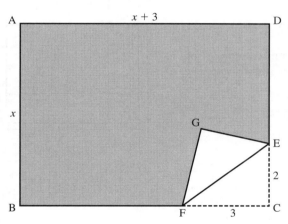

All lengths are in cm.

(a) Calculate the area of △ECF.

(b) Find an expression for the shaded area ABFGED in terms of x.

Given that the shaded area is 20 cm², show that $x(x + 3) = 26$

Solve this equation, giving your answer correct to one decimal place.

9. In the rectangle PQRS, PQ = x cm and QR = 1 cm. The line LM is drawn so that PLMS is a square.

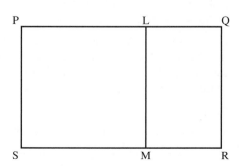

(a) Write down, in terms of x, the length LQ.

(b) If $\dfrac{PQ}{QR} = \dfrac{QR}{LQ}$, obtain an equation in x.

Hence find x correct to two decimal places.

10. Find the number whose cube is 100 times as large as its square root. Give your answer correct to 1 decimal place.

$$\square^3 = 100\sqrt{\square}$$

11. In triangle ABC, angle ABC $= x°$, BC $= x$ cm and
AB $= (x + 10)$ cm.

Find the value of x, correct to the nearest 0.5.

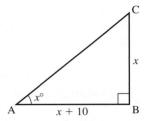

12. A rectangle is drawn inside a semicircle of radius 1 unit.
(a) Find an expression, in terms of x, for the area of the rectangle.
(b) Use trial and improvement, possibly using a spreadsheet
program, to find the value of x, correct to 2 d.p., which gives
the largest value for the area of the rectangle.

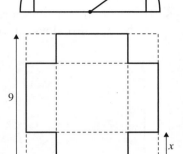

13. The net for an open box is made by cutting corners from a square
card as shown.
(a) Show that the volume of the box is given by the formula
$V = x(9 - 2x)(9 - 2x)$.
(b) Find the value of x which gives the maximum volume of the
box. Again you may find it helpful to use a spreadsheet.

3.4 Finding a rule

Here is a sequence of 'houses' made from matches.

 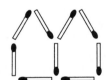  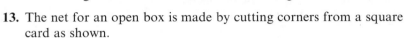

The table on the right records the number of houses h and the number
of matches m.

If the number in the h column goes up one at a time, look at the
number in the m column. If it goes up (or down) by the same number
each time, the function connecting m and h is linear. This means that
there are no terms in h^2 or anything more complicated.

h		m
1		5
2		9
3		13
4		17

In this case, the numbers in the m column go up by 4 each time. This
suggests that a column for $4h$ might help.

Now it is fairly clear that m is one more than $4h$.
So the formula linking m and h is: $m = 4h + 1$

h	$4h$	m
1	4	5
2	8	9
3	12	13
4	16	17

The table shows how r changes with n. What is the formula linking r with n?

n		r
2		3
3		8
4		13
5		18

Because r goes up by 5 each time, write another column for $5n$. The table shows that r is always 7 less than $5n$, so the formula linking r with n is $r = 5n - 7$

n	$5n$	r
2	10	3
3	15	8
4	20	13
5	25	18

The method only works when the first set of numbers goes up by one each time.

Exercise 16

1. Below is a sequence of diagrams showing black tiles b and white tiles w with the related table.

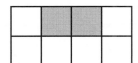

b		w
1		5
2		6
3		7
4		8

What is the formula for w in terms of b? Write it as $w = \ldots$

2. This is a different sequence with black tiles b and white tiles w and the related table.

b		w
2		10
3		12
4		14
5		16

What is the formula? Write it as $w = \ldots$

3. Here is a sequence of Is.

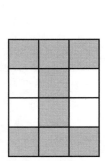 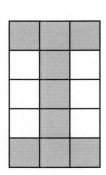 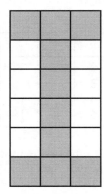

Make your own table for black tiles b and white tiles w. What is the formula for w in terms of b?

4. This sequence shows matches m arranged in triangles t.

Make a table for t and m starting like this:
Continue the table and find a formula for m in terms of t. Write it as $m = \ldots$

t		m
1		3
2		5
⋮		⋮

5. Here is a different sequence of matches and triangles.

Make a table and find a formula connecting m and t.

6. In this sequence, there are triangles t and squares s around the outside.

 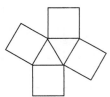 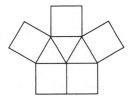 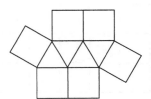

What is the formula connecting t and s?

7. Look at the tables below. In each case, find a formula connecting the two letters.

(a)

n	p
1	3
2	8
3	13
4	18

(b)

n	k
2	17
3	24
4	31
5	38

(c)

n	w
3	17
4	19
5	21
6	23

8. This is one member of a sequence of cubes c made from matches m.

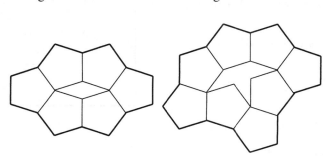

Find a formula connecting m and c.

9. In these tables, the numbers on the left do not go up by one each time. Try to find a formula in each case.

(a)

n	y
1	4
3	10
7	22
8	25

(b)

n	h
2	5
3	9
6	21
10	37

(c)

n	k
3	14
7	26
9	32
12	41

10. Some attractive loops can be made by fitting pentagon tiles together.

Diagram $n = 1$ Diagram $n = 2$ Diagram $n = 3$

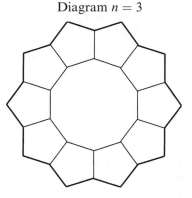

6 tiles t
14 external edges e

8 tiles t
17 external edges e

10 tiles t
20 external edges e

Make your own tables involving n, t and e.
(a) Find a formula connecting n and t.
(b) Find a formula connecting n and e.
(c) Find a formula connecting t and e.

11. When a group of workers were made redundant, the management
used a formula to calculate the redundancy money paid.

Here are the details.

Number of years worked N	Redundancy money £R
1	£5400
2	£5800
3	£6200
4	£6600
10	£9000

   (a) Sam has worked for 7 years. How much redundancy money will
       he receive?

   (b) Write down the formula connecting the amount of
       redundancy paid R and the number of years worked N.

12. In each diagram there are black squares
along one diagonal.
Find a formula for the number of white
squares in the $n \times n$ square.

  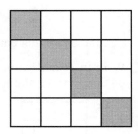

The nth term in a sequence

● For the sequence 3, 6, 9, 12, 15 ... the rule is 'add 3'.

Here is the *mapping diagram* for the sequence.

The terms are found by multiplying the term number by 3.
So the 10th term is 30, the 33rd term is 99.

A *general* term in the sequence is the nth term, where n stands
for any number.
The nth term of this sequence is $3n$.

Term number (n)		Term
1	→	3
2	→	6
3	→	9
⋮		⋮
10	→	30
⋮		
n	→	$3n$

● Here is a more difficult sequence: 3, 7, 11, 15, ...

The rule is 'add 4' so, in the mapping diagram
below, we have written a column for 4 times
the term number [i.e. $4n$].

We see that each term is 1 less than $4n$.
So, the 10th term is $(4 \times 10) - 1 = 39$
the nth term is $(4 \times n) - 1 = 4n - 1$

Term number (n)		$4n$		Term
1	→	4	→	3
2	→	8	→	7
3	→	12	→	11
4	→	16	→	15

Exercise 17

1. Write down each sequence and select the correct expression for the
*n*th term from the list given.
(a) 10, 20, 30, 40, ...
(b) 5, 10, 15, 20, ...
(c) 3, 4, 5, 6, ...
(d) 3, 5, 7, 9, 11, ...
(e) 30, 60, 90, 120, ...
(f) 5, 11, 17, 23, ...
(g) 1, 4, 7, 10, 13, ...

$n + 2$ $5n$

$6n - 1$ $2n + 1$

$10n$ $30n$

$3n - 2$

2. Copy and complete the mapping diagrams. Notice that an extra
column has been written.

(a)

Term number (*n*)		2*n*		Term
1	→	2	→	5
2	→	4	→	7
3	→	6	→	9
4	→	8	→	11
⋮				
10	→	☐	→	☐
⋮				
n	→	☐	→	☐

(b)

Term number (*n*)		3*n*		Term
1	→	3	→	4
2	→	6	→	7
3	→	9	→	10
⋮				
20	→	☐	→	☐
⋮				
n	→	☐	→	☐

3. Here are the first five terms of a sequence: 6, 11, 16, 21, 26, ...
(a) Draw a mapping diagram similar to those above.
[Hint: The rule for the sequence is 'add 5' so write a column for
'5*n*' in the diagram.]
(b) Write down: (i) the 10th term.
(ii) the *n*th term.

4. Here are three sequences:
A: 1, 4, 7, 10, 13, ...
B: 6, 10, 14, 18, 22, ...
C: 5, 12, 19, 26, 33, ...

For each sequence: (a) draw a mapping diagram
(b) find the *n*th term.

5. Look at the sequences A1, A2,
A3, A4 ... and B1, B2, B3,
B4 ...
Write down
(a) A10   (b) B10
(c) the *n*th term in the 'A'
sequence.
(d) the *n*th term in the 'B'
sequence.

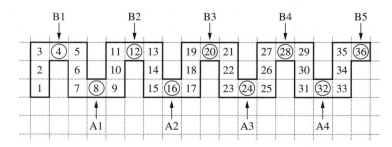

6.

(a) Find the nth term in the 'P' sequence.
(b) Find the nth term in the 'Q' sequence.

7. Here are two more difficult sequences. Copy and complete the mapping diagrams.

(a)

Term number (n)		Term
1	$\rightarrow \quad 1 \times 2 \quad \rightarrow$	2
2	$\rightarrow \quad 2 \times 3 \quad \rightarrow$	6
3	$\rightarrow \quad 3 \times 4 \quad \rightarrow$	12
4	$\rightarrow \quad 4 \times 5 \quad \rightarrow$	20
$\vdots$		
10	$\rightarrow \quad \square \quad \rightarrow$	$\square$
$\vdots$		
n	$\rightarrow \quad \square \quad \rightarrow$	$\square$

(b)

Term number (n)		Term
1	$\rightarrow \quad 1^2 + 1 \quad \rightarrow$	2
2	$\rightarrow \quad 2^2 + 1 \quad \rightarrow$	5
3	$\rightarrow \quad 3^2 + 1 \quad \rightarrow$	10
4	$\rightarrow \quad 4^2 + 1 \quad \rightarrow$	17
$\vdots$		
10	$\rightarrow \quad \square \quad \rightarrow$	$\square$
$\vdots$		
n	$\rightarrow \quad \square \quad \rightarrow$	$\square$

8. Look at the sequence: 5, 8, 13, 20, ...

Decide which of the following is the correct expression for the nth term of the sequence.

$$\boxed{4n + 1} \qquad \boxed{3n + 2} \qquad \boxed{n^2 + 4}$$

In Questions **9** to **13** look carefully at how each sequence is formed. Write down:

(a) the 10th term
(b) the nth term.

9. $1^2, 2^2, 3^2, 4^2, \ldots$

10. $(1 \times 2), (2 \times 3), (3 \times 4), (4 \times 5), \ldots$

11. $\frac{1}{2}, \frac{2}{3}, \frac{3}{4}, \frac{4}{5}, \ldots$

12. $\frac{5}{1^2}, \frac{5}{2^2}, \frac{5}{3^2}, \frac{5}{4^2}, \ldots$

13. $(1 \times 3), (2 \times 4), (3 \times 5), (4 \times 6), \ldots$

14. Here are three sequences and three expressions. Write down each sequence and select the correct expression for its nth term.

A: $-1, 2, 5, 8, 11, \ldots$
B: $-1, 2, 7, 14, 23, \ldots$
C: $3, 8, 15, 24, 35, \ldots$

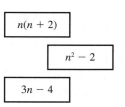

$$n(n + 2)$$

$$n^2 - 2$$

$$3n - 4$$

15. The nth term of a sequence is t_n.

If $t_n = n^3$: the first term, $t_1 = 1^3$
the second term, $t_2 = 2^3$
the third term, $t_3 = 3^3$ etc.

Write down the first 3 terms of the sequences where the nth term is given.

(a) $t_n = 2n + 1$ (b) $t_n = 5n - 3$

(c) $t_n = 100 - 2n$ (d) $t_n = \dfrac{n}{n + 2}$

(e) $t_n = 10^n$ (f) $t_n = n(n + 1)$

16. Two pupils, Anna and Dipika, did the same investigation and got the same results. The patterns they noticed were different.

Anna's table

n	t	pattern
1	3	1×3
2	8	2×4
3	15	3×5
4	24	4×6

Dipika's table

n	t	pattern
1	3	$1^2 + 2 \times 1$
2	8	$2^2 + 2 \times 2$
3	15	$3^2 + 2 \times 3$
4	24	$4^2 + 2 \times 4$

(a) What formula do you think Anna got ($t = \ldots$)?
(b) What formula do you think Dipika got ($t = \ldots$)?
(c) Are they in fact the same formula, written differently? Explain your answer.

†**17.** Find an expression for the nth term of each sequence.

(a) $1, \frac{1}{4}, \frac{1}{9}, \frac{1}{16}, \ldots$ (b) $2, 4, 8, 16, 32, \ldots$
(c) $(3 \times 2), (3 \times 4), (3 \times 8), (3 \times 16), \ldots$ (d) $9, 99, 999, 9999, \ldots$
(e) $(2 \times 2), (4 \times 3), (6 \times 4), (8 \times 5), \ldots$ (f) $\frac{1}{3}, \frac{4}{9}, \frac{9}{27}, \frac{16}{81}, \ldots$

18. Diagonals are drawn in a sequence of polygons.

Draw polygons with 6, 7, 8 ... sides and use your results to find an expression for the number of diagonals in a polygon with n sides.

19. (a) The nth term of the sequence 6, 14, 24, 36, ... is $n^2 + kn$.
   Find k.
   (b) The nth term of the sequence 4, 5, 8, 13, ... is $n^2 + an + b$.
   Find a and b.
   (c) The nth term of the sequence 8, 17, 32, 53, ... is $cn^2 + d$.
   Find c and d.

3.5 Linear graphs, $y = mx + c$

A *linear* graph is a graph which is a straight line. The graphs of
$y = 3x - 1$, $y = 7 - 2x$, $y = \frac{1}{2}x + 100$ are all linear.
Notice that on the right hand side there are only x terms and number
terms. There are *no* terms involving x^2, x^3, $1/x$, etc.
Linear graphs can be used to represent a wide variety of practical
situations like those in the first exercise below.
We begin by drawing linear graphs *accurately*.

Exercise 18

Draw the following graphs, using a scale of 2 cm to 1 unit on the
x-axis and 1 cm to 1 unit on the y-axis.

1. $y = 2x + 1$ for $-3 \leqslant x \leqslant 3$ **2.** $y = 3x - 4$ for $-3 \leqslant x \leqslant 3$

3. $y = 8 - x$ for $-2 \leqslant x \leqslant 4$ **4.** $y = 10 - 2x$ for $-2 \leqslant x \leqslant 4$

5. $y = \dfrac{x + 5}{2}$ for $-3 \leqslant x \leqslant 3$ **6.** $y = 3(x - 2)$ for $-3 \leqslant x \leqslant 3$

7. $y = \frac{1}{2}x + 4$ for $-3 \leqslant x \leqslant 3$ **8.** $v = 2t - 3$ for $-2 \leqslant t \leqslant 4$

9. $z = 12 - 3t$ for $-2 \leqslant t \leqslant 4$

10. Kendal Motors hires out vans.

KENDAL £35 MOTORS
plus 20 pence per mile *(including VAT!)*

Copy and complete the table where x is the number of miles
travelled and C is the total cost in pounds.

x	0	50	100	150	200	250	300
C	35			65			95

Draw a graph of C against x, using scales of 2 cm for 50 miles on
the x-axis and 1 cm for £10 on the C-axis.

(a) Use the graph to find the number of miles travelled when the
   total cost was £71.
(b) What is the formula connecting C and x?

11. Jeff sets up his own business as a plumber.

Copy and complete the table where C stands for his total charge and h stands for the number of hours he works.

h	0	1	2	3
C		33		

Draw a graph with h across the page and C up the page. Use scales of 2 cm to 1 hour for h and 2 cm to £10 for C.

(a) Use your graph to find how long he worked if his charge was £55.50.

(b) What is the equation connecting C and h?

12. The equation connecting the annual mileage, M miles, of a certain car and the annual running cost, £C is $C = \dfrac{M}{20} + 200$.

Draw the graph for $0 \leqslant M \leqslant 10\,000$ using scales of 1 cm for 1000 miles for M and 2 cm for £100 for C.

(a) From the graph find:
  (i) the cost when the annual mileage is 7200 miles
  (ii) the annual mileage corresponding to a cost of £320.

(b) There is a '200' in the formula for C. Explain where this figure might come from.

13. Some drivers try to estimate their annual cost of repairs £c in relation to their average speed of driving s km/h using the equation $c = 6s + 50$.

Draw the graph for $0 \leqslant s \leqslant 160$. From the graph find

(a) the estimated repair bill for a man who drives at an average speed of 23 km/h.

(b) the average speed at which a motorist drives if his annual repair bill is £1000.

Line segment

The graph shows a line segment from A(1, 2) to B(7, 4).

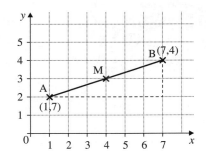

The coordinates of the mid-point M are the mean of the coordinates A

and B i.e. M has coordinates $\left(\dfrac{1+7}{2}, \dfrac{2+4}{2}\right) = (4, 3)$

The length of the line segment is found using Pythagoras.

$$AB = \sqrt{(7-1)^2 + (4-2)^2}$$
$$= \sqrt{40}\ \text{units.}$$

Exercise 19

1. Find the coordinates of the mid-point of the line joining the following pairs of points.
   (a) A (3, 7), B (5, 13)
   (b) C (0, 10), D (12, 4)
   (c) P (−3, 6), Q (5, 0)
   (d) X (4, 3), Y (−2, 3)

2. Find the length of the line segment joining the following pairs of points.
   (a) E (7, 4), F (10, 8)
   (b) G (0, −2), H (5, 10)
   (c) S (3, −1), T (10, 4)

3. The line $2x + y = 12$ cuts the axes at the points A and B.

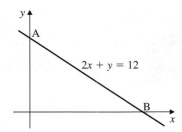

Find the coordinates of A and B and hence find:
(a) the coordinates of the mid-point of AB
(b) the length of the line AB.

4. Find the coordinates of the mid-point of the line joining $P(a, 5a)$ and $Q(7a, -3a)$.

The general straight line, $y = mx + c$

Gradient

The gradient of a straight line is a measure of how steep it is.

Gradient of line AB $= \dfrac{3-1}{6-1} = \dfrac{2}{5}$.

Gradient of line AC $= \dfrac{6-1}{3-1} = \dfrac{5}{2}$.

Gradient of line BC $= \dfrac{6-3}{3-6} = -1$.

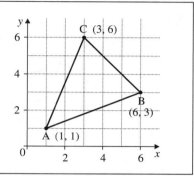

A line which slopes upwards to the right has a *positive* gradient.
A line which slopes upwards to the left has a *negative* gradient.

- Gradient $= \dfrac{\text{difference in } y\text{-coordinates}}{\text{difference in } x\text{-coordinates}}$

Exercise 20

1. Find the gradient of AB, BC, AC.

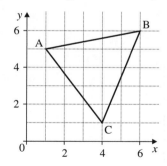

2. Find the gradient of PQ, PR, QR.

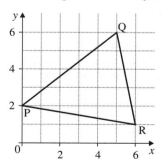

3. Find the gradient of the lines joining the following pairs of points:
   (a) $(5, 2) \rightarrow (7, 8)$
   (b) $(-1, 3) \rightarrow (1, 6)$
   (c) $(\frac{1}{2}, 1) \rightarrow (\frac{3}{4}, 2)$
   (d) $(3{\cdot}1, 2) \rightarrow (3{\cdot}2, 2{\cdot}5)$

4. Find the value of a if the line joining the points $(3a, 4)$ and $(a, -3)$ has a gradient of 1.

5. (a) Write down the gradient of the line joining the points $(2m, n)$
   and $(3, -4)$.
   (b) Find the value of n if the line is parallel to the x-axis
   (c) Find the value of m if the line is parallel to the y-axis.

6. On the grid 1 square $= 1$ unit both across and up the page.
   Three pairs of intersecting lines are drawn. In each case AB is
   perpendicular to CD.

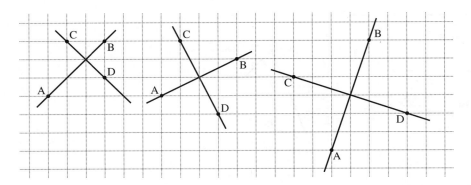

   (a) In each case find the gradient of AB and the gradient of CD.
   (b) In each case work out the product,

   $$(\text{gradient of AB}) \times (\text{gradient of CD})$$

   (c) Copy and complete the following sentence:
   'The product of the gradients of lines which are perpendicular is
   always ☐.'
   (d) A more straightforward result is true for *parallel* lines.
   Copy and complete this sentence:
   'Parallel lines have _____ gradient.'

7. Copy and complete the table

Gradient of line 1	Gradient of line 2	Information
5	☐	Lines are parallel
2	☐	Lines are perpendicular
☐	$-\frac{2}{3}$	Lines are parallel
-5	☐	Lines are perpendicular
☐	$\frac{3}{4}$	Lines are perpendicular

8. PQR is a right-angled triangle with vertices at $P(2, 1)$, $Q(4, 3)$, and
   $R(8, t)$. Given that angle PQR is $90°$, find t.

Intercept constant c

Here are two straight lines.

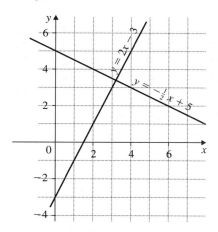

For $y = 2x - 3$, the gradient is 2 and the y-intercept is -3.

For $y = -\frac{1}{2}x + 5$ the gradient is $-\frac{1}{2}$ and the y-intercept is 5.

These two lines illustrate a general rule.
When the equation of a straight line is written in the form

$y = mx + c$

the gradient of the line is m and the intercept on the y-axis is c.

Draw the line $y = 2x + 3$ on a *sketch* graph.

The word 'sketch' implies that we do not plot a series of points but simply show the position and slope of the line.

The line $y = 2x + 3$ has a gradient of 2 and cuts the y-axis at $(0, 3)$.

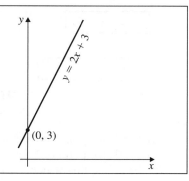

Draw the line $x + 2y - 6 = 0$ on a sketch graph.
(a) Rearrange the equation to make y the subject.
$$x + 2y - 6 = 0$$
$$2y = -x + 6$$
$$y = -\frac{1}{2}x + 3.$$

(b) The line has a gradient of $-\frac{1}{2}$ and cuts the y-axis at $(0, 3)$.

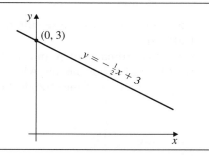

Exercise 21

In Questions **1** to **20**, find the gradient of the line and the intercept on the *y*-axis. Hence draw a small sketch graph of each line.

1. $y = x + 3$ **2.** $y = x - 2$ **3.** $y = 2x + 1$ **4.** $y = 2x - 5$
5. $y = 3x + 4$ **6.** $y = \frac{1}{2}x + 6$ **7.** $y = 3x - 2$ **8.** $y = 2x$
9. $y = \frac{1}{4}x - 4$ **10.** $y = -x + 3$ **11.** $y = 6 - 2x$ **12.** $y = 2 - x$
13. $y + 2x = 3$ **14.** $3x + y + 4 = 0$ **15.** $2y - x = 6$ **16.** $3y + x - 9 = 0$
17. $4x - y = 5$ **18.** $3x - 2y = 8$ **19.** $10x - y = 0$ **20.** $y - 4 = 0$

21. Find the equations of the lines A and B.

22. Find the equations of the lines C and D.

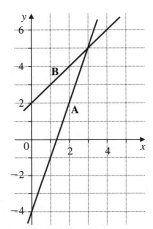

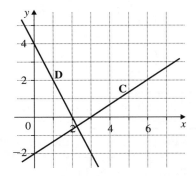

23. Here are the equations of several straight lines

A $y = -x + 7$ B $2y = x - 7$ C $4x + 12y = 5$

D $x + 3y = 10$ E $y = x - 3$ F $y = 5 - 5x$ G $5x + y = 12$

H $y = 3 - 2x$ I $y = 2x + 4$

(a) Find *two* pairs of lines which are parallel.
(b) Find *two* pairs of lines which are perpendicular.
(c) Find *one* line which is neither parallel nor perpendicular to any other line.

24. The sketch represents a section of the curve $y = x^2 - 2x - 8$. Calculate:
(a) the coordinates of A and of B,
(b) the gradient of the line AB,
(c) the equation of the straight line AB.

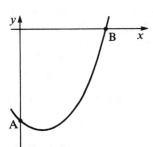

Finding the equation of a line

Find the equation of the straight line which passes through
(1, 3) and (3, 7).

(a) Let the equation of the line take the form $y = mx + c$.

The gradient, $m = \dfrac{7 - 3}{3 - 1} = 2$

so we may write the equation as

$y = 2x + c$ 　　　　　　　　　　　...[1]

(b) Since the line passes through (1, 3), substitute 3 for y and
1 for x in [1].

$\therefore$ 　　$3 = 2 \times 1 + c$
　　　　$1 = c$

The equation of the line is $y = 2x + 1$

Exercise 22

In Questions **1** to **10** find the equation of the line which:

1. passes through (0, 7) at a gradient of 3

2. passes through (0, −9) at a gradient of 2

3. passes through (0, 5) at a gradient of −1

4. passes through (2, 3) at a gradient of 2

5. passes through (2, 11) at a gradient of 3

6. passes through (4, 3) at a gradient of −1

7. passes through (6, 0) at a gradient of $\frac{1}{2}$

8. passes through (2, 1) and (4, 5)

9. passes through (5, 4) and (6, 7)

10. passes through (0, 5) and (3, 2).

Finding a law

A ball is released from the top of a building. Its position is recorded
every second using an electronic photoflash camera.
Here are the results, giving distance fallen d and time taken t.

t	0	1	2	3
d	0	4·9	19·6	44·1

By plotting d against t, a curve is obtained.

It is not easy to find the equation (the law) connecting d and t from a
curve.

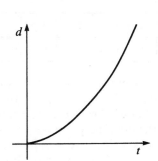

Try plotting d against t^2.

t	0	1	2	3
t^2	0	1	4	9
d	0	4·9	19·6	44·1

This is a straight line with gradient 4·9 and intercept 0 on the d-axis.

$\therefore$ the law connecting d and t is
$$d = 4.9t^2 + 0$$
or $d = 4.9t^2$

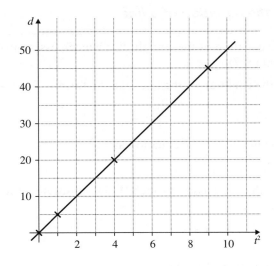

Note: We can only find the law from a straight line graph. It is not easy to discover an equation from a curve.

Exercise 23

1. Two variables Z and X are thought to be connected by an equation of the form $Z = aX + c$. Here are some values of Z and X.

X	1	2	3·6	4·2	6·4
Z	4	6·5	10·5	12	17·5

Draw a graph with X on the horizontal axis using a scale of 2 cm to 1 unit and Z on the vertical axis with a scale of 1 cm to 1 unit.
Use your graphs to find a and c and hence write down the equation relating Z and X.

2. In an experiment, the following measurements of the variables q and t were taken.

q	0·5	1·0	1·5	2·0	2·5	3·0
t	3·85	5·0	6·1	7·0	7·75	9·1

A scientist suspects that q and t are related by an equation of the form $t = mq + c$, (m and c constants). Plot the values obtained from the experiment and draw the line of best fit through the points. Plot q on the horizontal axis with a scale of 4 cm to 1 unit, and t on the vertical axis with a scale of 2 cm to 1 unit. Find the gradient and intercept on the t-axis and hence estimate the values of m and c.

3. In an experiment, the following measurements of p and z were taken:

z	1·2	2·0	2·4	3·2	3·8	4·6
p	11·5	10·2	8·8	7·0	6·0	3·5

Plot the points on a graph with z on the horizontal axis and draw the line of best fit through the points. Hence estimate the values of n and k if the equation relating p and z is of the form $p = nz + k$.

4. In an experiment the following measurements of t and z were taken:

t	1·41	2·12	2·55	3·0	3·39	3·74
z	3·4	3·85	4·35	4·8	5·3	5·75

Draw a graph, plotting t^2 on the horizontal axis and z on the vertical axis, and hence confirm that the equation connecting t and z is of the form $z = mt^2 + c$. Find approximate values for m and c.

5. Two variables T and V are thought to be related by an equation which is either

$$V = a\sqrt{T} + b \text{ or } V = \frac{a}{T} + b.$$

Here are some values of T and V.

T	1	2	4	10
V	15·5	9·5	6·5	4·7

Draw a suitable graph (or graphs) to determine which equation is the correct one.
What are the values of a and b?

6. Two engineers measured the power P (kilowatts) required for a car to travel at various speeds V (km/h). Here are their results.

V	40	60	80	100	120	140
P	13·5	30·5	54	84·5	122	166

John thinks the formula connecting P and V is of the form $P = mV + c$.
Steve thinks the formula is $\sqrt{P} = kV$.
Steve says that John must be wrong, just by looking at the results.
Explain why John's formula cannot be correct.
Draw a graph with V on the horizontal axis ($1\,\text{cm} = 20\,\text{units}$) and $\sqrt{P}$ on the vertical axis ($1\,\text{cm} = 1\,\text{unit}$).
Draw a line of best fit to confirm that Steve's formula is correct and estimate the value of the constant k.

3.6 Real-life graphs

Travel graphs

Exercise 24

1. The graph shows the journeys made by
   a van and a car starting at York,
   travelling to Durham and returning to
   York.
   (a) For how long was the van
       stationary during the journey?
   (b) At what time did the car first
       overtake the van?
   (c) At what speed was the van
       travelling between 09:30 and 10:00?
   (d) What was the greatest speed
       attained by the car during the entire
       journey?
   (e) What was the average speed of the
       car over its entire journey?

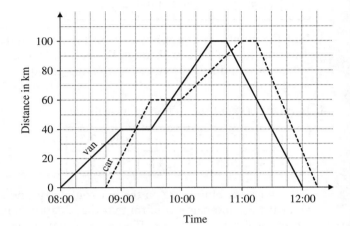

2. The graph shows the journeys of a bus
   and a car along the same road. The bus
   goes from Leeds to Darlington and back
   to Leeds. The car goes from Darlington to
   Leeds and back to Darlington.
   (a) When did the bus and the car meet for
       the second time?
   (b) At what speed did the car travel from
       Darlington to Leeds?
   (c) What was the average speed of the bus
       over its entire journey?
   (d) Approximately how far apart were the
       bus and the car at 09:45?
   (e) What was the greatest speed attained
       by the car during its entire journey?

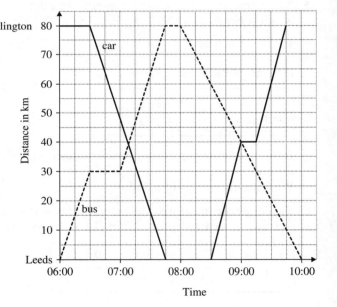

In Questions **3, 4, 5,** draw a travel graph to illustrate the journey
described. Draw axes with the same scales as in Question **2**.

3. Mrs Chuong leaves home at 08:00 and drives at a speed of 50 km/h.
   After $\frac{1}{2}$ hour she reduces her speed to 40 km/h and continues at this
   speed until 09:30. She stops from 09:30 until 10:00 and then returns
   home at a speed of 60 km/h.
   Use a graph to find the approximate time at which she arrives
   home.

4. Mr Coe leaves home at 09:00 and drives at a speed of 20 km/h. After $\frac{3}{4}$ hour he increases his speed to 45 km/h and continues at this speed until 10:45. He stops from 10:45 until 11:30 and then returns home at a speed of 50 km/h.
   Use the graph to find the approximate time at which he arrives home.

5. At 10:00 Akram leaves home and cycles to his grandparents' house which is 70 km away. He cycles at a speed of 20 km/h until 11:15, at which time he stops for $\frac{1}{2}$ hour. He then completes the journey at a speed of 30 km/h. At 11:45 Akram's sister, Hameeda, leaves home and drives her car at 60 km/h. Hameeda also goes to her grandparents' house and uses the same road as Akram.
   At approximately what time does Hameeda overtake Akram?

6. A boat can travel at a speed of 20 km/h in still water. The current in a river flows at 5 km/h so that downstream the boat travels at 25 km/h and upstream it travels at only 15 km/h.

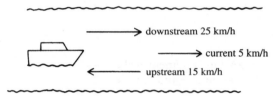

The boat has only enough fuel to last 3 hours. The boat leaves its base and travels downstream. Draw a distance–time graph and draw lines to indicate the outward and return journeys. After what time must the boat turn round so that it can get back to base without running out of fuel?

7. The boat in Question **6** sails in a river where the current is 10 km/h and it has fuel for four hours. After what time must the boat turn round this time if it is not to run out of fuel?

Sketch graphs

Exercise 25

1. Which of the graphs A to D below best fits the following statement:
   'Unemployment is still rising but by less each month.'?

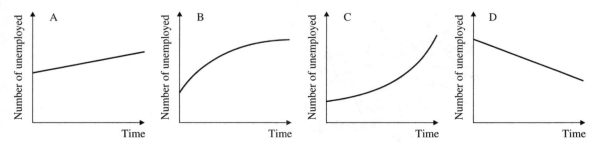

2. Which of the graphs A to D below best fits each of the following statements?

(a) The birthrate was falling but is now steady.

(b) Unemployment, which rose slowly until 1990, is now rising rapidly.

(c) Inflation, which has been rising steadily, is now beginning to fall.

(d) The price of gold has fallen steadily over the last year.

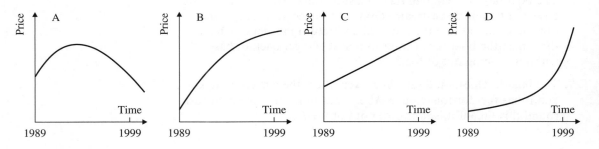

3. Which of the graphs A to D best fits the following statement: 'The price of oil was rising more rapidly in 1999 than at any time in the previous ten years.'?

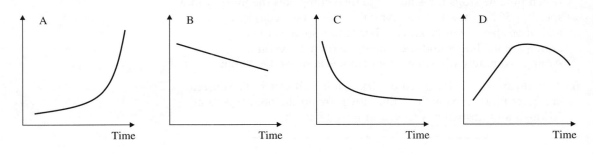

4. A car hire firm makes the charges shown. Draw a sketch graph showing 'miles travelled' across the page and 'total cost' up the page. Assume the car is hired for just one day.

5. Here are two car racing circuits.

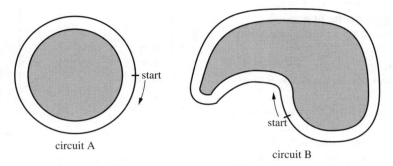

circuit A

circuit B

Sketch a speed–time graph to show the speed of a racing car as it
goes around one lap of each circuit.
[Not the first lap. Why?]

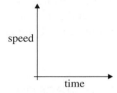

speed

time

6. Water is poured at a constant rate into each
of the containers A, B and C.

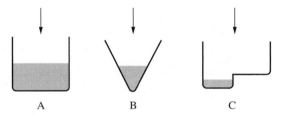

A   B   C

The graphs X, Y and Z show how the water level rises.
Decide which graph fits each container.

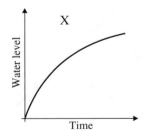

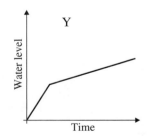

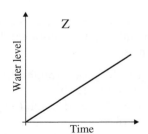

7. Mr Gibson organises a trip for pupils from his school. He hires a
coach for £100 and decides to divide the cost equally between the
pupils on the trip. Sketch a graph showing the 'number of pupils'
across the page and the 'cost per pupil' up the page.
Is it a linear graph?

8. A car travels along a motorway and the amount of petrol in its tank is monitored as shown on the graph.
  (a) How much petrol was bought at the first stop?
  (b) What was the petrol consumption in miles per gallon:
     (i) before the first stop,
     (ii) between the two stops?
  (c) What was the average petrol consumption over the 200 miles?

After it leaves the second service station the car encounters road works and slow traffic for the next 20 miles. Its petrol consumption is reduced to 20 m.p.g. After that, the road clears and the car travels a further 75 miles during which time the consumption is 30 m.p.g. Draw the graph above and extend it to show the next 95 miles. How much petrol is in the tank at the end of the journey?

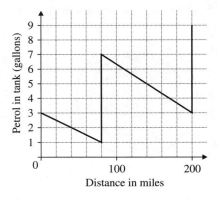

3.7 Simultaneous linear equations

Graphical solution

Louise and Philip are two children; Louise is 5 years older than Philip. The sum of their ages is 12 years. How old is each child?

Let Louise be x years old and Philip be y years old.
The sum of their ages is 12,

so $x + y = 12$

The difference of their ages is 5,

so $x - y = 5$

Because both equations relate to the same information, both can be plotted to the same axes.

$x + y = 12$ goes through $(0,12)$, $(2,10)$, $(6,6)$, $(12,0)$.
$x - y = 5$ goes through $(5,0)$, $(7,2)$, $(10,5)$.
The values of x and y are found from the point where the two lines intersect. The point $(8 \cdot 5, 3 \cdot 5)$ lies on both lines.

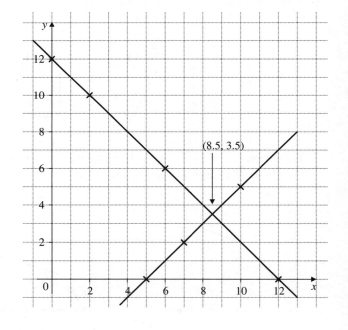

The solution is $x = 8 \cdot 5$, $y = 3 \cdot 5$

So Louise is $8\frac{1}{2}$ years old and Philip is $3\frac{1}{2}$ years old.

Note that only three points are required to define a straight line.

Exercise 26

1. Use the graphs below to solve the equations.

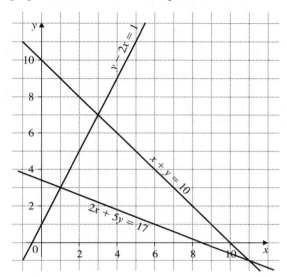

Solve:

(a) $x + y = 10$
$y - 2x = 1$

(b) $2x + 5y = 17$
$y - 2x = 1$

(c) $x + y = 10$
$2x + 5y = 17$

In Questions **2** to **6**, solve the simultaneous equations by first drawing graphs.

2. $x + y = 6$
$2x + y = 8$
Draw axes with x and y from 0 to 8.

3. $x + 2y = 8$
$3x + y = 9$
Draw axes with x and y from 0 to 9.

4. $x + 3y = 6$
$x - y = 2$
Draw axes with x from 0 to 8 and y from -2 to 4.

5. $5x + y = 10$
$x - y = -4$
Draw axes with x from -4 to 4 and y from 0 to 10.

6. $a + 2b = 11$
$2a + b = 13$
Here, the unknowns are a and b. Draw the a axis across the page from 0 to 13 and the b axis up the page also from 0 to 13.

7. There are four lines drawn here.
Write down the solutions to the following:

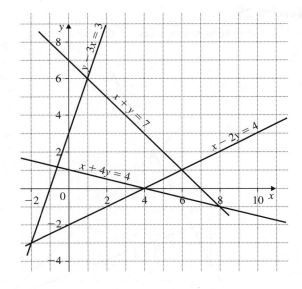

(a) $x - 2y = 4$
$\quad x + 4y = 4$

(b) $\quad x + y = 7$
$\quad y - 3x = 3$

(c) $y - 3x = 3$
$\quad x - 2y = 4$

(d) $x + 4y = 4$
$\quad x + y = 7$

(e) $x + 4y = 4$
$\quad y - 3x = 3$

[Here, x and y should be given correct to 1 d.p.]

Algebraic solution

Simultaneous equations can be solved without drawing graphs. There are two methods: substitution and elimination. Choose the one which seems suitable.

(a) Substitution method

This method is used when one equation contains a single x or y as in equation [2] of the example below.

$$3x - 2y = 0 \qquad \ldots [1]$$
$$2x + y = 7 \qquad \ldots [2]$$

(a) Label the equations so that the working is made clear.
(b) In *this* case, write y in terms of x from equation [2].
(c) Substitute this expression for y in equation [1] and solve to find x.
(d) Find y from equation [2] using this value of x.

$$2x + y = 7$$
$$\quad y = 7 - 2x$$

Substituting in [1]
$$3x - 2(7 - 2x) = 0$$
$$3x - 14 + 4x = 0$$
$$7x = 14$$
$$x = 2$$

Substituting in [2]
$$2 \times 2 + y = 7$$
$$y = 3$$

The solutions are $x = 2$, $y = 3$.

Exercise 27

Use the substitution method to solve the following:

1. $2x + y = 5$
   $x + 3y = 5$

2. $x + 2y = 8$
   $2x + 3y = 14$

3. $3x + y = 10$
   $x - y = 2$

4. $2x + y = -3$
   $x - y = 2$

5. $4x + y = 14$
   $x + 5y = 13$

6. $x + 2y = 1$
   $2x + 3y = 4$

7. $2x + y = 5$
   $3x - 2y = 4$

8. $2x + y = 13$
   $5x - 4y = 13$

9. $7x + 2y = 19$
   $x - y = 4$

10. $b - a = -5$
    $a + b = -1$

11. $a + 4b = 6$
    $8b - a = -3$

12. $a + b = 4$
    $2a + b = 5$

13. $3m = 2n - 6\frac{1}{2}$
    $4m + n = 6$

14. $2w + 3x - 13 = 0$
    $x + 5w - 13 = 0$

15. $x + 2(y - 6) = 0$
    $3x + 4y = 30$

16. $2x = 4 + z$
    $6x - 5z = 18$

17. $3m - n = 5$
    $2m + 5n = 7$

18. $5c - d - 11 = 0$
    $4d + 3c = -5$

(b) Elimination method

Use this method when the first method is unsuitable (some prefer to use it for every question).

$$2x + 3y = 5 \qquad \ldots [1]$$
$$5x - 2y = -16 \qquad \ldots [2]$$

$[1] \times 5 \qquad 10x + 15y = 25 \qquad \ldots [3]$
$[2] \times 2 \qquad 10x - 4y = -32 \qquad \ldots [4]$
$[3] - [4] \qquad 15y - (-4y) = 25 - (-32)$
$$19y = 57$$
$$y = 3$$

Substitute in [1] $2x + 3 \times 3 = 5$
$$2x = 5 - 9 = -4$$
$$x = -2$$

The solutions are $x = -2$, $y = 3$.

Exercise 28

Use the elimination method to solve the following:

1. $2x + 5y = 24$
   $4x + 3y = 20$

2. $5x + 2y = 13$
   $2x + 6y = 26$

3. $3x + y = 11$
   $9x + 2y = 28$

4. $x + 2y = 17$
   $8x + 3y = 45$

5. $3x + 2y = 19$
   $x + 8y = 21$

6. $2a + 3b = 9$
   $4a + b = 13$

7. $2x + 7y = 17$
   $5x + 3y = -1$

8. $5x + 3y = 23$
   $2x + 4y = 12$

9. $3x + 2y = 11$
   $2x - y = -3$

10. $3x + 2y = 7$
    $2x - 3y = -4$

11. $x - 2y = -4$
    $3x + y = 9$

12. $5x - 7y = 27$
    $3x - 4y = 16$

13. $x + 3y - 7 = 0$
    $2y - x - 3 = 0$

14. $3a - b = 9$
    $2a + 2b = 14$

15. $2x - y = 5$
    $\dfrac{x}{4} + \dfrac{y}{3} = 2$

16. $3x - y = 17$
    $\dfrac{x}{5} + \dfrac{y}{2} = 0$

17. $4x - 0{\cdot}5y = 12{\cdot}5$
    $3x + 0{\cdot}8y = 8{\cdot}2$

18. $0{\cdot}4x + 3y = 2{\cdot}6$
    $x - 2y = 4{\cdot}6$

Solving problems using simultaneous equations

As with linear equations, solving problems involves four steps.

(a) Let the two unknown quantities be x and y.
(b) Write the problem in the form of two equations.
(c) Solve the equations and give the answers in words.
(d) Check your solution using the problem and not your equations.

A motorist buys 24 litres of petrol and 5 litres of oil for £26·75, while another motorist buys 18 litres of petrol and 10 litres of oil for £31. Find the cost of 1 litre of petrol and 1 litre of oil at this garage.

(a) Let the cost of 1 litre of petrol be x pence and the cost of 1 litre of oil be y pence.

(b) $24x + 5y = 2675$ ... [1]
        $18x + 10y = 3100$ ... [2]

(c) Solve the equations: $x = 75$, $y = 175$

   1 litre of petrol costs 75 pence.
   1 litre of oil costs 175 pence.

(d) Check: $24 \times 75 + 5 \times 175 = 2675\text{p} = £26·75$

             $18 \times 75 + 10 \times 175 = 3100\text{p} = £31$ ✓

Exercise 29

Solve each problem by forming a pair of simultaneous equations.

1. Find two numbers with a sum of 15 and a difference of 4.

2. Twice one number added to three times another gives 21. Find the numbers, if the difference between them is 3.

3. The average of two numbers is 7, and three times the difference between them is 18. Find the numbers.

4. Here is a puzzle from a newspaper. The ? and * stand for numbers which are to be found. The totals for the rows and columns are given.
   Write down two equations involving ? and * and solve them to find the values of ? and *

?	*	?	*	36
?	*	*	?	36
*	?	*	*	33
?	*	?	*	36
39	33	36	33	

5. The line, with equation $y + ax = c$, passes through the points (1, 5) and (3, 1). Find a and c.
Hint: For the point (1, 5) put $x = 1$ and $y = 5$ into $y + ax = c$, etc.

6. The line $y = mx + c$ passes through (2, 5) and (4, 13). Find m and c.

7. A stone is thrown into the air and its height, h metres above the ground, is given by the equation

$h = at - bt^2$.

From an experiment we know that $h = 40$ when $t = 2$ and that $h = 45$ when $t = 3$.
Show that, $a - 2b = 20$
       and $a - 3b = 15$.
Solve these equations to find a and b.

8. A television addict can buy either two televisions and three video-recorders for £1750 or four televisions and one video-recorder for £1250. Find the cost of one of each.

9. A pigeon can lay either white or brown eggs. Three white eggs and two brown eggs weigh 13 ounces, while five white eggs and four brown eggs weigh 24 ounces. Find the weight of a brown egg, b, and of a white egg, w.

10. A tortoise makes a journey in two parts; it can either walk at 2 m/min or crawl at 1 m/min.

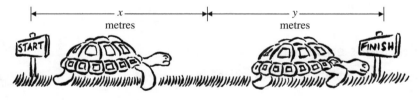

walks at 2 m/min   crawls at 1 m/min

If the tortoise walks the first part and crawls the second, the journey takes 110 minutes.
If it crawls the first part and walks the second, the journey takes 100 minutes.
Let x metres be the length of the first part and y metres be the length of the second part.
Write down two simultaneous equations and solve them to find the lengths of the two parts of the journey.
[Use the formula, Time = Distance ÷ Speed]

11. A cyclist completes a journey of 500 m in 22 seconds, part of the way at 10 m/s and the remainder at 50 m/s. How far does she travel at each speed?

12. A bag contains forty coins, all of them either 2p or 5p coins. If the value of the money in the bag is £1·55, find the number of each kind.

13. A slot machine takes only 10p and 50p coins and contains a total of twenty-one coins altogether. If the value of the coins is £4·90, find the number of coins of each value.

14. Thirty tickets were sold for a concert, some at 60p and the rest at £1. If the total raised was £22, how many had the cheaper tickets?

15. A kipper can swim at $14\,\text{m/s}$ with the current and at $6\,\text{m/s}$ against it. Find the speed of the current and the speed of the kipper in still water.

16. If the numerator and denominator of a fraction are both decreased by one the fraction becomes $\frac{2}{3}$. If the numerator and denominator are both increased by one the fraction becomes $\frac{3}{4}$. Find the original fraction.

17. In three years' time a pet mouse will be as old as his owner was four years ago. Their present ages total 13 years. Find the age of each now.

18. The curve $y = ax^2 + bx + c$ passes through the points (1, 8), (0, 5) and (3, 20). Find the values of a, b and c and hence the equation of the curve.

19. The curve $y = ax^2 + bx + c$ passes through (1, 8), (−1, 2) and (2, 14). Find the equation of the curve.

Linear/quadratic simultaneous equations

A pair of simultaneous equations can contain equations which are linear, quadratic, cubic and so on.
Consider the pair of simultaneous equations

$$y = 2x + 2$$
$$\text{and } y = x^2 - 1$$

The diagram shows that the line and the curve intersect at (−1, 0) and (3, 8).

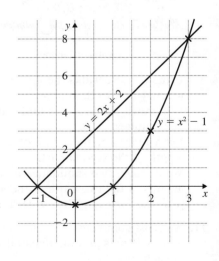

These solutions can be obtained algebraically using the method of substitution as follows:

$$y = 2x + 2 \quad \dots [A]$$
$$y = x^2 - 1 \quad \dots [B]$$

Substitute $y = 2x + 2$ into equation [B]

$$2x + 2 = x^2 - 1$$
$$\Rightarrow \quad x^2 - 2x - 3 = 0$$
$$(x - 3)(x + 1) = 0$$
$$x = 3 \text{ or } x = -1$$

when $x = 3$, $y = 8$
when $x = -1$, $y = 0$ $\Big\}$ These are the solutions.

Exercise 30

This work requires a knowledge of the solution of quadratic equations.

1. Solve the simultaneous equations:
   (a) $y = x^2 - 2x$
       $y = x + 4$
   (b) $y = 7x - 8$
       $y = x^2 - x + 7$

2. Solve the simultaneous equations:
   (a) $y = x^2 - 3x + 7$
       $5x - y = 8$
   (b) $y = 9x - 4$
       $y = 2x^2$

3. A circle with centre (0, 0) and radius r has equation $x^2 + y^2 = r^2$. Find the coordinates of the points of intersection of the line $y = x + 1$ and the circle $x^2 + y^2 = 13$.

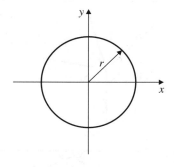

4. Solve the simultaneous equations: $x^2 + y^2 = 20$, $y = x - 2$

5. Find the coordinates of the two points where the line $y = 4x - 8$ intersects the curve $y^2 = 16x$.

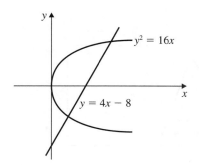

6. Explain why there are no solutions to the simultaneous equations $x^2 + y^2 = 1$ and $y = x + 10$.

7. (a) Draw accurately the graph of the circle $x^2 + y^2 = 25$. Take values of x from -5 to $+5$ and use a scale of 1 cm to 1 unit on both axes.
   (b) Draw the graph of $y = x + 1$ and find the coordinates of the two points of intersection.

†8. Find the coordinates of the point of intersection of the curve $y = x^3$ and the line $x + y = 10$.

4 SHAPE, SPACE AND MEASURES 1

4.1 Angle facts

● The angles at a point add up to 360°.
The angles on a straight line add up to 180°.

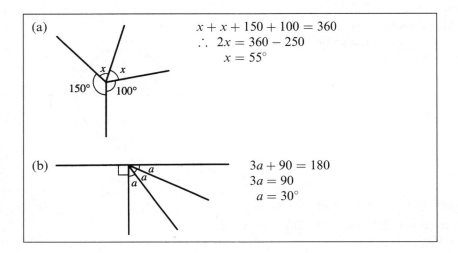

(a)

$$x + x + 150 + 100 = 360$$
$$\therefore \ 2x = 360 - 250$$
$$x = 55°$$

$150°$ $100°$

(b)

$$3a + 90 = 180$$
$$3a = 90$$
$$a = 30°$$

● The angles in a triangle add up to 180°.

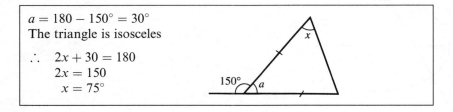

$$a = 180 - 150° = 30°$$
The triangle is isosceles

$$\therefore \ 2x + 30 = 180$$
$$2x = 150$$
$$x = 75°$$

$150°$ a x

● When a line cuts a pair of parallel lines all the acute angles are equal
and all the obtuse angles are equal.

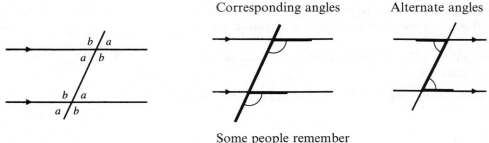

Corresponding angles

Alternate angles

b a
a b

b a
a b

Some people remember
'F angles'

and 'Z angles'

Exercise 1

Find the angles marked with letters. The lines AB and CD are straight.

1.

2.

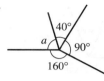

3.

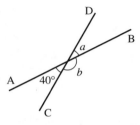

4.

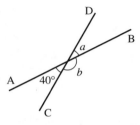

5.

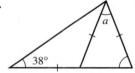

6.

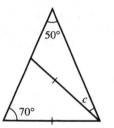

7.

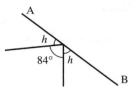

8.

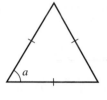

9.

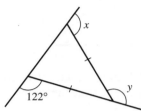

10.

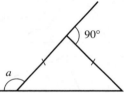

11.

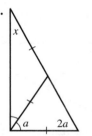

12.

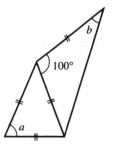

13.

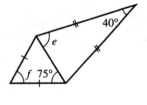

14.

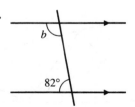

15.

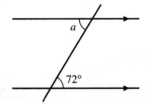

16.

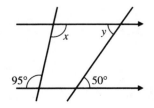

17.

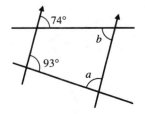

18.

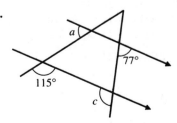

19.

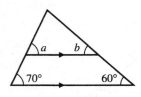

20.

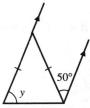

A line is drawn
parallel to one side.

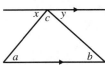

21.

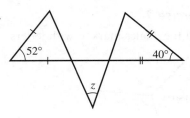

22. Here is a triangle.

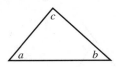

(a) What is angle x? What is angle y?
(b) Prove that the sum of the angles in a triangle is 180°.

23. (a) Write down the sum of the angles p, q and r.
(b) Write down the sum of the angles q and x.
(c) Hence show that the exterior angle of a triangle is equal to the sum of the interior angles at the other two vertices.

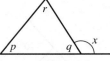

24. The diagram shows two equal squares joined to a triangle. Find the angle x.

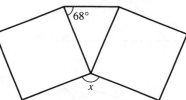

25. Find the angle a between the diagonals of the parallelogram.

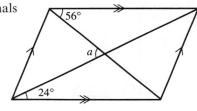

26. The diagram shows a series of isosceles triangles drawn between two lines. Find the value of x.

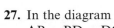

27. In the diagram
$AB = BD = DC$ and
$A\hat{D}C - D\hat{A}C = 70°$

Find $A\hat{D}B$.

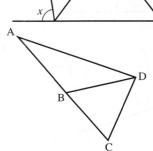

Quadrilaterals

Square: Four equal sides;
All angles 90°;
Four lines of symmetry.
Rotational symmetry of order 4

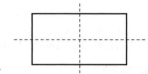

Rectangle (not square): Two pairs of equal
and parallel sides;
All angles 90°;
Two lines of symmetry;
Rotational symmetry of order 2.

Rhombus: Four equal sides;
Opposite sides parallel;
Diagonals bisect at right angles;
Diagonals bisect angles of rhombus;
Two lines of symmetry;
Rotational symmetry of order 4.

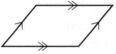

Parallelogram: Two pairs of equal and parallel sides;
Opposite angles equal;
No lines of symmetry (in general);
Rotational symmetry of order 2.

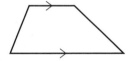

Trapezium: One pair of parallel sides;
Rotational symmetry of order 1.

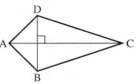

Kite: AB = AD, CB = CD;
Diagonals meet at 90°;
One line of symmetry;
Rotational symmetry of order 1.

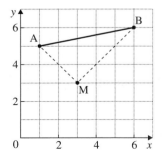

● For all quadrilaterals the sum of the interior angles is 360°.

Exercise 2

1. ABCD is a rhombus whose diagonals intersect at M.
   Find the coordinates of C and D.

2.

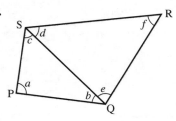

Quadrilateral PQRS is split into two triangles.
(a) Write down the sum of the angles a, b and c.
(b) Write down the sum of the angles d, e and f.
(c) Show that the sum of the angles in the quadrilateral is 360°.

In Questions **3** to **11** begin by drawing a diagram and remember to put the letters around the shape in alphabetical order.

3. In a parallelogram WXYZ, $W\hat{X}Y = 72°$, $Z\hat{W}Y = 80°$.
Calculate:
(a) $W\hat{Z}Y$ (b) $X\hat{W}Z$ (c) $W\hat{Y}X$

4. In a kite ABCD, AB = AD, BC = CD, $C\hat{A}D = 40°$ and $C\hat{B}D = 60°$. Calculate:
(a) $B\hat{A}C$ (b) $B\hat{C}A$ (c) $A\hat{D}C$

5. In a rhombus ABCD, $A\hat{B}C = 64°$. Calculate:
(a) $B\hat{C}D$ (b) $A\hat{D}B$ (c) $B\hat{A}C$

6. In the parallelogram PQRS line QA bisects (cuts in half) angle PQR.
Calculate the size of angle RAQ.

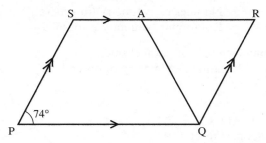

7. In a trapezium ABCD, AB is parallel to DC, AB = AD, BD = DC and $B\hat{A}D = 128°$. Find:
(a) $A\hat{B}D$ (b) $B\hat{D}C$ (c) $B\hat{C}D$

8.

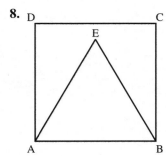

ABE is an equilateral triangle drawn inside square ABCD. Calculate the size of angle DEC.

9. In a kite PQRS with PQ = PS and RQ = RS, $Q\hat{R}S = 40°$ and $Q\hat{P}S = 100°$. Find $P\hat{Q}R$.

10. In a rhombus PQRS, $R\hat{P}Q = 54°$. Find:
(a) $P\hat{R}Q$ (b) $P\hat{S}R$ (c) $R\hat{Q}S$

11. In a kite PQRS, $R\widehat{P}S = 2\,P\widehat{R}S$, PQ = QS = PS and QR = RS.
Find:

    (a) $Q\widehat{P}S$ (b) $P\widehat{R}S$ (c) $Q\widehat{S}R$

Angles in polygons

Exterior angles of a polygon

The exterior angle of a polygon is the angle between a produced side
and the adjacent side of the polygon. The word 'produced' in this
context means 'extended'.

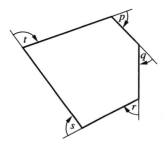

If we put all the exterior angles together we can see that the sum of the
angles is 360°. This is true for any polygon.

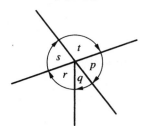

- The sum of the exterior angles of a polygon = 360°

Note:

(a) In a *regular* polygon all exterior angles are equal.

(b) For a regular polygon with n sides, each exterior angle $= \dfrac{360}{n}$.

The diagram shows a regular octagon (8 sides).

(a) Calculate the size of each exterior angle (marked e)

(b) Calculate the size of each interior angle (marked i)

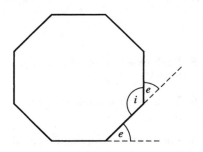

(a) There are 8 exterior angles and the sum of this angle is 360°

   ∴   angle $e = \frac{360}{8} = 45°$

(b) $e + i = 180°$ (angles on a straight line)

   ∴ $i = 135°$

Exercise 3

1. Look at the polygon shown.
   (a) Calculate each exterior angle
   (b) Check that the total of the exterior angles is 360°.

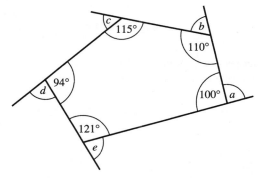

2. The diagram shows a regular decagon.
   (a) Calculate the angle *a*.
   (b) Calculate the interior angle of a regular decagon.

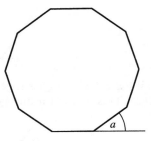

3. Find: (a) the exterior angle
   (b) the interior angle of a regular polygon with
   (i) 9 sides (ii) 18 sides (iii) 45 sides (iv) 60 sides.

4. Find the angles marked with letters.

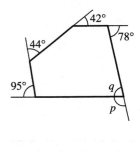
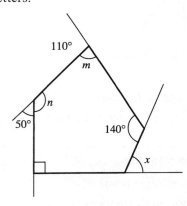

5. Each exterior angle of a regular polygon is 15°. How many sides has the polygon?

6. Each interior angle of a regular polygon is 140°. How many sides has the polygon?

7. Each exterior angle of a regular polygon is 18°. How many sides has the polygon?

Sum of interior angles

pentagon, 5 sides

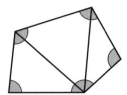

There are 3 triangles.
Sum of interior angles = $3 \times 180°$
$= (5 - 2) \times 180°$

hexagon, 6 sides

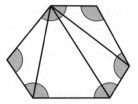

There are 4 triangles.
Sum of interior angles = $4 \times 180°$
$= (6 - 2) \times 180°$

- In general: The sum of the interior angles of a polygon with n sides is $(n - 2) \times 180°$.

Exercise 4

1. Use the formula above to find the sum of the interior angles in:
   (a) an octagon (8 sides)
   (b) a decagon (10 sides).

2. (a) Work out the sum of the interior angles in a polygon with 20 sides.
   (b) What is the size of each interior angle in a *regular* polygon with 20 sides?

3. Find the angles marked with letters.
   (a)   (b)   (c)

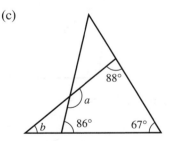

4. A regular dodecagon has 12 sides.
   (a) Calculate the size of each interior angle, i.
   (b) Use your answer to find the size of each exterior angle, e.

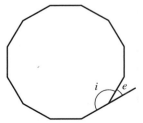

5. The sum of the interior angles of a polygon with n sides is 3600°.
Find the value of n.

6. The sides of a regular polygon subtend angles of 18° at the centre of
the polygon. How many sides has the polygon?

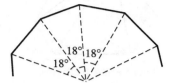

4.2 Congruent triangles

Two plane figures are congruent if one fits exactly on the other.
The four types of congruence for triangles are as follows:

(a) Two sides and the included angle (S.A.S)

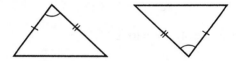

(b) Two angles and a corresponding side (A.A.S)

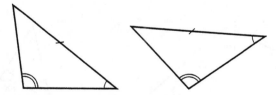

(c) Three sides (S.S.S)

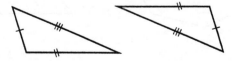

(d) Right angle, hypotenuse and one other side (R.H.S)

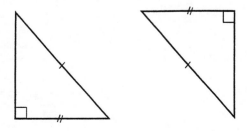

PQRS is a square. M is the mid-point of PQ, N is the mid-point of RQ.
Prove that SN = RM.

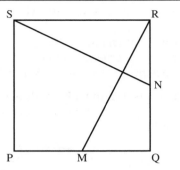

$\widehat{SRQ} = \widehat{PQR}$ (Both right angles)
   $SR = RQ$ (Sides of a square)
   $RN = MQ$ (Given)

∴ Triangle SRN is congruent to triangle RQM. (S.A.S)
   We can write △SRN ≡ △RQM
   The sign ≡ means 'is identically equal to'.
   [Note that the order of letters shows corresponding vertices in the triangles.]

Since the triangles are congruent, it follows that SN = RM.

Exercise 5

For Questions **1** to **6**, decide whether the pair of triangles are congruent. If they are congruent, state which conditions for congruency are satisfied.

1.

2.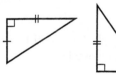

3.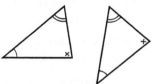

4.

5.

6.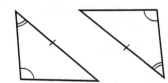

7.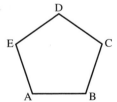

ABCDE is a regular pentagon.
List all the triangles which are congruent to triangle BCE.

8. By construction show that it is possible to draw two *different* triangles with the angle and sides shown.

This shows that 'A.S.S' is not a condition for triangles to be congruent.

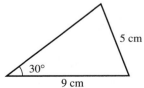

5 cm

30°

9 cm

9. Draw triangle ABC with AB = BC and point D at the mid-point of AC.
   Prove that triangles ABD and CBD are congruent and state which case of congruency applies.
   Hence prove that angles A and C are equal.

10. Tangents TA and TB are drawn to touch the circle with centre O. Use congruent triangles to prove that the two tangents are the same length.

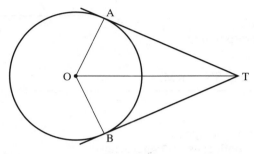

11. Triangle LMN is isosceles with LM = LN; X and Y are points on LM, LN respectively such that LX = LY. Prove that triangles LMY and LNX are congruent.

12. ABCD is a quadrilateral and a line through A parallel to BC meets DC at X. If $\widehat{D} = \widehat{C}$, prove that $\triangle ADX$ is isosceles.

13. XYZ is a triangle with XY = XZ. The bisectors of angles Y and Z meet the opposite sides in M and N respectively. Prove that YM = ZN.

14. In the diagram, DX = XC, DV = ZC and the lines AB and DC are parallel. Prove that:
    (a) AX = BX
    (b) AC = BD
    (c) triangles DBZ and CAV are congruent.

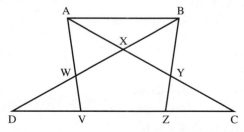

15. Draw a parallelogram ABCD and prove that its diagonals bisect each other.

16. PQRS is a square and equilateral triangles PQA and RQB are drawn with A inside and B outside the square.

    By considering triangles PQR and AQB, prove that AB is equal to PR.

4.3 Locus

In mathematics, the word *locus* describes the position of points which obey a certain rule. The locus can be the path traced out by a moving point.

Three important loci

(a) Circle

The locus of points which are equidistant from a fixed point O is shown. It is a **circle** with centre O.

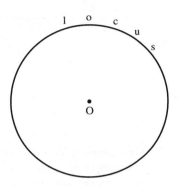

(b) Perpendicular bisector

The locus of points which are equidistant from two fixed points A and B is shown.

The locus is the **perpendicular bisector** of the line AB. Use compasses to draw arcs, as shown, or use a ruler and a protractor.

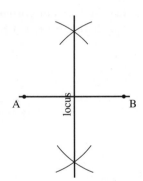

(c) Angle bisector

The locus of points which are equidistant from two fixed lines AB and AC is shown.

The locus is the line which bisects the angle BAC. Use compasses to draw arcs or use a protractor to construct the locus.

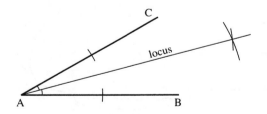

Exercise 6

1. Draw the locus of a point P which moves so that it is always 3 cm from a fixed point X.

   •X

2. Mark two points P and Q which are 10 cm apart. Draw the locus of points which are equidistant from P and Q.

3. Draw two lines AB and AC of length 8 cm, where $B\widehat{A}C = 40°$.
Draw the locus of points which are equidistant from AB and
AC.

4. A sphere rolls along a surface from A to B. Sketch the locus of the
centre of the sphere in each case.

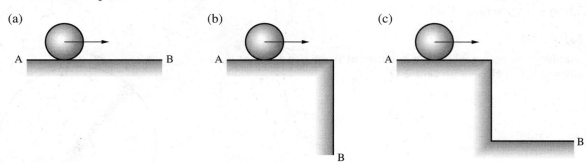

(a)   (b)   (c)

5. A rectangular slab is rotated around
corner B from position 1 to position 2.
Draw a diagram, on squared paper, to
show:
(a) the locus of corner A
(b) the locus of corner C.

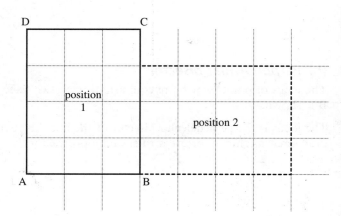

6.

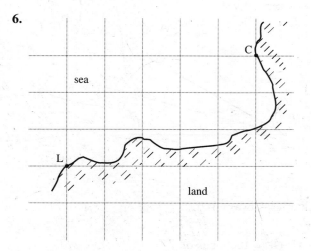

Scale 1 cm : 10 km

The diagram shows a section of coastline with
a lighthouse L and coast guard C. A sinking
ship sends a distress signal. The ship appears
to be up to 40 km from L and up to 20 km
from C.

Copy the diagram and show the region in
which the sinking ship could be.

7. (a) Draw the triangle LMN full size.
   (b) Draw the locus of the points which are:
      (i) equidistant from L and N
      (ii) equidistant from LN and LM
      (iii) 4 cm from M

[Draw the three loci in different colours].

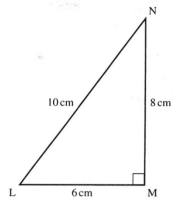

8. Draw a line AB of length 6 cm. Draw the locus of a point P so that angle ABP = 90°.

9. The diagram shows a garden with a fence on two sides and trees at two corners. A sand pit is to be placed so that it is:

(a) equidistant from the 2 fences
(b) equidistant from the 2 trees.

Make a scale drawing (1 cm = 1 m) and mark where the sand pit goes.

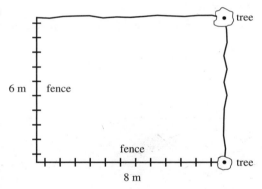

10. Channel 9 in Australia is planning the position of a new TV satellite to send pictures all over the country. The new satellite is to be placed:

(a) an equal distance from Darwin and Adelaide.
(b) not more than 2000 km from Perth.
(c) not more than 1800 km from Brisbane.

Make a copy of the map on squared paper using the grid lines as reference.

Show clearly where the satellite could be placed so that it satisfies the conditions (a), (b), (c) above.

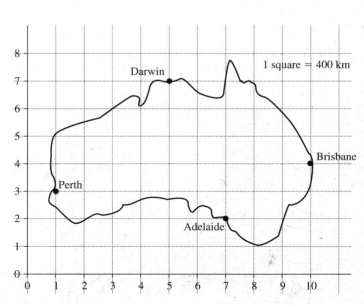

11. A tortoise is tethered to a post P in gardens A and B.
The rope is 3 m long in both gardens.
Show the locus of points which the tortoise can reach.

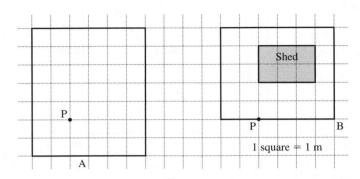

1 square = 1 m

Exercise 7

1. The perpendicular from a point to a line is obtained as follows.

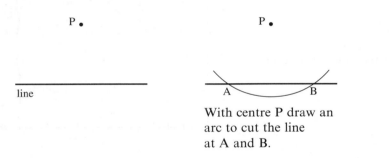

With centre P draw an
arc to cut the line
at A and B.

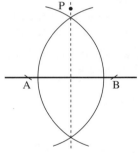

Construct the perpendicular
of bisector of AB.

Draw a line and a point P about 5 cm away.
Construct the line which passes through P and which is
perpendicular to the line.

2. Draw a straight line and mark a point A on the line. Construct the
perpendicular at A.

3. A circle, centre O, radius 5 cm, is inscribed inside a square ABCD.
Point P moves so that OP $\leqslant$ 5 cm and BP $\leqslant$ 5 cm. Shade the set of
points indicating where P can be.

4. Construct a triangle ABC where AB = 9 cm, BC = 7 cm and
AC = 5 cm.
(a) Sketch and describe the locus of points within the triangle which
are equidistant from AB and AC.
(b) Shade the set of points within the triangle which are less than
5 cm from B and are also nearer to AC than to AB.

5. A goat is tied to one corner on the outside of a barn.
The diagram shows a plan view.
Sketch two plan views of the barn and show the locus of points
where the goat can graze if:
(a) the rope is 4 m long
(b) the rope is 7 m long.

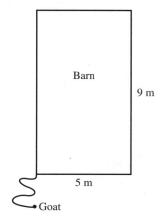

6. Rod AB rotates about the fixed point A. B is joined to C which
slides along a fixed rod. AB = 15 cm and BC = 25 cm.

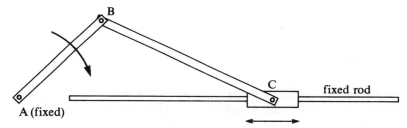

Describe the locus of C as AB rotates clockwise about A.
Find the smallest and the largest distance between C and A.

7. Draw two points M and N 16 cm apart. Draw the locus of a point
P which moves so that the area of triangle MNP is 80 cm^2.

8. Describe the locus of a point which moves in three dimensional
space and is equidistant from two fixed points.

9. Draw two points A and B 10 cm apart.

Place the corner of a piece of paper (or a set square) so that the
edges of the paper pass through A and B.
Mark the position of corner C.
Slide the paper around so the edge still passes through
A and B and mark the new position of C. Repeat several times and
describe the locus of the point C which moves so that angle ACB is
always 90°.

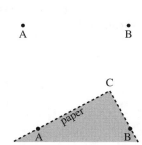

10. Draw any triangle ABC and construct the bisectors of angles B and C to meet at point Y.

With centre at Y draw a circle which just touches the sides of the triangle.
This is the *inscribed* circle of the triangle.

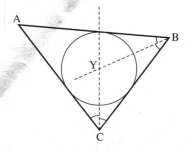

11. Copy the diagram shown. OC = 4 cm.
Sketch the locus of P which moves so that PC is equal to the perpendicular distance from P to the line AB. This locus is called a *parabola*.

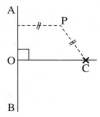

12. A rod OA of length 60 cm rotates about O at a constant rate of 1 revolution per minute. An ant, with good balance, walks along the rod at a speed of 1 cm per second. Sketch the locus of the ant for 1 minute after it leaves O.

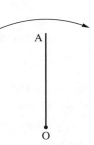

13. Here is a circle with point A on its circumference. The centre of the circle is unknown.
Using ruler and compasses, construct a diameter to the circle through the point A.

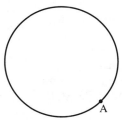

4.4 Pythagoras' theorem

In a right-angled triangle the square on the hypotenuse is equal to the sum of the squares on the other two sides.

$$a^2 + b^2 = c^2$$

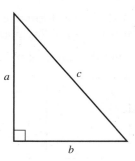

Exercise 8

In Questions **1** to **8**, find x. All the lengths are in cm.

1.
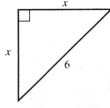

2.
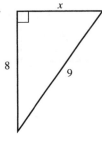

3.
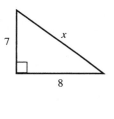

4.
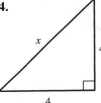

5.
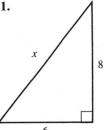

6.
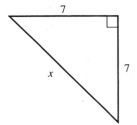

7.
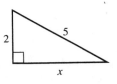

8.
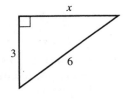

9. Find the length of a diagonal of a rectangle of length 9 cm and width 4 cm.

10. An isosceles triangle has sides 10 cm, 10 cm and 4 cm. Find the height of the triangle.

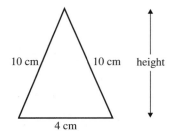

11. A 4 m ladder rests against a vertical wall with its foot 2 m from the wall. How far up the wall does the ladder reach?

12. A ship sails 20 km due North and then 35 km due East. How far is it from its starting point?

13. Find the length of a diagonal of a square of side 9 cm.

14.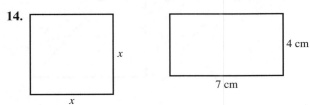

The square and the rectangle have the same length diagonal. Find x.

15. (a) Find the height of the triangle, h.
(b) Find the area of the triangle ABC.

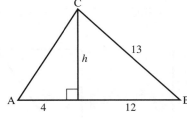

16. A thin wire of length 18 cm is bent in the shape shown.

Calculate the length from A to B.

17. A paint tin is a cylinder of radius 12 cm and height 22 cm. Leonardo, the painter, drops his stirring stick into the tin and it disappears.
Work out the maximum length of the stick.

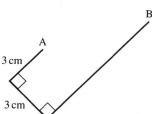

18. A square of side 10 cm is drawn inside a circle.
(a) Find the diameter of the circle.
(b) Find the shaded area.

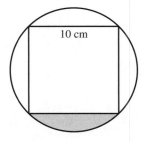

Exercise 9

1. In the diagram A is (1, 2) and B is (6, 4).
Work out the length AB. (First find the length of AN and BN).

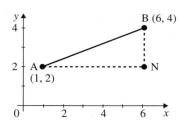

2. On squared paper plot P(1, 3), Q(6, 0), R(6, 6). Find the lengths of
the sides of triangle PQR. Is the triangle isosceles?

In Questions **3** to **8** find x.

3.

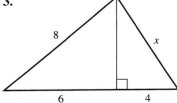

4.

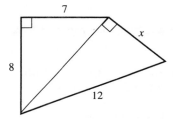

5.

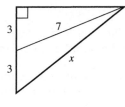

6.

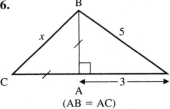

(AB = AC)

7.

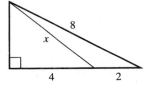

8.

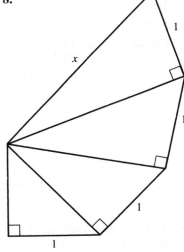

9. The diagram shows a rectangular block.
Calculate: (a) AC   (b) AY.

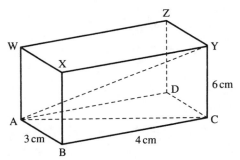

10. The diagram shows a cuboid 5 cm by 4 cm by 12 cm.

Calculate: (a) AC   (b) AD.

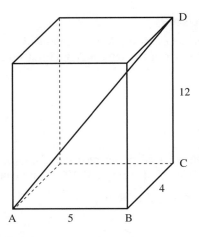

11. Find the length of a diagonal of a rectangular room of length 5 m, width 3 m and height 2·5 m.

12. Find the height of a rectangular box of length 8 cm, width 6 cm where the length of a diagonal is 11 cm.

13. TC is a vertical pole whose base lies at a corner of the horizontal rectangle ABCD.
The top of the pole T is connected by straight wires to points A, B and D.

Calculate: (a) TC
             (b) TD
             (c) (harder) TA.

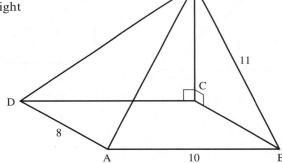

14. The diagonal of a rectangle exceeds the length by 2 cm. If the width of the rectangle is 10 cm, find the length.

15. A ladder reaches H when held vertically against a wall.
When the base is 6 feet from the wall, the top of the ladder is 2 feet lower than H.
How long is the ladder?

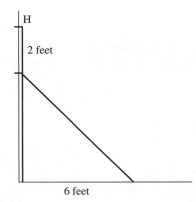

16. (a) Sketch axes in three dimensions as shown.
(b) Find the length of the line from O (0, 0, 0) to A (3, 4, 7).
(c) Find the length of the line from O (0, 0, 0) to B (1, 10, 13).

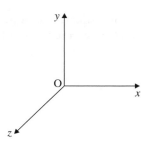

17. The diagram represents the starting position (AB) and the finishing position (CD) of a ladder as it slips. The ladder is leaning against a vertical wall.

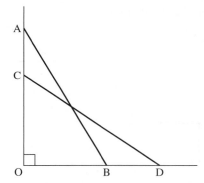

Given: $AC = x$, $OC = 4AC$, $BD = 2AC$ and $OB = 5$ m.
Form an equation in x, find x and hence find the length of the ladder.

18. The most well known right-angled triangle is the 3, 4, 5 triangle $[3^2 + 4^2 = 5^2]$.
It is interesting to look at other right-angled triangles where all the sides are whole numbers.

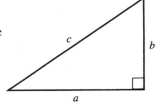

(a) (i) Find c if $a = 5$, $b = 12$.
   (ii) Find c if $a = 7$, $b = 24$.
   (iii) Find a if $c = 41$, $b = 40$.

(b) Write the results in a table.

a	b	c
3	4	5
5	12	?
7	24	?
?	40	41

(c) Look at the sequences in the 'a' column and in the 'b' column. Also write down the connection between b and c for each triangle.

(d) Predict the next three sets of values of a, b, c. Check to see if they really do form right-angled triangles.

19. The net of a square-based pyramid is shown. The base is 8 cm × 8 cm and the vertical height is 10 cm. Find x.

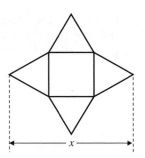

4.5 Area

For a **triangle,**   area $= \dfrac{1}{2} \times$ base $\times$ height *or* area $= \dfrac{1}{2} ab \sin C$

Use the second formula when you know two sides and the included angle.

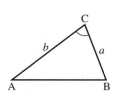

For a **rectangle,** area $=$ base $\times$ height

Trapezium

A trapezium has two parallel sides.

area $= \dfrac{1}{2}(a + b) \times h$

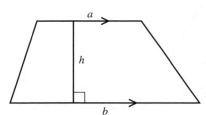

Parallelogram

area $= b \times h$

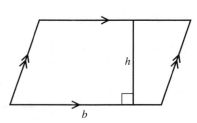

Exercise 10

For Questions **1** to **6,** find the area of each shape. All lengths are in cm.

1.

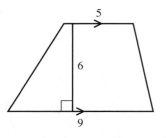

2.

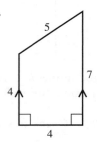

3.

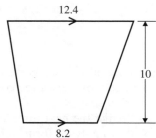

4.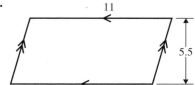

5.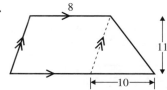

6.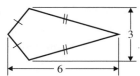

7. A rectangle has an area of $117\,\text{m}^2$ and a width of $9\,\text{m}$. Find its length.

8. A trapezium of area $105\,\text{cm}^2$ has parallel sides of length $5\,\text{cm}$ and $9\,\text{cm}$. How far apart are the parallel sides?

9. A floor $5\,\text{m}$ by $20\,\text{m}$ is covered by square tiles of side $20\,\text{cm}$. How many tiles are needed?

10. Find the area of each triangle. All lengths are in cm.

(a)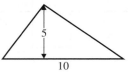

(b)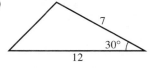

(c)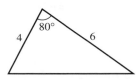

(d)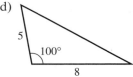

(e)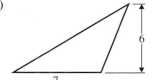

(f)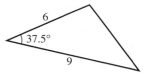

In Questions **11** to **14**, find the area shaded.

11.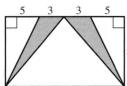

12.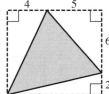

13.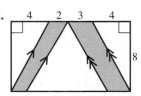

14.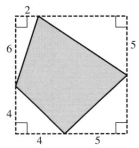

15.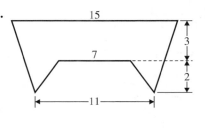

16.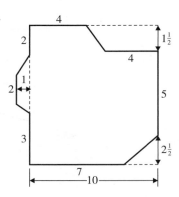

Exercise 11

1. A rectangular pond measuring 10 m by 6 m is surrounded by a path which is 2 m wide. Find the area of the path.

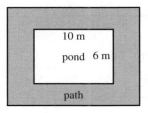

2. This shape is made with centimetre squares
(a) Write down its perimeter.
(b) Write down its area.
(c) Draw a shape of your own design with area 8 cm² and perimeter 14 cm.

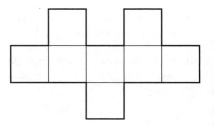

3. Find the length x. The area is shown inside the shape.
(a) 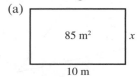 (b) 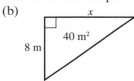 (c)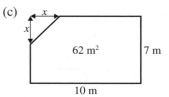

4. Work out the area of this shape. All lengths are in cm.

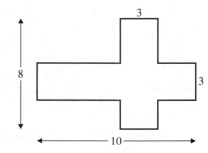

5. The arrowhead has an area of 3·6 cm². Find the length x.

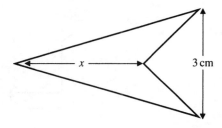

6. The diagram shows a square ABCD in which
DX = XY = YC = AW. The area of the square is 45 cm².

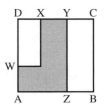

  (a) What is the fraction $\dfrac{DX}{DC}$?

  (b) What fraction of the square is shaded?

  (c) Find the area of the unshaded part.

7. On squared paper draw a 7 × 7 square. Design a pattern which
divides it up into nine smaller squares.

8. A rectangular field, 400 m long, has an area of 6 hectares. Calculate
the perimeter of the field. [1 hectare = 10 000 m²].

9. On squared paper draw the triangle with vertices at (1, 1), (5, 3),
(3, 5). Find the area of the triangle.

10. Draw the quadrilateral with vertices at (1, 1), (6, 2), (5, 5), (3, 6).
Find the area of the quadrilateral.

11. A square wall is covered with square tiles. There are 85 tiles
altogether along the two diagonals. How many tiles are there on the
whole wall?

12. The side of the small square is half the length of the side of the large
square. The L-shape has an area of 75 cm². Find the side length of the
large square.

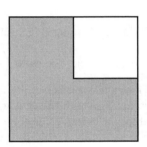

13. A regular hexagon is circumscribed by a circle of radius 3 cm with
centre O.

  (a) What is angle EOD?

  (b) Find the area of triangle EOD and hence find the area of the
hexagon ABCDEF.

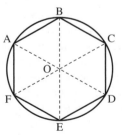

14. Find the area of a parallelogram ABCD with AB = 7 m,
AD = 20 m and BÂD = 62°.

15. In the diagram if AE = $\frac{1}{3}$AB, find the area shaded.

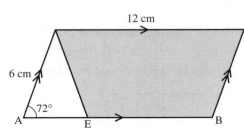

16. The area of an equilateral triangle ABC is $50\,\text{cm}^2$. Find AB.

17. The area of a triangle XYZ is $11\,\text{m}^2$. Given $YZ = 7\,\text{m}$ and $X\hat{Y}Z = 130°$, find XY.

18. Find the length of a side of an equilateral triangle of area $10{\cdot}2\,\text{m}^2$.

19. A rhombus has an area of $40\,\text{cm}^2$ and adjacent angles of $50°$ and $130°$. Find the length of a side of the rhombus.

†**20.** The diagram shows a part of the perimeter of a regular polygon with n sides.
   The centre of the polygon is at O and $OA = OB = 1$ unit.
   (a) What is the angle AOB in terms of n?
   (b) Work out an expression in terms of n for:
       (i) the area of triangle OAB
       (ii) the area of the whole polygon.
   (c) Find the area of the polygons where $n = 100$ and $n = 1000$. What do you notice?

†**21.** The area of a regular pentagon is $600\,\text{cm}^2$.
   Calculate the length of one side of the pentagon.

4.6 Circles, arcs and sectors

For any circle, the ratio $\left(\dfrac{\text{circumference}}{\text{diameter}}\right)$ is equal to π.

The value of π is usually taken to be $3{\cdot}14$, but this is not an exact value. Through the centuries, mathematicians have been trying to obtain a better value for π.

For example, in the third century A.D., the Chinese mathematician Liu Hui obtained the value $3{\cdot}14159$ by considering a regular polygon having 3072 sides! Ludolph van Ceulen (1540–1610) worked even harder to produce a value correct to 35 significant figures. He was so proud of his work that he had this value of π engraved on his tombstone. Electronic computers are now able to calculate the value of π to many millions of figures, but its value is still not exact. It was shown in 1761 that π is an *irrational number* which, like $\sqrt{2}$ or $\sqrt{3}$ cannot be expressed exactly as a fraction.

The first fifteen significant figures of π can be remembered from the number of letters in each word of the following sentence.

   How I need a drink, cherryade of course, after the silly lectures involving Italian kangaroos.

There remain a lot of unanswered questions concerning π, and many mathematicians today are still working on them.

The following formulae should be memorised.

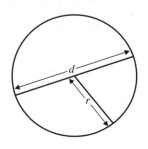

- circumference $= \pi d$
                $= 2\pi r$
-       area $= \pi r^2$

Find the circumference and area of a circle of diameter 8 cm. Take π from a calculator.

$$\text{Circumference} = \pi d$$
$$= \pi \times 8$$
$$= 25 \cdot 1 \text{ cm (3 s.f.)}$$

$$\text{Area} = \pi r^2$$
$$= \pi \times 4^2$$
$$= 50 \cdot 3 \text{ cm}^2 \text{ (3 s.f.)}$$

Find the perimeter and area of the shape below.

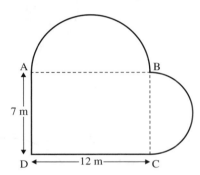

Perimeter $= \text{arc AB} + \text{arc BC} + \text{CD} + \text{DA}$

$$= \left(\frac{\pi \times 12}{2}\right) + \left(\frac{\pi \times 7}{2}\right) + 12 + 7$$

$$= 9\tfrac{1}{2}\pi + 19$$

$$= 48 \cdot 8 \text{ m (3 s.f.)}$$

- Note: Do not approximate *early* in questions of this type. If your final answer is given to 3 s.f., work to 5 s.f. in the working. Better still, leave the working in the calculator memory.

Area $=$ large semicircle $+$ small semicircle $+$ rectangle

$$= \left(\frac{\pi \times 6^2}{2}\right) + \left(\frac{\pi \times 3 \cdot 5^2}{2}\right) + (12 \times 7)$$

$$= 160 \text{ m}^2 \text{ (3 s.f.)}$$

Exercise 12

For each shape find (a) the perimeter, (b) the area. All lengths are in cm unless otherwise stated. All the arcs are either semicircles or quarter circles.

1.
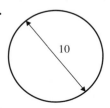
10

2.

3

3.

50 m

4.
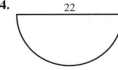
22

5.

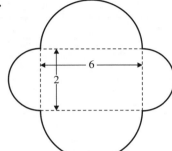

6.

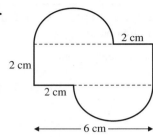

7.

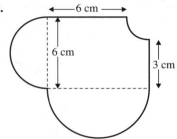

8.

9.

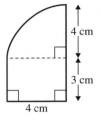

10.

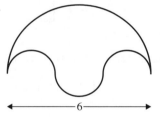

11.

12.

13.

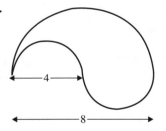

(a) A circle has a circumference of 20 m. Find the radius of the circle.

Let the radius of the circle be r m.

$$\text{Circumference} = 2\pi r$$
$$\therefore \quad 2\pi r = 20$$
$$\therefore \quad r = \frac{20}{2\pi}$$
$$r = 3 \cdot 18$$

The radius of the circle is $3 \cdot 18$ m (3 s.f.).

(b) A circle has an area of 45 cm^2. Find the radius of the circle.

Let the radius of the circle be r cm.

$$\pi r^2 = 45$$
$$r^2 = \frac{45}{\pi}$$
$$r = \sqrt{\left(\frac{45}{\pi}\right)} = 3 \cdot 78$$

The radius of the circle is $3 \cdot 78$ cm (3 s.f.).

Exercise 13

1. A circle has an area of 15 cm^2. Find its radius.

2. An odometer is a wheel used by surveyors to measure distances along roads. The circumference of the wheel is one metre. Find the diameter of the wheel.

3. Find the radius of a circle of area 22 km^2.

4. Find the radius of a circle of circumference $58 \cdot 6 \text{ cm}$.

5. The handle of a paint tin is a semicircle of wire which is 28 cm long. Calculate the diameter of the tin.

6. A circle has an area of 16 mm^2. Find its circumference.

7. A circle has a circumference of 2500 km. Find its area.

8. A circle of radius 5 cm is inscribed inside a square as shown. Find the area shaded.

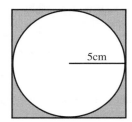

9. Discs of radius 4 cm are cut from a rectangular plastic sheet of length 84 cm and width 24 cm.
How many complete discs can be cut out?
Find:
(a) the total area of the discs cut
(b) the area of the sheet wasted.

10. The tyre of a car wheel has an outer diameter of 30 cm. How many times will the wheel rotate on a journey of 5 km?

11. A golf ball of diameter 1·68 inches rolls a distance of 4 m in a straight line. How many times does the ball rotate completely?
(1 inch = 2·54 cm)

12. A circular pond of radius 6 m is surrounded by a path of width 1 m.
(a) Find the area of the path.
(b) The path is resurfaced with astroturf which is bought in packs each containing enough to cover an area of 7 m^2.
How many packs are required?

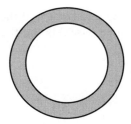

13. A rectangular metal plate has a length of 65 cm and a width of 35 cm. It is melted down and recast into circular discs of the same thickness. How many complete discs can be formed if:
(a) the radius of each disc is 3 cm?
(b) the radius of each disc is 10 cm?

14. Calculate the radius of a circle whose area is equal to the sum of the areas of three circles of radii 2 cm, 3 cm and 4 cm respectively.

15. The diagram below shows a lawn (unshaded) surrounded by a path of uniform width (shaded). The curved end of the lawn is a semicircle of diameter 10 m.

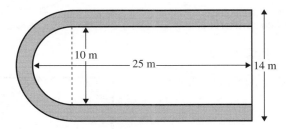

Calculate the total area of the path.

16. The diameter of a circle is given as 10 cm, correct to the nearest cm. Calculate:
(a) the maximum possible circumference
(b) the minimum possible area of the circle consistent with this data.

17. A square is inscribed in a circle of radius 7 cm. Find:
(a) the area of the square
(b) the area shaded.

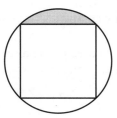

18. The governor of a prison has 100 m of wire fencing. What area can he enclose if he makes a circular compound?

19. In 'equable shapes' the numerical value of the area is equal to the numerical value of the perimeter.
Find the dimensions of the following equable shapes:
(a) square   (b) circle   (c) equilateral triangle.

†**20.** The semicircle and the isosceles triangle have the same base AB and the same area. Find the angle x.

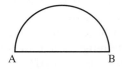

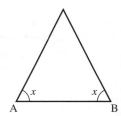

†**21.** Mr Gibson decided to measure the circumference of the earth using a very long tape measure. For reasons best known to himself he held the tape measure 1 m from the surface of the (perfectly spherical) earth all the way round. When he had finished Mrs Gibson told him that his measurement gave too large an answer. She suggested taking off 6 m. Was she correct? [Take the radius of the earth to be 6400 km (if you need it).]

†**22.** The large circle has a radius of 10 cm. Find the radius of the largest circle which will fit in the middle.

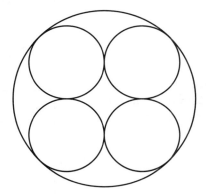

Mixed questions

Exercise 14

1. The sloping line divides the area of the square in the ratio 1 : 5. What is the ratio $a : b$?

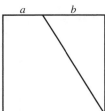

2. ABCD is a square of side 14 cm. P, Q, R, S are the mid-points of the sides. Semicircles are drawn with centres P, Q, R, S as shown on the diagram. Find the shaded area, using $\pi = \dfrac{22}{7}$.

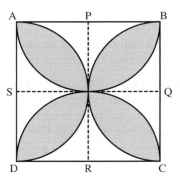

3. What fraction of the area of a circle of radius 4 cm is more than 3 cm from the centre?

† **4.** Two equal circles are cut from a square piece of paper of side 2 m. Calculate the radius of the largest possible circles. Give your answer correct to 3 significant figures.

5. These circles have radii 2, 3, 5 cm. Express the shaded area as a percentage of the area of the largest circle.

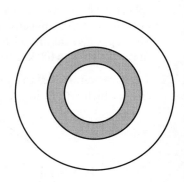

6. A square of side x cm is drawn in the middle of a square of side 10 cm. The corners of the smaller square are 2 cm from the corners of the larger square and are on the diagonals of the larger square. Find x correct to 4 s.f.

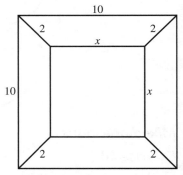

7. ABCDEFGH is a regular octagon. Calculate angle ACD.

8. The diagram shows two diagonals drawn on the faces of a cube. Calculate the angle between the diagonals.

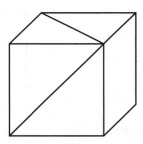

† **9.** The radius of the circle inscribed in a regular hexagon has length 4 cm. Find the area of the hexagon, correct to 3 s.f.

†**10.** In △ABC, AC = BC = 12 cm. CD is perpendicular to AB.
MD = 1 cm.
M is equidistant from A, B and C.
Calculate the length CM.

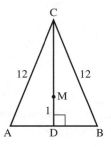

†11. The circumcircle and the inscribed circle of an equilateral triangle are shown. Calculate the ratio of the area of the circumcircle to the area of the inscribed circle.

†12. ABCDEFGH is a regular octagon of side 1 unit.
   (a) Calculate the length AF
   (b) Calculate the length AC
   (c) (Much harder) You probably used a calculator to work out AF. Can you calculate the *exact* value of AF? Leave your answer in a form involving a square root.

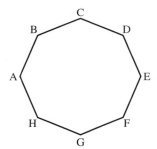

Arcs and sectors

Arc length

- Arc length, $l = \dfrac{\theta}{360} \times 2\pi r$

  We take a fraction of the whole circumference depending on the angle at the centre of the circle.

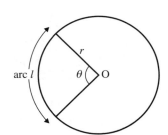

Sector area

- Sector area, $A = \dfrac{\theta}{360} \times \pi r^2$

  We take a fraction of the whole area depending on the angle at the centre of the circle.

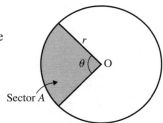

(a) Find the length of an arc which subtends an angle of 140° at the centre of a circle of radius 12 cm. (Take $\pi = \frac{22}{7}$.)

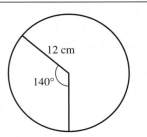

$$\text{Arc length} = \frac{140}{360} \times 2 \times \frac{22}{7} \times 12$$

$$= \frac{88}{3}$$

$$= 29\frac{1}{3} \text{ cm.}$$

(b) A sector of a circle of radius 10 cm has an area of 25 cm². Find the angle at the centre of the circle.

Let the angle at the centre of the circle be θ.

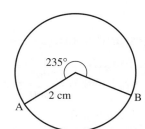

$$\frac{\theta}{360} \times \pi \times 10^2 = 25$$

$$\therefore \quad \theta = \frac{25 \times 360}{\pi \times 100}$$

$$\theta = 28\cdot6° \text{ (1 d.p.)}$$

The angle at the centre of the circle is 28·6°

Exercise 15

1. Find the length of the arc AB.

(a)

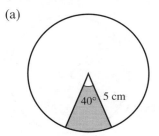

(b)

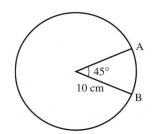

(c)

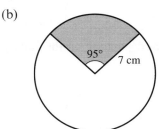

2. A pendulum of length 55 cm swings through an angle of 16°. Through what distance does the tip of the pendulum swing?

3. Find the area of the sector shaded.

(a)

(b)

(c)

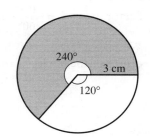

4. Find the shaded area.

(a)

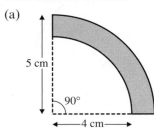

(b)

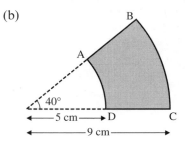

5. Find the length of the perimeter of the shaded area in question **4** part (b).

6. A sector is cut from a round cheese as shown. Calculate the volume of the piece of cheese.

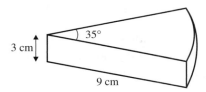

7. Work out the shaded area.

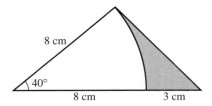

8. The lengths of the minor and major arcs of a circle are 5·2 cm and 19·8 cm respectively. Find:

(a) the radius of the circle
(b) the angle subtended at the centre by the minor arc.

9. The length of the minor arc AB of a circle, centre O, is 2π cm and the length of the major arc is 22π cm. Find

(a) the radius of the circle,
(b) the acute angle AOB.

10. A wheel of radius 10 cm is turning at a rate of 5 revolutions per minute. Calculate

(a) the angle through which the wheel turns in 1 second
(b) the distance moved by a point on the rim in 2 seconds.

11. In the diagram the arc length is l and the sector area is A.
 (a) Find θ, when $r = 5\,\text{cm}$ and $l = 7{\cdot}5\,\text{cm}$
 (b) Find θ, when $r = 2\,\text{m}$ and $A = 2\,\text{m}^2$
 (c) Find r, when $\theta = 55°$ and $l = 6\,\text{cm}$.

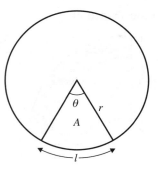

12. Two parallel lines are drawn 2 cm from the centre of a circle of radius 4 cm. Calculate the area shaded.

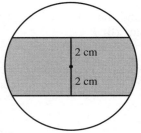

13. Find the angle θ so that the arc length is equal to the radius. The angle you have found is called a radian. Your calculator may have a $\boxed{\text{RAD}}$ mode, where $\boxed{\text{RAD}}$ stands for radian.

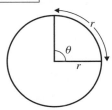

14. (a) ABCDE is a regular pentagon. Calculate the size of angle x.

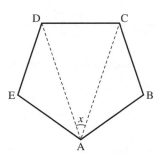

 (b) In this diagram, arcs are drawn with centres at the opposite corners of the pentagon, e.g. arc CD has centre at A. The radius of each arc is 5 cm. Show that the perimeter of the shape is 5π cm.

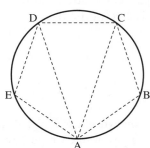

15. In the diagram, the arc length is l and the sector area is A.

   (a) Find l, when $\theta = 72°$ and $A = 15\,\text{cm}^2$
   (b) Find l, when $\theta = 135°$ and $A = 162\,\text{m}^2$
   (c) Find A, when $l = 11\,\text{cm}$ and $r = 5\cdot2\,\text{cm}$

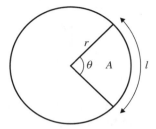

16. An arc of length one nautical mile subtends an angle of one minute at the centre of the circle around the equator of the Earth. Calculate the length of a nautical mile in metres, given that 60 minutes = 1 degree and the radius of the Earth is about 6370 km.

†**17.** The length of an arc of a circle is 12 cm. The corresponding sector area is $108\,\text{cm}^2$. Find:

   (a) the radius of the circle
   (b) the angle subtended at the centre of the circle by the arc.

†**18.** The length of an arc of a circle is $7\cdot5$ cm. The corresponding sector area is $37\cdot5\,\text{cm}^2$. Find:

   (a) the radius of the circle
   (b) the angle subtended at the centre of the circle by the arc.

†**19.** In the diagram, **AB** is a tangent to the circle at A. Straight lines BCD and ACE pass through the centre of the circle C. Angle ACB $= x°$ and the radius of the circle is 1 unit.

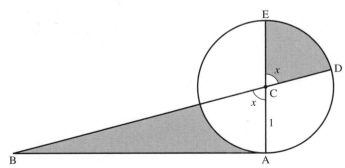

The two shaded areas are equal.
   (a) Show that x satisfies the equation $x = \left(\dfrac{90}{\pi}\right)\tan x$.

   (b) Use trial and improvement to find a solution for x correct to 1 decimal place.

†**20.** (Hard, unless you can find a neat solution!)
In the diagram:
BC is a diameter of the semicircle,
$A\widehat{B}C = 90°$,
shaded area ① = shaded area ② and angle ACB = x.

Show that $\tan x = \dfrac{\pi}{4}$.

Hence find angle x.

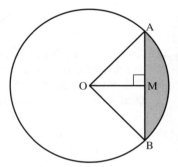

Segments

The line AB is a chord. The area of a circle cut off by a chord is called
a *segment*. In the diagram the *minor* segment is shaded and the *major*
segment is unshaded.

(a) The line from the centre of a circle to the mid-point M of a chord
bisects the chord at *right angles*.
(b) The line from the centre of a circle to the mid-point of a chord
bisects the angle subtended by the chord at the centre of the circle.

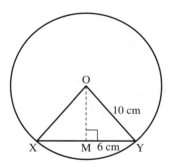

XY is a chord of length 12 cm of a circle of radius 10 cm,
centre O. Calculate

(a) the angle XOY
(b) the area of the minor segment cut off by the chord XY.

Let the mid-point of XY be M

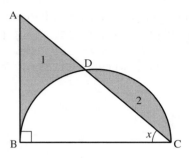

$$\therefore \quad MY = 6\,\text{cm}$$

$$\sin M\widehat{O}Y = \frac{6}{10}$$

$$\therefore \quad M\widehat{O}Y = 36\cdot87°$$

$$\therefore \quad X\widehat{O}Y = 2 \times 36\cdot87$$
$$= 73\cdot74°$$

area of minor segment = area of sector XOY − area of △XOY

$$\text{area of sector XOY} = \frac{73\cdot74}{360} \times \pi \times 10^2$$
$$= 64\cdot32\,\text{cm}^2$$

$$\text{area of △XOY} = \tfrac{1}{2} \times 10 \times 10 \times \sin 73\cdot74°$$
$$= 48\cdot00\,\text{cm}^2$$

$$\therefore \quad \text{Area of minor segment} = 64\cdot32 - 48\cdot00$$
$$= 16\cdot3\,\text{cm}^2 \text{ (3 s.f.)}$$

Exercise 16

Use the π button on a calculator.

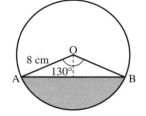

1. The chord AB subtends an angle of 130° at the centre O.
   The radius of the circle is 8 cm. Find:
   (a) the length of AB,
   (b) the area of sector OAB,
   (c) the area of triangle OAB,
   (d) the area of the minor
       segment (shown shaded).

2. Find the shaded area when:
   (a) $r = 6$ cm, $\theta = 70°$
   (b) $r = 14$ cm, $\theta = 104°$
   (c) $r = 5$ cm, $\theta = 80°$

3. Find θ and hence the shaded area when:
   (a) AB $= 10$ cm, $r = 10$ cm
   (b) AB $= 8$ cm, $r = 5$ cm

4. How far is a chord of length 8 cm from the centre of a circle of
   radius 5 cm?

5. How far is a chord of length 9 cm from the centre of a circle of
   radius 6 cm?

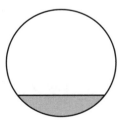

6. The diagram shows the cross section of a cylindrical pipe with
   water lying in the bottom.
   (a) If the maximum depth of the water is 2 cm and the radius of
       the pipe is 7 cm, find the area shaded.
   (b) What is the *volume* of water in a length of 30 cm?

7. An equilateral triangle is inscribed in a circle of radius 10 cm. Find:
   (a) the area of the triangle,
   (b) the area shaded.

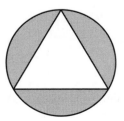

8. A regular hexagon is circumscribed by a circle of radius 6 cm. Find the area shaded.

9. A regular octagon is circumscribed by a circle of radius r cm. Find the area enclosed between the circle and the octagon. (Give the answer in terms of r.)

†**10.** Find the radius of the circle:
   (a) when $\theta = 90°$, $A = 20\,\text{cm}^2$
   (b) when $\theta = 30°$, $A = 35\,\text{cm}^2$
   (c) when $\theta = 150°$, $A = 114\,\text{cm}^2$

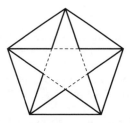

†**11.** The diagram shows a regular pentagon of side 10 cm with a star inside. Calculate the area of the star.

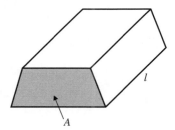

4.7 Volume and surface area

A **prism** is an object with the same cross section throughout its length.

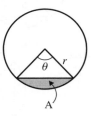

- Volume of prism = (area of cross section) × length
  $$= A \times l.$$

A **cuboid** is a prism whose six faces are all rectangles. A cube is a special case of a cuboid in which all six faces are squares.

A **cylinder** is a prism whose cross section is a circle.

radius $= r$
height $= h$

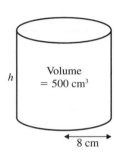

- Volume of cylinder = area of cross section × length
    Volume $= \pi r^2 h$

Calculate the height of a cylinder of volume 500 cm³ and base radius 8 cm.

Let the height of the cylinder be h cm.

$$\pi r^2 h = 500$$
$$3{\cdot}14 \times 8^2 \times h = 500$$
$$h = \frac{500}{3{\cdot}14 \times 64}$$
$$h = 2{\cdot}49 \ (3\,\text{s.f.})$$

h Volume = 500 cm³

The height of the cylinder is 2·49 cm.

8 cm

Exercise 17

1. Calculate the volume of the prisms. All lengths are in cm.

(a)

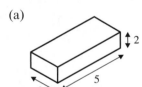

(b)

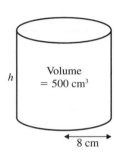

(c)

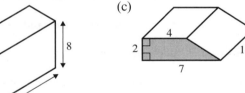

(d)

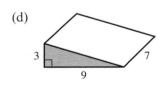

(e)

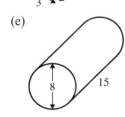

(f)

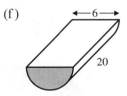

2. Calculate the volume of the following cylinders:
   (a) $r = 4$ cm, $h = 10$ cm
   (b) $r = 11$ m, $h = 2$ m

3. A gas cylinder has diameter 18 cm and length 40 cm. Calculate the capacity of the cylinder, correct to the nearest litre.

4. A solid cylinder of radius 5 cm and length 15 cm is made from material of density 6 g/cm³. Calculate the mass of the cylinder.

5. The two solid cylinders shown have the same mass. Calculate the density, $x\,\text{g/cm}^3$, of cylinder B.

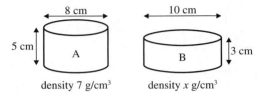

6. The diagram shows a view of the water in a swimming pool.

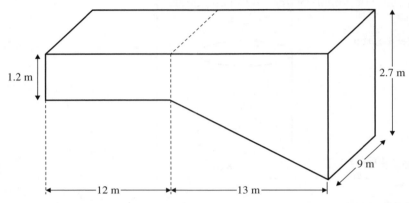

Calculate the volume of water in the pool.

7. Find the height of a cylinder of volume $200\,\text{cm}^3$ and radius $4\,\text{cm}$.

8. Find the length of a cylinder of volume 2 litres and radius $10\,\text{cm}$.

9. Find the radius of a cylinder of volume $45\,\text{cm}^3$ and length $4\,\text{cm}$.

10. When 3 litres of oil is removed from an upright cylindrical can, the level falls by $10\,\text{cm}$. Find the radius r of the can.

11. A solid cylinder of radius $4\,\text{cm}$ and length $8\,\text{cm}$ is melted down and recast into a solid cube. Find the side of the cube.

12. A solid rectangular block of copper $5\,\text{cm}$ by $4\,\text{cm}$ by $2\,\text{cm}$ is drawn out to make a cylindrical wire of diameter $2\,\text{mm}$. Calculate the length of the wire.

13. Water flows through a circular pipe of internal diameter 3 cm at a speed of 10 cm/s. If the pipe is full, how much water issues from the pipe in one minute? (answer in litres)

14. Water issues from a hose-pipe of internal diameter 1 cm at a rate of 5 litres per minute. At what speed is the water flowing through the pipe?

15. A cylindrical metal pipe has external diameter of 6 cm and internal diameter of 4 cm. Calculate the volume of metal in a pipe of length 1 m. If the density of the metal is 8 g/cm^3, find the weight of the pipe.

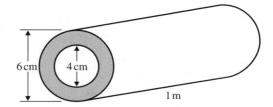

16. Mr Gibson decided to build a garage and began by calculating the number of bricks required. The garage was to be 6 m by 4 m and 2·5 m in height. Each brick measures 22 cm by 10 cm by 7 cm. Mr Gibson estimated that he would need about 40 000 bricks. Is this a reasonable estimate?

17. A cylindrical can of internal radius 20 cm stands upright on a flat surface. It contains water to a depth of 20 cm. Calculate the rise h in the level of the water when a brick of volume 1500 cm^3 is immersed in the water.

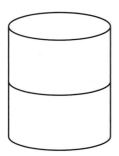
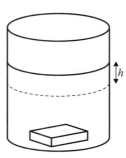

18. A cylindrical tin of height 15 cm and radius 4 cm is filled with sand from a rectangular box. How many times can the tin be filled if the dimensions of the box are 50 cm by 40 cm by 20 cm?

19. Rain which falls onto a flat rectangular surface of length 6 m and width 4 m is collected in a cylinder of internal radius 20 cm. What is the depth of water in the cylinder after a storm in which 1 cm of rain fell?

Pyramid, sphere, cone

Pyramid

Volume $= \frac{1}{3}$(base area) $\times$ height.

Sphere

Volume $= \frac{4}{3}\pi r^3$

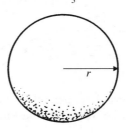

Cone

Volume $= \frac{1}{3}\pi r^2 h$

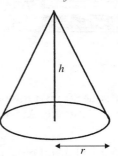

● The formula for the volume of a pyramid is demonstrated below:

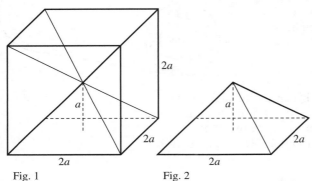

Fig. 1   Fig. 2

Figure 1 shows a cube of side $2a$ broken down into six pyramids of height a as shown in figure 2.

If the volume of each pyramid is V,

then $\quad 6V = 2a \times 2a \times 2a$

$\qquad V = \frac{1}{6} \times (2a)^2 \times 2a$

so $\qquad V = \frac{1}{3} \times (2a)^2 \times a$

$\qquad V = \frac{1}{3}$(base area) $\times$ height.

(a) Calculate the volume of the cone.

Volume $= \frac{1}{3} \times \pi \times 2^2 \times 5$

$\qquad = \dfrac{\pi \times 20}{3}$

$\qquad = 20 \cdot 9 \, \text{cm}^3 \quad$ (3 S.F.)

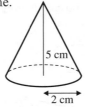

5 cm

2 cm

(b) A solid sphere of radius 4 cm is made of metal of density $8 \, \text{g/cm}^3$. Calculate the mass of the sphere.

Volume $= \frac{4}{3} \times \pi \times 4^3 \, \text{cm}^3$

Mass $= \frac{4}{3} \times \pi \times 4^3 \times 8 = 2140 \, \text{g} \quad$ (3 s.f.)

Exercise 18

In Questions **1** to **6** find the volume of each object.
All lengths are in cm.

1.

4

5

6

2.

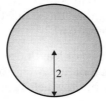

2

3.

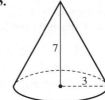

7

3

4.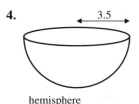
hemisphere

5.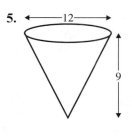

6.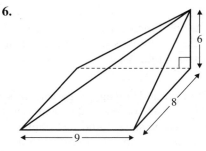

7. A solid sphere is made of metal of density $9\,\text{g/cm}^3$. Calculate the mass of the sphere if the radius is $5\,\text{cm}$.

8. Find the volume of a hemisphere of radius $5\,\text{cm}$.

9. Calculate the volume of a cone of height and radius $7\,\text{m}$.

10. A cone is attached to a hemisphere of radius $4\,\text{cm}$. If the total height of the object is $10\,\text{cm}$, find its volume.

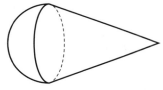

11. Find the height of a pyramid of volume $20\,\text{m}^3$ and base area $12\,\text{m}^2$.

12. A single drop of oil is a sphere of radius $3\,\text{mm}$. The drop of oil falls on water to produce a thin circular film of radius $100\,\text{mm}$. Calculate the thickness of this film in mm.

13. Gold is sold in solid spherical balls. Which is worth more: 10 balls of radius $2\,\text{cm}$ or 1 ball of radius $4\,\text{cm}$?

r = 2 cm
r = 4 cm

14. A toy consists of a cylinder of diameter $6\,\text{cm}$ 'sandwiched' between a hemisphere and a cone of the same diameter. If the cone is of height $8\,\text{cm}$ and the cylinder is of height $10\,\text{cm}$, find the total volume of the toy.

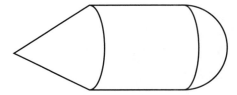

15. Water is flowing into an inverted cone, of diameter and height $30\,\text{cm}$, at a rate of 4 litres per minute. How long, in seconds, will it take to fill the cone?

16. Metal spheres of radius 2 cm are packed into a rectangular box of internal dimensions 16 cm × 8 cm × 8 cm. When 16 spheres are packed the box is filled with a preservative liquid. Find the volume of this liquid.

17. The Pyramids in Egypt are seriously large! One Pyramid has a square base of side 100 m and height 144 m. An average slave could just manage to carry a brick measuring 40 cm by 20 cm by 16 cm. Assuming no spaces were left for mummies or treasure, work out how many bricks would be needed to make this Pyramid.

18. The cylindrical end of a pencil is sharpened to produce a perfect cone at the end with no overall loss of length. If the diameter of the pencil is 1 cm, and the cone is of length 2 cm, calculate the volume of the shavings.

19. One corner of a solid cube of side 8 cm is removed by cutting through the mid-points of three adjacent sides. Calculate the volume of the piece removed.

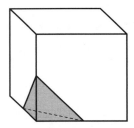

20. A hollow spherical vessel has internal and external radii of 6 cm and 6·4 cm respectively. Calculate the weight of the vessel if it is made of metal of density 10 g/cm^3.

Calculate the radius of a sphere of volume 500 cm^3.

Let the radius of the sphere be r cm

$\frac{4}{3}\pi r^3 = 500$

$r^3 = \dfrac{3 \times 500}{4\pi}$

$r = \sqrt[3]{\left(\dfrac{3 \times 500}{4\pi}\right)} = 4.92 \text{ (3 s.f.)}$

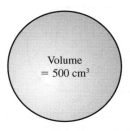

Volume
= 500 cm^3

The radius of the sphere is 4·92 cm.

Exercise 19

1. Calculate the radius of each sphere.

   (a)   (b)   (c)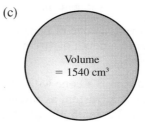

2. A solid metal sphere of radius 4 cm is recast into a solid cube. Find the length of a side of the cube.

3. Find the height of a cone of volume 2500 cm^3 and radius 10 cm.

4. A solid metal sphere is recast into many smaller spheres. Calculate the number of the smaller spheres if the initial and final radii are as follows:
   (a) initial radius = 10 cm, final radius = 2 cm
   (b) initial radius = 1 m,   final radius = $\frac{1}{3}$ cm.

5. A spherical ball is immersed in water contained in a vertical cylinder.
   Assuming the water covers the ball, calculate the rise in the water level if:
   (a) sphere radius = 3 cm, cylinder radius = 10 cm
   (b) sphere radius = 2 cm, cylinder radius = 5 cm.

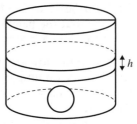

6. A spherical ball is immersed in water contained in a vertical cylinder. The rise in water level is measured in order to calculate the radius of the spherical ball. Calculate the radius of the ball in the following cases:
   (a) cylinder of radius 10 cm, water level rises 4 cm
   (b) cylinder of radius 100 cm, water level rises 8 cm.

7. The diagram shows the cross section of an inverted cone of height MC = 12 cm. If AB = 6 cm and XY = 2 cm, use similar triangles to find the length NC. Hence find the volume of the cone of height NC.

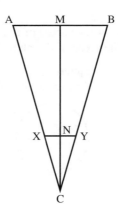

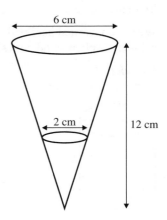

8. An inverted cone of height 10 cm and base radius 6·4 cm contains water to a depth of 5 cm, measured from the vertex. Calculate the volume of water in the cone.

9. An inverted cone of height 15 cm and base radius 4 cm contains water to a depth of 10 cm. Calculate the volume of water in the cone.

10. A frustum is a cone with 'the end chopped off'. A bucket in the shape of a frustum as shown has diameters of 10 cm and 4 cm at its ends and a depth of 3 cm. Calculate the volume of the bucket.

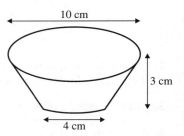

11. Find the volume of a frustum with end diameters of 60 cm and 20 cm and a depth of 40 cm.

12. Builders' sand is tipped into the shape of a cone where the height and radius are both 1·4 m.
    (a) Calculate the volume of the sand, correct to 1 d.p.
    (b) A further 2 m^3 is added to the pile. This now forms a larger cone, but the height still equals the radius. Calculate the height of this larger pile correct to 1 d.p.

13. The diagram shows a sector of a circle of radius 10 cm.
    (a) Find, as a multiple of π, the arc length of the sector.
    The straight edges are brought together to make a cone.
    Calculate
    (b) the radius of the base of the cone.
    (c) the vertical height of the cone.

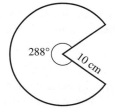

†14. A sphere passes through the eight vertices of a cube of side 10 cm. Find the volume of the sphere.

Surface area

We are concerned here with the surface areas of the *curved* parts of cylinders, spheres and cones. The areas of the plane faces are easier to find.

Cylinder
Curved surface area $= 2\pi rh$

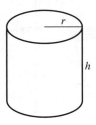

Sphere
Surface area $= 4\pi r^2$

Cone
Curved surface area $= \pi rl$, where l is the slant height

Calculate the *total* surface area of a solid cylinder of radius 3 cm and height 8 cm.

Curved surface area $= 2\pi r h$

$$= 2 \times \pi \times 3 \times 8$$

$$= 48\pi \, \text{cm}^2$$

Area of two ends $= 2 \times \pi r^2$

$$= 2 \times \pi \times 3^2$$

$$= 18\pi \, \text{cm}^2$$

Total surface area $= (48\pi + 18\pi) \, \text{cm}^2$

$$= 207 \, \text{cm}^2 \quad \text{(3 s.f.)}$$

Exercise 20

1. Work out the *curved* surface area of these objects. Leave π in your answers. All lengths are in cm.

   (a)

   $r = 3$

   (b)

   $r = 4$
   $h = 5$

   (c)

   $l = 10$

   $r = 6$

   (d)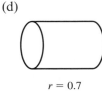

   $r = 0.7$
   $h = 1$

2. Work out the *total* surface area of these objects. Leave π in your answers. All lengths are in cm.

   (a)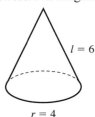

   $l = 6$

   $r = 4$

   (b)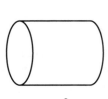

   $r = 2$
   $h = 5$

   (c)

   hemisphere
   $r = 4$

   (d)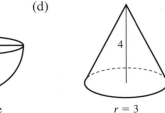

   4

   $r = 3$
   vertical height $= 4$

3. A solid cylinder of height 10 cm and radius 4 cm is to be plated with material costing £11 per cm^2. Find the cost of the plating.

4. A tin of paint covers a surface area of 60 m^2 and costs £4·50. Find the cost of painting the outside surface of a cylindrical gas holder of height 30 m and radius 18 m. The top of the gas holder is a flat circle.

5. A solid wooden cylinder of height 8 cm and radius 3 cm is cut in two along a vertical axis of symmetry. Calculate the total surface area of the two pieces.

6. Calculate the total surface area of the combined cone/cylinder/hemisphere.

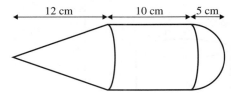

7. A man is determined to spray the entire surface of the earth (including the oceans) with a revolutionary new weed killer. If it takes him 10 seconds to spray $1\,\mathrm{m}^2$, how long will it take to spray the whole world?
(radius of the earth $= 6370\,\mathrm{km}$; ignore leap years)

8. A rectangular piece of card $10\,\mathrm{cm}$ by $30\,\mathrm{cm}$ is rolled up to make a tube (with no overlap). Find the radius of the tube if:
(a) the long sides are joined,
(b) the short sides are joined.

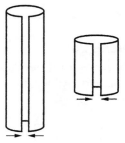

9. Find the radius of a sphere of surface area $34\,\mathrm{cm}^2$.

10. Find the slant height of a cone of curved surface area $20\,\mathrm{cm}^2$ and radius $3\,\mathrm{cm}$.

11. Find the height of a solid cylinder of radius $1\,\mathrm{cm}$ and *total* surface area $28\,\mathrm{cm}^2$.

12. Find the volume of a sphere of surface area $100\,\mathrm{cm}^2$.

13. Find the surface area of a sphere of volume $28\,\mathrm{cm}^3$.

14. An inverted cone of vertical height $12\,\mathrm{cm}$ and base radius $9\,\mathrm{cm}$ contains water to a depth of $4\,\mathrm{cm}$. Find the area of the interior surface of the cone not in contact with the water.

15. A circular paper of radius $20\,\mathrm{cm}$ is cut in half and each half is made into a hollow cone by joining the straight edges. Find the slant height and base radius of each cone.

16. A solid metal cube of side $6\,\mathrm{cm}$ is recast into a solid sphere.
(a) Find the radius of the sphere.
(b) By how much is the surface area of the original cube greater than the surface area of the new sphere?

17. A solid cuboid, measuring $5\,\text{cm} \times 5\,\text{cm} \times 3\,\text{cm}$, has a hole of radius $2\,\text{cm}$ drilled right through.
Calculate the surface area of the object.

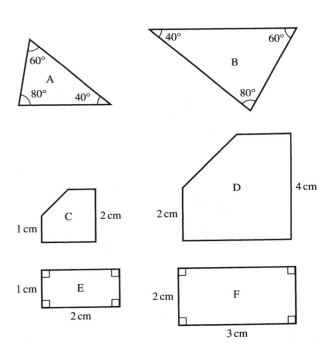

18. A cone of radius $6\,\text{cm}$ has a *total* surface area of $300\,\text{cm}^2$. Calculate the volume of the cone.

4.8 Similar shapes

If one shape is an enlargement of another, the two shapes are mathematically *similar*.

The two triangles A and B are similar if they have the same angles.

For other shapes to be similar, not only must corresponding angles be equal, but also corresponding edges must be in the same proportion.

The two quadrilaterals C and D are similar. All the edges of shape D are twice as long as the edges of shape C.

The two rectangles E and F are not similar even though they have the same angles.

(a) The triangles below are similar. Find x.

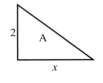
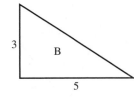

(b) Triangle B is an enlargement of triangle A. Corresponding sides are in the same ratio.

$$\frac{x}{5} = \frac{2}{3}$$

$$x = \tfrac{2}{3} \times 5$$

$$x = 3\tfrac{1}{3}$$

Exercise 21

1. Which of the shapes B, C, D is/are similar to shape A?

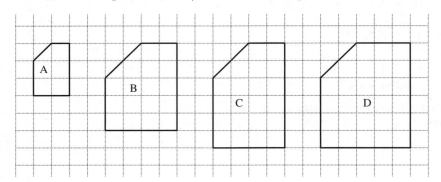

In Questions **2** to **7**, find the sides marked with letters; all lengths are given in cm. The pairs of shapes are similar.

2.

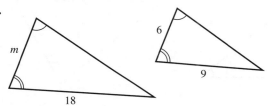

3.

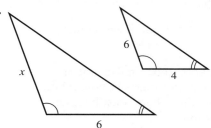

4.

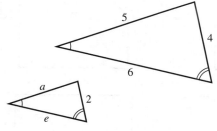

5.

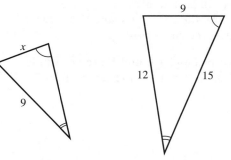

6.

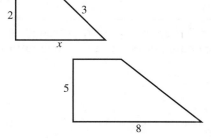

7.

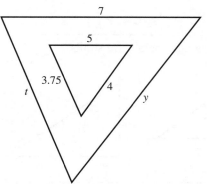

8. Picture B is an enlargement of picture A. Calculate the length x.

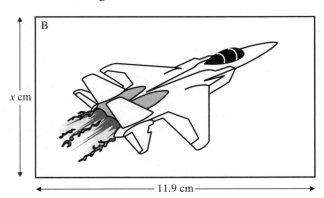

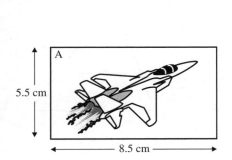

9. The drawing shows a rectangular picture 16 cm × 8 cm surrounded
by a border of width 4 cm.
Are the two rectangles similar?

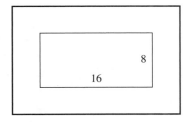

10. Which of the following *must* be similar to each other:
   (a) two equilateral triangles.   (b) two rectangles.
   (c) two isosceles triangles.   (d) two squares.
   (e) two regular pentagons.   (f) two kites.
   (g) two rhombuses.   (h) two circles.

11. (a) Explain why triangles ABC and EBD are similar.
   (b) Given that EB = 7 cm, calculate the length AB.
   (c) Write down the length AE.

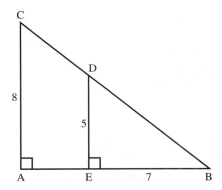

In Questions **12**, **13**, **14** use similar triangles to find the sides marked with letters. All lengths are in cm.

12.

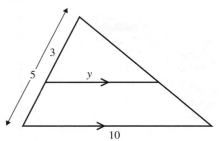

13.

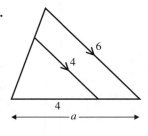

14.

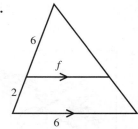

Exercise 22

1. A tree of height 4 m casts a shadow of length 6·5 m. Find the height of a house casting a shadow 26 m long.

2. A small cone is cut from a larger cone. Find the radius of the smaller cone.

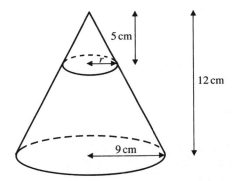

3. The diagram shows the side view of a swimming pool being filled with water. Calculate the length x.

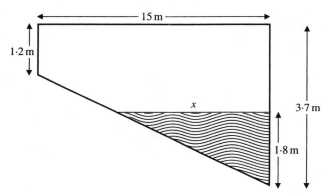

4. The diagonals of a trapezium ABCD intersect at O. AB is parallel to DC, AB = 3 cm and DC = 6 cm. Show that triangles ABO and CDO are similar. If CO = 4 cm and OB = 3 cm, find AO and DO.

In Questions **5** and **6** find the sides marked with letters.

5. $\hat{BAC} = \hat{DBC}$

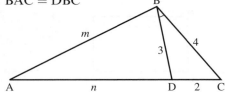

6.

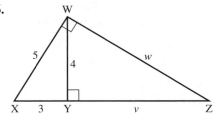

7. A rectangle 11 cm by 6 cm is similar to a rectangle 2 cm by x cm. Find the two possible values of x.

8. From the rectangle ABCD a square is cut off to leave rectangle BCEF.
Rectangle BCEF is similar to ABCD. Find x and hence state the ratio of the sides of rectangle ABCD. ABCD is called the Golden Rectangle and is an important shape in architecture.

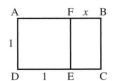

9. Triangles ABC and EBD are similar but DE is *not* parallel to AC. Work out the length x.

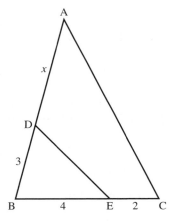

10. In the diagram $\hat{ABC} = \hat{ADB} = 90°$, $AD = p$ and $DC = q$.

(a) Use similar triangles ABD and ABC to show that $x^2 = pz$.

(b) Find a similar expression for y^2

(c) Add the expressions for x^2 and y^2 and hence prove Pythagoras' theorem.

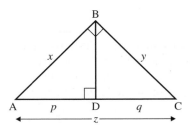

†**11.** In a triangle ABC, a line is drawn parallel to BC to meet AB at D and AC at E. DC and BE meet at X. Prove that:

(a) the triangles ADE and ABC are similar

(b) the triangles DXE and BXC are similar

(c) $\dfrac{AD}{AB} = \dfrac{EX}{XB}$

Areas of similar shapes

The two rectangles shown are similar.
The ratio of their corresponding sides is k.

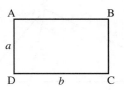

area of $\mathbf{ABCD} = ab$
area of $\mathbf{WXYZ} = ka \times kb = k^2ab$

$$\therefore \quad \frac{\text{area of WXYZ}}{\text{area ABCD}} = \frac{k^2ab}{ab} = k^2$$

This illustrates an important general
rule for all similar shapes:

If two figures are similar and the ratio of corresponding
sides is k, then the ratio of their areas is k^2.

Note. k may be called the *linear scale factor*.

This result also applies for the surface areas of similar
three-dimensional objects.

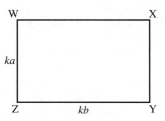

In triangle $\mathbf{ABC}$, XY is parallel to BC

and $\dfrac{\mathbf{AB}}{\mathbf{AX}} = \frac{3}{2}$. If the area of triangle AXY is

$4\,\text{cm}^2$, find the area of triangle ABC.
The triangles ABC and AXY are similar.

Ratio of corresponding sides $(k) = \frac{3}{2}$

$\therefore \qquad\qquad$ Ratio of areas $(k^2) = \frac{9}{4}$

$\therefore \qquad\qquad$ Area of $\triangle\text{ABC} \quad = \frac{9}{4} \times (\text{area of } \triangle\text{AXY})$

$\qquad\qquad\qquad\qquad\qquad = \frac{9}{4} \times (4) = 9\,\text{cm}^2$

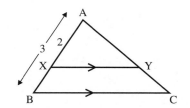

Two similar triangles have areas of
$18\,\text{cm}^2$ and $32\,\text{cm}^2$ respectively. If the
base of the smaller triangle is $6\,\text{cm}$, find
the base of the larger triangle.

Ratio of areas $(k^2) = \dfrac{32}{18} = \dfrac{16}{9}$

$\therefore \quad$ Ratio of corresponding sides $(k) = \sqrt{\left(\dfrac{16}{9}\right)} = \dfrac{4}{3}$

$\therefore \quad$ Base of larger triangle $= 6 \times \dfrac{4}{3} = 8\,\text{cm}$

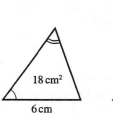

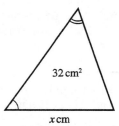

Exercise 23

In this exercise, a number written inside a figure represents the area of the shape in cm^2. Numbers on the outside give linear dimensions in cm. In each case the shapes are similar. In Questions **1** to **6**, find the unknown area A.

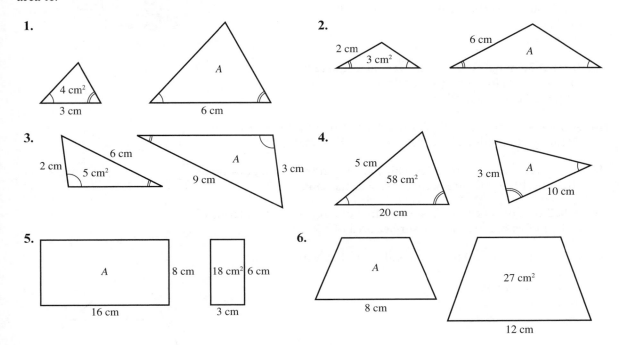

1.

4 cm^2
3 cm

A
6 cm

2.

2 cm
3 cm^2

6 cm
A

3.

2 cm
6 cm
5 cm^2

A
9 cm
3 cm

4.

5 cm
58 cm^2
20 cm

3 cm
A
10 cm

5.

A
8 cm
16 cm

18 cm^2 6 cm
3 cm

6.

A
8 cm

27 cm^2
12 cm

In Questions **7** to **10**, find the lengths marked for each pair of similar shapes.

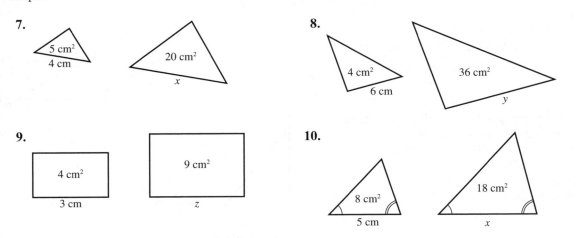

7.

5 cm^2
4 cm

20 cm^2
x

8.

4 cm^2
6 cm

36 cm^2
y

9.

4 cm^2
3 cm

9 cm^2
z

10.

8 cm^2
5 cm

18 cm^2
x

11. A triangle with sides 3, 4 and 5 cm has an area of 6 cm^2. A similar triangle has an area of 54 cm^2. Find the sides of the larger triangle.

12. A giant ball is made to promote the sales of a
    new make of golf ball. The surface area of an
    ordinary ball is 50 cm².
    The diameter of the giant ball is 100 times
    as great as a normal ball.
    Work out the surface area of the giant ball:
    (a) in cm²
    (b) in m².

13. It takes 30 minutes to cut the grass in a square field
    of side 20 m. How long will it take to cut the grass in
    a square field of side 60 m?

14. A floor is covered by 600 tiles which are 10 cm by 10 cm. How many
    20 cm by 20 cm tiles are needed to cover the same floor?

15. A wall is covered by 160 tiles which are 15 cm by 15 cm. How many
    10 cm by 10 cm tiles are needed to cover the same wall?

16. A ball of radius r cm has a surface area of 20 cm². Find, in terms of
    r, the radius of a ball with a surface area of 500 cm².

17. Given AD = 3 cm, AB = 5 cm and area of △ADE = 6 cm².
    Find:
    (a) area of △ABC
    (b) area of DECB.

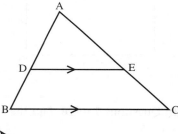

18. Given XY = 5 cm, MY = 2 cm and area of △MYN = 4 cm².
    Find:
    (a) area of △XYZ
    (b) area of MNZX.

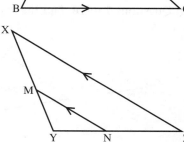

19. The triangles ABC and EBD are similar
    (AC and DE are *not* parallel).
    If AB = 8 cm, BE = 4 cm and the
    area of △DBE = 6 cm²,
    and the area of △BDE = 6 cm²,
    find the area of △ABC.

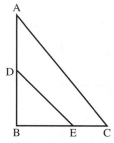

20. When potatoes are peeled do you lose more peel or less when big
    potatoes are used as opposed to small ones?

21. A supermarket offers cartons of cheese with 10% extra for the same price.

The old carton has a radius of 6 cm and a thickness of 1·2 cm. The new carton has the same thickness. Calculate its radius.

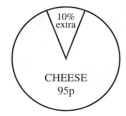

22.

An A3 sheet of paper can be cut into two sheets of A4. The A3 and A4 sheets are mathematically similar.

Find the scale factor: $\left(\dfrac{\text{long side of A3 sheet}}{\text{long side of A4 sheet}}\right)\left[\text{i.e. } \dfrac{x}{y}\right]$

†**23.** Rectangle ABCD is inscribed in △PQR. The length of AD is one third the length of the perpendicular from Q to PR.

Calculate the ratio $\left(\dfrac{\text{area ABCD}}{\text{area PQR}}\right)$.

Volumes of similar objects

When solid objects are similar, one is an accurate enlargement of the other. If two objects are similar and the ratio of corresponding sides is k, then the ratio of their volumes is k^3.

A line has one dimension, and the scale factor is used once.
An area has two dimensions, and the scale factor is used twice.
A volume has three dimensions, and the scale factor is used three times.

Two similar cylinders have heights of 3 cm and 6 cm respectively. If the volume of the smaller cylinder is 30 cm³, find the volume of the larger cylinder.

$$\text{ratio of heights } (k) = \tfrac{6}{3} \text{ (linear scale factor)}$$
$$= 2$$
$$\therefore \quad \text{ratio of volumes } (k^3) = 2^3$$
$$= 8$$
$$\text{and volume of larger cylinder} = 8 \times 30$$
$$= 240 \text{ cm}^3$$

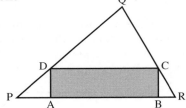

Two similar spheres made of the same material have weights of 32 kg and 108 kg respectively. If the radius of the larger sphere is 9 cm, find the radius of the smaller sphere.

We may take the ratio of weights to be the same as the ratio of volumes.

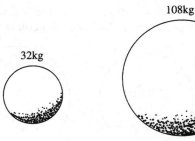

32kg

108kg

$$\text{ratio of volumes } (k^3) = \frac{32}{108} = \frac{8}{27}$$

ratio or corresponding

$$\text{lengths } (k) = \sqrt[3]{\left(\frac{8}{27}\right)} = \frac{2}{3}$$

$\therefore$ radius of smaller sphere $= \dfrac{2}{3} \times 9 = 6$ cm

Exercise 24

In this exercise, the objects are similar and a number written inside a figure represents the volume of the object in cm^3.
Numbers on the outside give linear dimensions in cm. In Questions **1** to **8**, find the unknown volume V.

1.

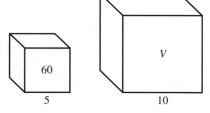

60

5

V

10

2.

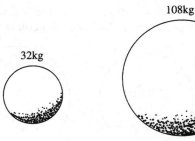

5   20   15 V

3.

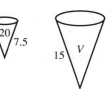

20 7.5

15 V

4.

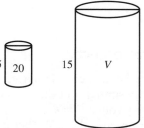

4.5

radius = 1.2 cm

V

radius = 12 cm

5.

24   6

9   V

6.

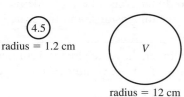

88

6.2

V

3.1

7.

8.

In Questions **9** to **14**, find the lengths marked by a letter.

9.
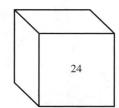

10.

11.

12.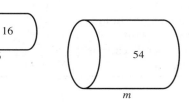

13.

14.

15. Two similar jugs have heights of 4 cm and 6 cm respectively. If the capacity of the smaller jug is 50 cm³, find the capacity of the larger jug.

16. Two similar cylindrical tins have base radii of 6 cm and 8 cm respectively. If the capacity of the larger tin is 252 cm³, find the capacity of the small tin.

17. Two solid metal spheres have masses of 5 kg and 135 kg respectively. If the radius of the smaller one is 4 cm, find the radius of the larger one.

18. Two similar cones have surface areas in the ratio 4:9. Find the ratio of:
   (a) their lengths,   (b) their volumes.

19. The area of the bases of two similar glasses are in the ratio 4:25. Find the ratio of their volumes.

20. A model of a treasure chest is 3 cm long and has a volume of 11 cm^3. Work out the length of the real treasure chest if its volume is 88 000 cm^3.

21. The ingredients of a standard Toblerone bar cost 20p. A giant-sized bar is similar in shape and four times as long. What should be the cost of the ingredients of the giant bar?

22. Sam has a model of a pirate ship which is made to a scale of 1 : 50.

Copy and complete the table using the given units.

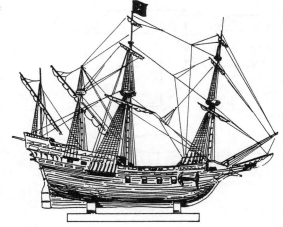

	On model	On actual ship
Length	42 cm	m
Capacity of hold	500 cm^3	m^3
Area of sails	cm^2	175 m^2
Number of cannon	12	
Deck area	370 cm^2	m^2

23. Two solid spheres have surface areas of 5 cm^2 and 45 cm^2 respectively and the mass of the smaller sphere is 2 kg. Find the mass of the larger sphere.

24. The masses of two similar objects are 24 kg and 81 kg respectively. If the surface area of the larger object is 540 cm^2, find the surface area of the smaller object.

25. A cylindrical can has a circumference of 40 cm and a capacity of 4·8 litres. Find the capacity of a similar cylinder of circumference 50 cm.

26. A container has a surface area of 5000 cm^2 and a capacity of 12·8 litres. Find the surface area of a similar container which has a capacity of 5·4 litres.

27. A full size snooker ball has a diameter of $2\frac{1}{16}$ inches and weighs 133·1 g. Calculate the weight of a snooker ball of diameter $1\frac{7}{8}$ inches, assuming that both balls are made of the same material.

5 ALGEBRA 2

5.1 Changing the subject of a formula

The operations involved in solving ordinary linear equations are exactly
the same as the operations required in changing the subject of a formula.

(a) Solve the equation $3x + 1 = 12$.

(b) Make x the subject of the formula $Mx + B = A$.

(c) Make y the subject of the formula $x(y - a) = e$.

(a) $3x + 1 = 12$
$3x = 12 - 1$
$x = \dfrac{12 - 1}{3} = \dfrac{11}{3}$

(b) $Mx + B = A$
$Mx = A - B$
$x = \dfrac{A - B}{M}$

(c) $x(y - a) = e$
$xy - xa = e$
$y = \dfrac{e + xa}{x}$

Exercise 1

Make x the subject of the following:

1. $Ax = B$
2. $Nx = T$
3. $Mx = K$
4. $xy = 4$
5. $9x = T + N$
6. $Ax = B - R$
7. $Cx = R + T$
8. $Lx = N - R^2$
9. $R - S^2 = Nx$
10. $x + 5 = 7$
11. $x + B = S$
12. $N = x + D$
13. $N^2 + x = T$
14. $L + x = N + M$
15. $x - R = A$
16. $x - A = E$
17. $F = x - B$
18. $F^2 = x - B^2$

Make y the subject of the following:

19. $L = y - B$
20. $N = y - T$
21. $3y + 1 = 7$
22. $2y - 4 = 5$
23. $Ay + C = N$
24. $By + D = L$
25. $Dy + E = F$
26. $Ny - F = H$
27. $Vy + m = Q$
28. $ty - m = n + a$
29. $V^2y + b = c$
30. $r = ny - 6$
31. $3(y - 1) = 5$
32. $A(y + B) = C$
33. $D(y + E) = F$
34. $h(y + n) = a$
35. $b(y - d) = q$
36. $n = r(y + t)$

(a) Solve the equation $\dfrac{3a + 1}{2} = 4$.

(b) Make a the subject of the formula $\dfrac{na + b}{m} = n$.

(c) Make a the subject of the formula $x - na = y$.

(a) $\dfrac{3a + 1}{2} = 4$
$3a + 1 = 8$
$3a = 7$
$a = \tfrac{7}{3}$

(b) $\dfrac{na + b}{m} = n$
$na + b = mn$
$na = mn - b$
$a = \dfrac{mn - b}{n}$

(c) $x - na = y$
Make the 'a' term positive
$x - y = na$
$\dfrac{x - y}{n} = a$

Exercise 2

Make a the subject.

1. $\dfrac{a}{4} = 3$

2. $\dfrac{a}{5} = 2$

3. $\dfrac{a}{D} = B$

4. $\dfrac{a}{B} = T$

5. $\dfrac{a}{N} = R$

6. $b = \dfrac{a}{m}$

7. $\dfrac{a-2}{4} = 6$

8. $\dfrac{a-A}{B} = T$

9. $\dfrac{a-D}{N} = A$

10. $\dfrac{a+Q}{N} = B^2$

11. $g = \dfrac{a-r}{e}$

12. $\dfrac{2a+1}{5} = 2$

13. $\dfrac{Aa+B}{C} = D$

14. $\dfrac{na+m}{p} = q$

15. $\dfrac{ra-t}{S} = v$

16. $\dfrac{za-m}{q} = t$

17. $\dfrac{m+Aa}{b} = c$

18. $A = \dfrac{Ba+D}{E}$

19. $n = \dfrac{ea-f}{h}$

20. $q = \dfrac{ga+b}{r}$

21. $6 - a = 2$

22. $7 - a = 9$

23. $5 = 7 - a$

24. $A - a = B$

25. $C - a = E$

26. $D - a = H$

27. $n - a = m$

28. $t = q - a$

29. $b = s - a$

30. $v = r - a$

31. $t = m - a$

32. $5 - 2a = 1$

33. $T - Xa = B$

34. $M - Na = Q$

35. $V - Ma = T$

36. $L = N - Ra$

37. $r = v^2 - ra$

38. $t^2 = w - na$

39. $n - qa = 2$

40. $\dfrac{3-4a}{2} = 1$

41. $\dfrac{5-7a}{3} = 2$

42. $\dfrac{B-Aa}{D} = E$

43. $\dfrac{D-Ea}{N} = B$

44. $\dfrac{h-fa}{b} = x$

45. $\dfrac{v^2-ha}{C} = d$

46. $\dfrac{M(a+B)}{N} = T$

47. $\dfrac{f(Na-e)}{m} = B$

48. $\dfrac{T(M-a)}{E} = F$

49. $\dfrac{y(x-a)}{z} = t$

50. $\dfrac{k^2(m-a)}{x} = x$

Formulas with fractions

(a) Solve the equation $\dfrac{4}{z} = 7$.

(b) Make z the subject of the formula $\dfrac{n}{z} = k$.

(c) Make t the subject of the formula $\dfrac{x}{t} + m = a$.

(a) $\dfrac{4}{z} = 7$

$4 = 7z$

$\dfrac{4}{7} = z$

(b) $\dfrac{n}{z} = k$

$n = kz$

$\dfrac{n}{k} = z$

(c) $\dfrac{x}{t} + m = a$

$\dfrac{x}{t} = a - m$

$x = (a - m)t$

$\dfrac{x}{(a-m)} = t$

Exercise 3

Make a the subject.

1. $\dfrac{7}{a} = 14$　　**2.** $\dfrac{5}{a} = 3$　　**3.** $\dfrac{B}{a} = C$　　**4.** $\dfrac{T}{a} = X$

5. $t = \dfrac{v}{a}$　　**6.** $\dfrac{n}{a} = \sin 20°$　　**7.** $\dfrac{7}{a} = \cos 30°$　　**8.** $\dfrac{B}{a} = x$

9. $\dfrac{v}{a} = \dfrac{m}{s}$　　**10.** $\dfrac{t}{b} = \dfrac{m}{a}$　　**11.** $\dfrac{B}{a+D} = C$　　**12.** $\dfrac{Q}{a-C} = T$

13. $\dfrac{V}{a-T} = D$　　**14.** $\dfrac{L}{Ma} = B$　　**15.** $\dfrac{N}{Ba} = C$　　**16.** $\dfrac{m}{ca} = d$

17. $t = \dfrac{b}{c-a}$　　**18.** $x = \dfrac{z}{y-a}$

Make x the subject.

19. $\dfrac{2}{x} + 1 = 3$　　**20.** $\dfrac{5}{x} - 2 = 4$　　**21.** $\dfrac{A}{x} + B = C$　　**22.** $\dfrac{V}{x} + G = H$

23. $\dfrac{r}{x} - t = n$　　**24.** $q = \dfrac{b}{x} + d$　　**25.** $t = \dfrac{m}{x} - n$　　**26.** $h = d - \dfrac{b}{x}$

27. $C - \dfrac{d}{x} = e$　　**28.** $r - \dfrac{m}{x} = e^2$　　**29.** $t^2 = b - \dfrac{n}{x}$　　**30.** $\dfrac{d}{x} + b = mn$

31. $3M = M + \dfrac{N}{P+x}$　　**32.** $A = \dfrac{B}{c+x} - 5A$　　**33.** $\dfrac{m^2}{x} - n = -p$　　**34.** $t = w - \dfrac{q}{x}$

Squares and square roots

Make x the subject of the formulas.

(a) $\sqrt{(x^2 + A)} = B$
$\qquad x^2 + A = B^2$ (square both sides)
$\qquad\quad x^2 = B^2 - A$
$\qquad\quad\ x = \pm\sqrt{(B^2 - A)}$

(b) $(Ax - B)^2 = M$
$\qquad Ax - B = \pm\sqrt{M}$ (square root both sides)
$\qquad\quad Ax = B \pm \sqrt{M}$
$\qquad\quad\ x = \dfrac{B \pm \sqrt{M}}{A}$

Exercise 4

Make x the subject.

1. $\sqrt{x} = 2$　　**2.** $\sqrt{(x - 2)} = 3$　　**3.** $\sqrt{(x + C)} = D$　　**4.** $\sqrt{(ax + b)} = c$
5. $b = \sqrt{(gx - t)}$　　**6.** $\sqrt{(d - x)} = t$　　**7.** $c = \sqrt{(n - x)}$　　**8.** $g = \sqrt{(c - x)}$
9. $\sqrt{(Ax + B)} = \sqrt{D}$　　**10.** $x^2 = g$　　**11.** $x^2 = B$　　**12.** $x^2 - A = M$

Make k the subject.

13. $C - k^2 = m$　　**14.** $mk^2 = n$　　**15.** $\dfrac{kz}{a} = t$　　**16.** $n = a - k^2$

17. $\sqrt{(k^2 - A)} = B$　　**18.** $t = \sqrt{(m + k^2)}$　　**19.** $A\sqrt{(k + B)} = M$　　**20.** $\sqrt{\left(\dfrac{N}{k}\right)} = B$

21. $\sqrt{(a^2 - k^2)} = t$　　**22.** $2\pi\sqrt{(k + t)} = 4$　　**23.** $\sqrt{(ak^2 - b)} = C$　　**24.** $k^2 + b = x^2$

Collecting from both sides

Make x the subject of the formulas.

(a) $Ax - B = Cx + D$
$Ax - Cx = D + B$ (x terms on one side)
$x(A - C) = D + B$ (factorise)
$$x = \frac{D + B}{A - C}$$

(b) $x + a = \dfrac{x + b}{c}$
$c(x + a) = x + b$
$cx + ca = x + b$
$cx - x = b - ca$ (x terms on one side)
$x(c - 1) = b - ca$ (factorise)
$$x = \frac{b - ca}{c - 1}$$

Exercise 5

Make y the subject.

1. $5(y - 1) = 2(y + 3)$ **2.** $7(y - 3) = 4(3 - y)$ **3.** $Ny + B = D - Ny$
4. $My - D = E - 2My$ **5.** $ay + b = 3b + by$ **6.** $my - c = e - ny$
7. $xy + 4 = 7 - ky$ **8.** $Ry + D = Ty + C$ **9.** $ay - x = z + by$

10. $m(y + a) = n(y + b)$ **11.** $x(y - b) = y + d$ **12.** $\dfrac{a - y}{a + y} = b$

13. $\dfrac{1 - y}{1 + y} = \dfrac{c}{d}$ **14.** $\dfrac{M - y}{M + y} = \dfrac{a}{b}$ **15.** $m(y + n) = n(n - y)$

16. $y + m = \dfrac{2y - 5}{m}$ **17.** $y - n = \dfrac{y + 2}{n}$ **18.** $y + b = \dfrac{ay + e}{b}$

19. $\dfrac{ay + x}{x} = 4 - y$ **20.** $c - dy = e - ay$ **21.** $y(a - c) = by + d$

22. $y(m + n) = a(y + b)$ **23.** $t - ay = s - by$ **24.** $\dfrac{y + x}{y - x} = 3$

25. $\dfrac{v - y}{v + y} = \dfrac{1}{2}$ **26.** $y(b - a) = a(y + b + c)$ **27.** $\sqrt{\left(\dfrac{y + x}{y - x}\right)} = 2$

28. $\sqrt{\left(\dfrac{z + y}{z - y}\right)} = \dfrac{1}{3}$ **29.** $\sqrt{\left[\dfrac{m(y + n)}{y}\right]} = p$ **30.** $n - y = \dfrac{4y - n}{m}$

Exercise 6   Mixed questions

Make the letter in square brackets the subject.

1. $ax + by + c = 0$ [x]   **2.** $\sqrt{\{a(y^2 - b)\}} = e$ [y]

3. $\dfrac{\sqrt{(k - m)}}{n} = \dfrac{1}{m}$ [k]   **4.** $a - bz = z + b$ [z]

5. $\dfrac{x + y}{x - y} = 2$ [x]   **6.** $\sqrt{\left(\dfrac{a}{z} - c\right)} = e$ [z]

7. $lm + mn + a = 0$ [n]   **8.** $t = 2\pi\sqrt{\left(\dfrac{d}{g}\right)}$ [d]

9. $t = 2\pi \sqrt{\left(\dfrac{d}{g}\right)}$ [g]

10. $\sqrt{(x^2 + a)} = 2x$ [x]

11. $\sqrt{\left\{\dfrac{b(m^2 + a)}{e}\right\}} = t$ [m]

12. $\sqrt{\left(\dfrac{x+1}{x}\right)} = a$ [x]

13. $a + b - mx = 0$ [m]

14. $\sqrt{(a^2 + b^2)} = x^2$ [a]

15. $\dfrac{a}{k} + b = \dfrac{c}{k}$ [k]

16. $a - y = \dfrac{b + y}{a}$ [y]

17. $G = 4\pi\sqrt{(x^2 + T^2)}$ [x]

18. $M(ax + by + c) = 0$ [y]

19. $x = \sqrt{\left(\dfrac{y-1}{y+1}\right)}$ [y]

20. $a\sqrt{\left(\dfrac{x^2 - n}{m}\right)} = \dfrac{a^2}{b}$ [x]

21. $\dfrac{M}{N} + E = \dfrac{P}{N}$ [N]

22. $\dfrac{Q}{P - x} = R$ [x]

23. $\sqrt{(z - ax)} = t$ [a]

24. $e + \sqrt{(x + f)} = g$ [x]

5.2 Inequalities and regions

Symbols:

- Here is the meaning of inequality symbols used.

  $x < 4$ means 'x is *less than* 4'
  $y > 7$ means 'y is *greater than* 7'
  $z \leqslant 10$ means 'z is *less than or equal to* 10'
  $t \geqslant -3$ means 't is *greater than or equal to* -3'

- With two symbols in one statement look at each part separately.
  For example, if n is an *integer* and $3 < n \leqslant 7$,
  n has to be greater than 3 but at the same time it has to be less than or equal to 7.

  So n could be 4, 5, 6 or 7 only.

Illustrate on a number line the range of values of x stated.

(a) $x > 1$

The circle at the left hand end of the range is open.
This means that 1 is not included.

(b) $x \leqslant -2$

The circle at -2 is filled in to indicate that -2 is included.

(c) $1 \leqslant x < 4$

Exercise 7

1. Write down each statement with either $>$ or $<$ in the box.

(a) 3 ☐ 7   (b) 0 ☐ −2   (c) 3·1 ☐ 3·01

(d) −3 ☐ −5   (e) 100 mm ☐ 1 m   (f) 1 kg ☐ 1 lb

2. Write down the inequality displayed. Use x for the variable.

(a)

2

(b)

5

(c)

100

(d)

−2   2

(e)

−6

(f)

3   8

3. Draw a number line to display these inequalities:

(a) $x \geqslant 7$ (b) $x < 2·5$ (c) $1 < x < 7$

(d) $0 \leqslant x \leqslant 4$ (e) $-1 < x \leqslant 5$

4. Write an inequality for each statement.

(a) You must be at least 16 to get married.
[Use A for age.]

(b) Vitamin J1 is not recommended for people over 70 or for children 3 years or under.

(c) To braise badger the oven temperature should be between 150°C and 175°C.
[Use T for temperature.]

(d) Applicants for training as paratroopers must be at least 1·75 m tall.
[Use h for height.]

5. Answer 'true' or 'false':

(a) n is an integer and $1 < n \leqslant 4$, so n can be 2, 3 or 4.

(b) x is an integer and $2 \leqslant x < 5$, so x can be 2, 3 or 4.

(c) p is an integer and $p \geqslant 10$, so p can be 10, 11, 12, 13 ...

6. Write two different inequalities involving x which are satisfied by the integers 3, 4, 5, 6, 7.

7. Write down one inequality to show the values of x which satisfy all three of the following inequalities.

| $x < 5$ | $0 < x < 6$ | $3 \leqslant x < 10$ |

Solving inequalities

We follow the same procedure used for solving equations except that when we multiply or divide by a *negative* number the inequality is *reversed*.

e.g. $4 > -2$ but multiplying by -2, $-8 < 4$

It is best to avoid dividing by a negative number as in the following example.

(a) $2x - 1 > 5$

$\qquad 2x > 5 + 1$ [add 1]

$\qquad x > \dfrac{6}{2}$ [divide by 2]

$\qquad x > 3$

(b) $x + 1 < 2x < 3x + 2$

$\qquad$ Solve the two inequalities separately:

$\qquad x + 1 < 2x \qquad 2x < 3x + 2$

$\qquad\quad 1 < x \qquad\qquad x < 2$

$\therefore$ The solution is $1 < x < 2$.

Exercise 8

Solve the following inequalities:

1. $x - 3 > 10$

2. $x + 1 < 0$

3. $5 > x - 7$

4. $2x + 1 \leqslant 6$

5. $3x - 4 > 5$

6. $10 \leqslant 2x - 6$

7. $5x < x + 1$

8. $2x \geqslant x - 3$

9. $4 + x < -4$

10. $3x + 1 < 2x + 5$

11. $2(x + 1) > x - 7$

12. $7 < 15 - x$

13. $9 > 12 - x$

14. $4 - 2x \leqslant 2$

15. $3(x - 1) < 2(1 - x)$

16. $7 - 3x < 0$

17. $\dfrac{x}{3} < -1$

18. $\dfrac{2x}{5} > 3$

19. $2x > 0$

20. $\dfrac{x}{4} < 0$

21. The height of the picture has to be greater than the width.

Find the range of possible values of x.

height
$2(x + 1)$

width
$(x + 7)$

(Hint: in Questions **22** to **27**, solve the two inequalities separately.)

22. $10 \leqslant 2x \leqslant x + 9$

23. $x < 3x + 2 < 2x + 6$

24. $10 \leqslant 2x - 1 \leqslant x + 5$

25. $3 < 3x - 1 < 2x + 7$

26. $x - 10 < 2(x - 1) < x$

27. $4x + 1 < 8x < 3(x + 2)$

28. Sumitra said 'I think of an integer.

$\qquad$ I subtract 14.

$\qquad$ I multiply the result by 5.

$\qquad$ I divide by 2.

$\qquad$ The answer is greater than the number I thought of.'

Write an inequality and solve it to find the smallest number Sumitra thought of.

● Squares and square roots in inequalities need care.

The equation $x^2 = 4$ becomes $x = \pm 2$, which is correct.

For the inequality $x^2 < 4$, we might wrongly write $x < \pm 2$.
Consider $x = -3$, say.
   -3 is less than -2 and is also less than $+2$.
   But $(-3)^2$ is not less than 4 and so
   $x = -3$ does not satisfy the inequality $x^2 < 4$.
The correct solution for $x^2 < 4$ is $-2 < x < 2$

Solve the inequality $2x^2 - 1 > 17$.
$$2x^2 - 1 > 17$$
$$2x^2 > 18$$
$$x^2 > 9$$
$$x > 3 \text{ or } x < -3$$

[Avoid the temptation to write $x > \pm 3$!]

Exercise 9

1. The area of the rectangle must be greater than the area of the triangle.
Find the range of possible values of x.

For Questions **2** to **8**, list the solutions which satisfy the given condition.

2. $3a + 1 < 20$; a is a positive integer.

3. $b - 1 \geqslant 6$; b is a prime number less than 20.

4. $1 < z < 50$; z is a square number.

5. $2x > -10$; x is a negative integer.

6. $x + 1 < 2x < x + 13$; x is an integer.

7. $0 \leqslant 2z - 3 \leqslant z + 8$; z is a prime number.

8. $\dfrac{a}{2} + 10 > a$; a is a positive even number.

9. Given that $4x > 1$ and $\dfrac{x}{3} \leqslant 1\frac{1}{3}$, list the possible integer values of x.

10. State the smallest integer n for which $4n > 19$.

11. Given that $-4 \leqslant a \leqslant 3$ and $-5 \leqslant b \leqslant 4$, find:

    (a) the largest possible value of a^2

    (b) the smallest possible value of ab

    (c) the largest possible value of ab

    (d) the value of b if $b^2 = 25$.

12. For any shape of triangle ABC, complete the statement
$\text{AB} + \text{BC} \ \square \ \text{AC}$, by writing $<, >$ or $=$ inside the box.

13. Find a simple fraction r such that $\frac{1}{3} < r < \frac{2}{3}$.

14. Find the largest prime number p such that $p^2 < 400$.

15 Find the integer n such that $n < \sqrt{300} < n + 1$.

16. If $f(x) = 2x - 1$ and $g(x) = 10 - x$ for what values of x is
$f(x) > g(x)$?

17. (a) The solution of $x^2 < 9$ is $-3 < x < 3$.
       [The square roots of 9 are -3 and 3.]

    (b) Copy and complete:

        (i) If $x^2 < 100$, then $\square < x < \square$

        (ii) If $x^2 < 81$, then $\square < x < \square$

        (iii) If $x^2 > 36$, then $x > \square$ or $x < \square$

Solve the inequalities.

18. $x^2 < 25$ **19.** $x^2 \leqslant 16$ **20.** $x^2 > 1$

21. $2x^2 \geqslant 72$ **22.** $3x^2 + 5 > 5$ **23.** $5x^2 - 2 < 18$

24. Given $2 \leqslant p \leqslant 10$ and $1 \leqslant q \leqslant 4$, find the range of values of:

    (a) pq (b) $\dfrac{p}{q}$ (c) $p - q$ (d) $p + q$

25. If $2^r > 100$, what is the smallest integer value of r?

26. Given $\left(\dfrac{1}{3}\right)^x < \dfrac{1}{200}$, what is the smallest integer value of x?

27. Find the smallest integer value of x which satisfies $x^x > 10\,000$.

28. What integer values of x satisfy
$100 < 5^x < 10\,000$?

†**29.** If x is an acute angle and $\sin x > \frac{1}{2}$, write down the range of values
that x can take.

†**30.** If x is an acute angle and $\cos x > \frac{1}{4}$, write down the range of values
that x can take.

Shading regions

It is useful to represent inequalities on a graph, particularly where two variables (x and y) are involved.

Draw a sketch graph and shade the area which represents the set of points that satisfy each of these inequalities.

(a) $x > 2$ 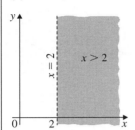 (b) $1 \leqslant y \leqslant 5$ 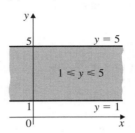 (c) $x + y \leqslant 8$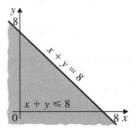

In each graph, the required region is shaded.

In (a), the line $x = 2$ is shown as a broken line to indicate that the points on the line are not included.

In (b) and (c) points on the line *are* included 'in the region' and the lines are drawn unbroken.

To decide which side to shade when the line is sloping, we take a *trial point*. This can be any point which is not actually on the line.

In (c) above, the trial point could be (1, 1).

Is (1, 1) in the region $x + y \leqslant 8$?
It satisfies $x + y < 8$ because $1 + 1 = 2$, which is less than 8.
So below the line is $x + y < 8$. We have shaded the required region.

Exercise 10

In Questions **1** to **6**, describe the region shaded.

1.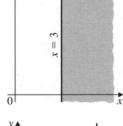

2.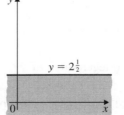

3.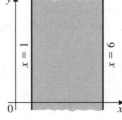

4.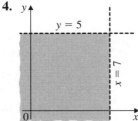

5.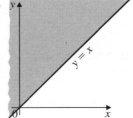

6.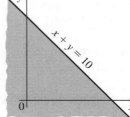

7. The point (1, 1), marked *, lies
in the shaded region. Use
this as a trial point to describe
the unshaded region.

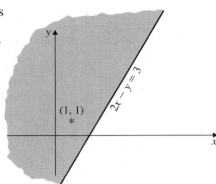

8. The point (3, 1), marked *, lies
in the shaded triangle. Use
this as a trial point to write
down the three inequalities which
describe the unshaded region.

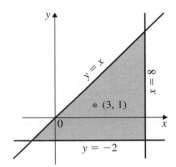

9. A trial point (1, 1) lies inside the shaded triangles. Write down the
three inequalities which describe each shaded region.

(a)

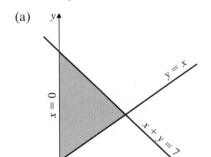

(b)

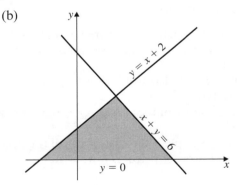

For Questions **10** to **25**, draw a sketch graph similar to those above and
indicate the set of points which satisfy the inequalities by shading the
required region.

10. $2 < x < 7$

11. $0 < y < 3\frac{1}{2}$

12. $-2 < x < 2$

13. $x < 6$ and $y < 4$

14. $0 < x < 5$ and $y < 3$

15. $1 < x < 6$ and $2 < y < 8$

16. $-3 < x < 0$ and $-4 < y < 2$

17. $y < x$

18. $x + y < 5$

19. $y > x + 2$ and $y < 7$

20. $x > 0$ and $y > 0$ and $x + y < 7$

21. $x > 0$ and $x + y < 10$ and $y > x$

22. $8 > y > 0$ and $x + y > 3$

23. $x + 2y < 10$ and $x > 0$ and $y > 0$

24. $3x + 2y < 18$ and $x > 0$ and $y > 0$

25. $x > 0$, $y > x - 2$, $x + y < 10$

26. $3x + 5y < 30$ and $y > \dfrac{x}{2}$

27. $y > \dfrac{x}{2}$, $y < 2x$ and $x + y < 8$

28. The two lines $y = x + 1$ and $x + y = 5$ divide the graph into four regions A, B, C, D. Write down the two inequalities which describe each of the regions A, B, C, D.

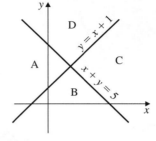

†**29.** Using the same axes, draw the graphs of $xy = 10$ and $x + y = 9$ for values of x from 1 to 10.

Hence find all pairs of positive integers whose product is greater than 10 and whose sum is less than 9.

5.3 Direct and inverse proportion

Direct proportion

(a) When you buy petrol, the more you buy the more money you have to pay. So if 2·2 litres costs 176p, then 4·4 litres will cost 352p.

We say the cost of petrol is *directly proportional* to the quantity bought.

To show that quantities are proportional, we use the symbol '$\propto$'.

So in our example if the cost of petrol is c pence and the number of litres of petrol is l, we write

$c \propto l$

The '$\propto$' sign can always be replaced by '$= k$' where k is a constant.
So $c = kl$

From above, if $c = 176$ when $l = 2·2$

then $176 = k \times 2·2$

$$k = \frac{176}{2·2} = 80$$

We can then write $c = 80l$, and this allows us to find the value of c for any value of l, and *vice versa*.

(b) If a quantity z is proportional to a quantity x, we have

$z \propto x$ or $z = kx$

Two other expressions are sometimes used when quantities are directly proportional. We could say

'z varies as x'

or 'z varies directly as x'.

The graph connecting z and x is a straight line which passes through the origin.

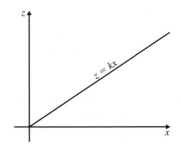

y varies as z, and $y = 2$ when $z = 5$; find:

(a) the value of y when $z = 6$
(b) the value of z when $y = 5$.

Because $y \propto z$, then $y = kz$ where k is a constant.

$$y = 2 \text{ when } z = 5$$
$$\therefore \quad 2 = k \times 5$$
$$k = \tfrac{2}{5}$$
So $\quad y = \tfrac{2}{5}z$

(a) When $z = 6$, $y = \tfrac{2}{5} \times 6 = 2\tfrac{2}{5}$.
(b) When $y = 5$, $5 = \tfrac{2}{5}z$; $z = \tfrac{25}{2} = 12\tfrac{1}{2}$.

The value V of a diamond is proportional to the square of its weight W. If a diamond weighing 10 grams is worth £200, find

(a) the value of a diamond weighing 30 grams
(b) the weight of a diamond worth £5000.

$$V \propto W^2$$
or $\quad V = kW^2$ where k is a constant.
$$V = 200 \text{ when } W = 10$$
$$\therefore \quad 200 = k \times 10^2$$
$$k = 2$$

So $\quad V = 2W^2$

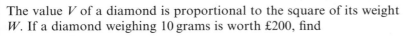

(a) When $W = 30$,
$$V = 2 \times 30^2 = 2 \times 900$$
$$V = £1800$$

So a diamond of weight 30 grams is worth £1800.

(b) When $\quad V = 5000$,
$$5000 = 2 \times W^2$$
$$W^2 = \frac{5000}{2} = 2500$$
$$W = \sqrt{2500} = 50$$

So a diamond of value £5000 weighs 50 grams.

Exercise 11

1. Rewrite the statement connecting each pair of variables using a constant k instead of '$\propto$'.
   (a) $S \propto e$ (b) $v \propto t$ (c) $x \propto z^2$
   (d) $y \propto \sqrt{x}$ (e) $T \propto \sqrt{L}$

2. y varies as t. If $y = 6$ when $t = 4$, calculate:
   (a) the value of y, when $t = 6$
   (b) the value of t, when $y = 4$.

3. z is proportional to m. If $z = 20$ when $m = 4$, calculate:
   (a) the value of z, when $m = 7$
   (b) the value of m, when $z = 55$.

4. A varies directly as r^2. If $A = 12$, when $r = 2$, calculate:
   (a) the value of A, when $r = 5$
   (b) the value of r, when $A = 48$.

5. Given that $z \propto x$, copy and complete the table.

x	1	3		$5\frac{1}{2}$
z	4		16	

6. Given that $V \propto r^3$, copy and complete the table.

r	1	2		$1\frac{1}{2}$
V	4		256	

7. The pressure of the water P at any point below the surface of the sea varies as the depth of the point below the surface d. If the pressure is 200 newtons/cm^2 at a depth of 3 m, calculate the pressure at a depth of 5 m.

8. The distance d through which a stone falls from rest is proportional to the square of the time taken t. If the stone falls 45 m in 3 seconds, how far will it fall in 6 seconds?
   How long will it take to fall 20 m?

9. The energy E stored in an elastic band is proportional to the square of the extension x. When the elastic is extended by 3 cm, the energy stored is 243 joules. What is the energy stored when the extension is 5 cm?
   What is the extension when the stored energy is 36 joules?

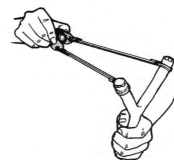

10. The resistance to motion of a car is proportional to the square of the speed of the car. If the resistance is 4000 newtons at a speed of 20 m/s, what is the resistance at a speed of 30 m/s?
    At what speed is the resistance 6250 newtons?

11. In an experiment, measurements of w and p were taken.

w	2	5	7
p	1·6	25	68·6

Which of these laws fits the results?

$p \propto w$, $p \propto w^2$, $p \propto w^3$.

12. A road research organisation recently claimed that the damage to road surfaces was proportional to the fourth power of the axle load. The axle load of a 44-ton HGV is about 15 times that of a car. Calculate the ratio of the damage to road surfaces made by a 44-ton HGV and a car.

Inverse proportion

If you travel a distance of 200 m at 10 m/s, the time taken is 20 s.
If you travel the same distance at 20 m/s, the time taken is 10 s.
As you *double* the speed, you *halve* the time taken.
For a fixed journey, the time taken is *inversely proportional* to the speed at which you travel.
If t is inversely proportional to s, we write

$$s \propto \frac{1}{t}$$

or $s = k \times \dfrac{1}{t}$

Notice that the product $s \times t$ is constant.
The graph connecting s and t is a curve.
The shape of the curve is the same as $y = \dfrac{1}{x}$.

Note: Sometimes we write 'x varies inversely as y'.
It means the same as 'x is inversely proportional to y'.

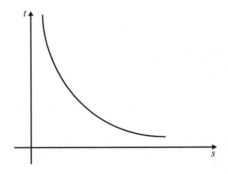

---

z is inversely proportional to t^2 and $z = 4$ when $t = 1$.
Calculate z when $t = 2$

We have $z \propto \dfrac{1}{t^2}$ or $z = k \times \dfrac{1}{t^2}$ (k is a constant)

$z = 4$ when $t = 1$,

$\therefore \quad 4 = k\left(\dfrac{1}{1^2}\right)$

so $k = 4$

$\therefore \quad z = 4 \times \dfrac{1}{t^2}$

When $t = 2$, $z = 4 \times \dfrac{1}{2^2} = 1$

Exercise 12

1. Rewrite the statements connecting the variables using a constant of variation, k.

(a) $x \propto \dfrac{1}{y}$
(b) $s \propto \dfrac{1}{t^2}$
(c) $t \propto \dfrac{1}{\sqrt{q}}$

(d) m varies inversely as w
(e) z is inversely proportional to t^2.

2. T is inversely proportional to m. If $T = 12$ when $m = 1$, find:
(a) T when $m = 2$
(b) T when $m = 24$.

3. L is inversely proportional to x. If $L = 24$ when $x = 2$, find:
(a) L when $x = 8$
(b) L when $x = 32$.

4. b varies inversely as e. If $b = 6$ when $e = 2$, calculate:
(a) the value of b when $e = 12$
(b) the value of e when $b = 3$.

5. x is inversely proportional to y^2. If $x = 4$ when $y = 3$, calculate:
(a) the value of x when $y = 1$
(b) the value of y when $x = 2\frac{1}{4}$.

6. p is inversely proportional to $\sqrt{y}$. If $p = 1\cdot2$ when $y = 100$, calculate:
(a) the value of p when $y = 4$
(b) the value of y when $p = 3$.

7. Given that $z \propto \dfrac{1}{y}$, copy and complete the table:

y	2	4		$\frac{1}{4}$
z	8		16	

8. Given that $v \propto \dfrac{1}{t^2}$, copy and complete the table:

t	2	5		10
v	25		$\frac{1}{4}$	

9. e varies inversely as $(y - 2)$. If $e = 12$ when $y = 4$, find:
(a) e when $y = 6$
(b) y when $e = \frac{1}{2}$.

10. The volume V of a given mass of gas varies inversely as the pressure P. When $V = 2\,\text{m}^3$, $P = 500\,\text{N/m}^2$. Find the volume when the pressure is $400\,\text{N/m}^2$. Find the pressure when the volume is $5\,\text{m}^3$.

11. The number of hours N required to dig a certain hole is inversely proportional to the number of men available x.

When 6 men are digging, the hole takes 4 hours. Find the time taken when 8 men are available. If it takes $\frac{1}{2}$ hour to dig the hole, how many men are there?

12. The force of attraction F between two magnets varies inversely as the square of the distance d between them. When the magnets are 2 cm apart, the force of attraction is 18 newtons. How far apart are they if the attractive force is 2 newtons?

13. The number of tiles, n, that can be pasted using one tin of tile paste is inversely proportional to the square of the side d, of the tile. One tin is enough for 180 tiles of side 10 cm. How many tiles of side 15 cm can be pasted using one tin?

14. The life expectancy L of a rat varies inversely as the square of the density d of poison distributed around his home. When the density of poison is $1 \, \text{g/m}^2$ the life expectancy is 50 days. How long will he survive if the density of poison is

(a) $5 \, \text{g/m}^2$?   (b) $\frac{1}{2} \, \text{g/m}^2$?

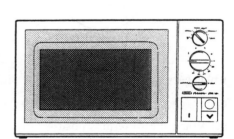

15. When cooking snacks in a microwave oven, a French chef assumes that the cooking time is inversely proportional to the power used. The five levels on his microwave have the powers shown in the table.

Level	Power used
Full	600 W
Roast	400 W
Simmer	200 W
Defrost	100 W
Warm	50 W

(a) Escargots de Bourgogne take 5 minutes on 'Simmer'. How long will they take on 'Warm'?

(b) Escargots à la Provençale are normally cooked on 'Roast' for 3 minutes. How long will they take on 'Full'?

16. Given $z = \dfrac{k}{x^n}$, find k and n, then copy and complete the table.

x	1	2	4	
z	100	$12\frac{1}{2}$		$\frac{1}{10}$

†**17.** Given $y = \dfrac{k}{\sqrt[n]{v}}$, find k and n, then copy and complete the table.

v	1	4	36	
y	12	6		$\frac{3}{25}$

[$\sqrt[n]{v}$ means the nth root of v]

5.4 Curved graphs

Common curves

It is helpful to know the general shape of some of the more common curves.

Quadratic curves have an x^2 term as the highest power of x.

e.g. $y = 2x^2 - 3x + 7$ and $y = 5 + 2x - x^2$

(a) When the x^2 term is positive, the curve is ∪-shaped.

(b) When the x^2 term is negative the curve is an inverted ∩.

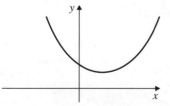

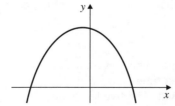

Cubic curves have an x^3 term as the highest power of x.

e.g. $y = x^3 + 7x - 4$ and $y = 8x - 4x^3$

(a) When the x^3 term is positive, the curve can be like one of the two shown below. Notice that as x gets larger, so does y.

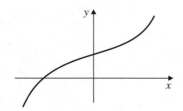

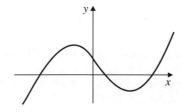

(b) When the x^3 term is negative, the curve can be like one of the two shown below. Notice that as x gets large, y is large but negative.

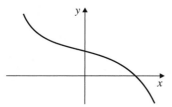
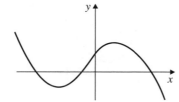

Reciprocal curves have a $\dfrac{1}{x}$ term.

e.g. $y = \dfrac{12}{x}$ and

$y = \dfrac{6}{x} + 5$

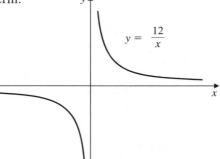

The curve has a break at $x = 0$. The x-axis and the y-axis are called *asymptotes* to the curve. The curve gets very near but never actually touches the asymptotes.

Exponential curves have a term involving a^x, where a is a constant.

e.g. $y = 3^x$
$\quad y = \left(\frac{1}{2}\right)^x$

The x-axis is an asymptote to the curve.

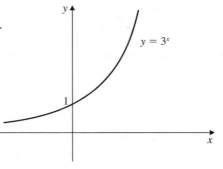

Exercise 13

1. What sort of curves are these? State as much information as you can.

(a)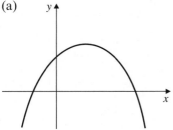

(b)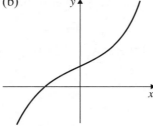

(c)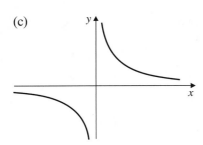

(d)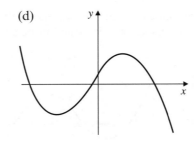

(e)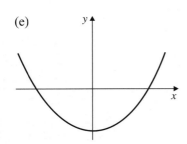

(f)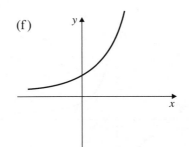

2. Draw the general shape of the following curves.

(a) $y = 3x^2 - 7x + 11$

(b) $y = 2^x$

(c) $y = \dfrac{100}{x}$

(d) $y = 8x - x^2$

(e) $y = 10x^3 + 7x - 2$

(f)† $y = \dfrac{1}{x^2}$

3. Here are the equations of the six curves in Question **1**, but not in the correct order.

(i) $y = \dfrac{8}{x}$

(ii) $y = 2x^3 + x + 2$

(iii) $y = 5 + 3x - x^2$

(iv) $y = x^2 - 6$

(v) $y = 5^x$

(vi) $y = 12 + 11x - 2x^2 - x^3$

Decide which equations fit the curves (a) to (f).

4. Sketch the two curves given and state the number of times the curves intersect.

(a) $y = x^3,$ (b) $y = x^2,$ (c) $y = 3^x$ (d) $y = x^3$
    $y = 10 - x$ $y = 10 - x^2$ $y = x^3$ $y = x$

Plotting curved graphs

Draw the graph of the function

$y = 2x^2 + x - 6$, for $-3 \leqslant x \leqslant 3$.

(a)

x	−3	−2	−1	0	1	2	3
$2x^2$	18	8	2	0	2	8	18
x	−3	−2	−1	0	1	2	3
−6	−6	−6	−6	−6	−6	−6	−6
y	9	0	−5	−6	−3	4	15

(b) Draw and label axes using suitable scales.
(c) Plot the points and draw a smooth curve through them with a pencil.
(d) Check any points which interrupt the smoothness of the curve.
(e) Label the curve with its equation.

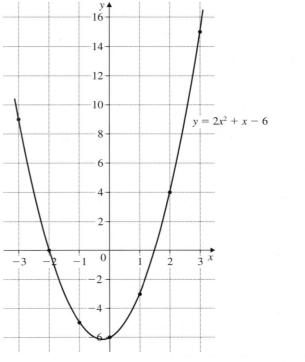

$y = 2x^2 + x - 6$

Common errors with graphs

Avoid the following.

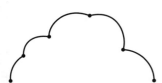

A series of 'mini curves'

Flat bottom

Wrong point

Exercise 14

Draw the graphs of the following functions using a scale of 2 cm for 1 unit on the x-axis and 1 cm for 1 unit on the y-axis.

1. $y = x^2 + 2x$, for $-3 \leqslant x \leqslant 3$

2. $y = x^2 + 4x$, for $-3 \leqslant x \leqslant 3$

3. $y = x^2 - 3x$, for $-3 \leqslant x \leqslant 3$

4. $y = x^2 + 2$, for $-3 \leqslant x \leqslant 3$

5. $y = x^2 - 7$, for $-3 \leqslant x \leqslant 3$

6. $y = x^2 + x - 2$, for $-3 \leqslant x \leqslant 3$

7. $y = x^2 + 3x - 9$, for $-4 \leqslant x \leqslant 3$

8. $y = x^2 - 3x - 4$, for $-2 \leqslant x \leqslant 4$

9. $y = x^2 - 5x + 7$, for $0 \leqslant x \leqslant 6$

10. $y = 2x^2 - 6x$, for $-1 \leqslant x \leqslant 5$

11. $y = 2x^2 + 3x - 6$, for $-4 \leqslant x \leqslant 2$

12. $y = 3x^2 - 6x + 5$, for $-1 \leqslant x \leqslant 3$

13. $y = 2 + x - x^2$, for $-3 \leqslant x \leqslant 3$

14. $f(x) = 1 - 3x - x^2$, for $-5 \leqslant x \leqslant 2$

15. $f(x) = 3 + 3x - x^2$, for $-2 \leqslant x \leqslant 5$

16. $f(x) = 7 - 3x - 2x^2$, for $-3 \leqslant x \leqslant 3$

17. $f(x) = 6 + x - 2x^2$, for $-3 \leqslant x \leqslant 3$

18. $y = 8 + 2x - 3x^2$, for $-2 \leqslant x \leqslant 3$

19. $y = x(x - 4)$, for $-1 \leqslant x \leqslant 6$

20. $y = (x + 1)(2x - 5)$, for $-3 \leqslant x \leqslant 3$.

Draw the graph of $y = \dfrac{12}{x} + x - 6$, for $1 \leqslant x \leqslant 8$.

Use the graph to find approximate values for

(a) the minimum value of $\dfrac{12}{x} + x - 6$

(b) the value of $\dfrac{12}{x} + x - 6$, when $x = 2 \cdot 25$.

Here is the table of values.

x	1	2	3	4	5	6	8
$\dfrac{12}{x}$	12	6	4	3	2·4	2	1·5
x	1	2	3	4	5	6	8
-6	-6	-6	-6	-6	-6	-6	-6
y	7	2	1	1	1·4	2	3·5

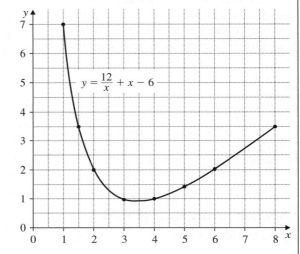

(a) From the graph, the minimum value of $\dfrac{12}{x} + x - 6$ (i.e. y) is approximately 0·9.

(b) At $x = 2 \cdot 25$, y is approximately 1·6.

Exercise 15

Draw the following curves. The scales given are for one unit of x and y.

1. $y = \dfrac{12}{x}$, for $1 \leqslant x \leqslant 10$. (Scales: 1 cm for x and y)

2. $y = \dfrac{9}{x}$, for $1 \leqslant x \leqslant 10$. (Scales: 1 cm for x and y)

3. $y = \dfrac{12}{x + 1}$, for $0 \leqslant x \leqslant 8$. (Scales: 2 cm for x, 1 cm for y)

4. $y = \dfrac{8}{x - 4}$, for $-4 \leqslant x \leqslant 3 \cdot 5$. (Scales: 2 cm for x, 1 cm for y)

5. $y = \dfrac{x}{x+4}$, for $-3.5 \leqslant x \leqslant 4$. (Scales: 2 cm for x and y)

6. $y = \dfrac{x+8}{x+1}$, for $0 \leqslant x \leqslant 8$. (Scales: 2 cm for x and y)

7. $y = \dfrac{10}{x} + x$, for $1 \leqslant x \leqslant 7$. (Scales: 2 cm for x, 1 cm for y)

8. $y = 3^x$, for $-3 \leqslant x \leqslant 3$. (Scales: 2 cm for x, $\frac{1}{2}$ cm for y)

9. $y = \left(\frac{1}{2}\right)^x$, for $-4 \leqslant x \leqslant 4$. (Scales: 2 cm for x, 1 cm for y)

10. $y = 5 + 3x - x^2$, for $-2 \leqslant x \leqslant 5$. (Scales: 2 cm for x, 1 cm for y)
Find
(a) the maximum value of the function $5 + 3x - x^2$,
(b) the two values of x for which $y = 2$.

11. $y = \dfrac{15}{x} + x - 7$, for $1 \leqslant x \leqslant 7$. (Scales: 2 cm for x and y)

From your graph find
(a) the minimum value of y,
(b) the y-value when $x = 5.5$.

12. $y = x^3 - 2x^2$, for $0 \leqslant x \leqslant 4$. (Scales: 2 cm for x, $\frac{1}{2}$ cm for y)
From your graph find
(a) the y-value at $x = 2.5$
(b) the x-value at $y = 15$.

13. $y = \frac{1}{10}(x^3 + 2x + 20)$, for $-3 \leqslant x \leqslant 3$. (Scales: 2 cm for x and y)
From your graph find
(a) the x-value where $x^3 + 2x + 20 = 0$,
(b) the x-value where $y = 3$.

†**14.** Draw the graph of
$$y = \dfrac{x}{x^2 + 1}, \text{ for } -6 \leqslant x \leqslant 6. \text{ (Scales: 1 cm for } x, \text{ 10 cm for } y)$$

†**15.** Draw the graph of
$$E = \dfrac{5000}{x} + 3x \text{ for } 10 \leqslant x \leqslant 80.$$

(Scales: 1 cm to 5 units for x and 1 cm to 25 units for E)
From the graph find,
(a) the minimum value of E,
(b) the value of x corresponding to this minimum value,
(c) the range of values of x for which E is less than 275.

Mixed questions

Exercise 16

1. A rectangle has a perimeter of 14 cm and length x cm. Show that the width of the rectangle is $(7 - x)$ cm and hence that the area A of the rectangle is given by the formula $A = x(7 - x)$. Draw the graph, plotting x on the horizontal axis with a scale of 2 cm to 1 unit, and A on the vertical axis with a scale of 1 cm to 1 unit. Take x from 0 to 7. From the graph find:
   (a) the area of the rectangle when $x = 2 \cdot 25$ cm,
   (b) the dimensions of the rectangle when its area is 9 cm^2,
   (c) the maximum area of the rectangle,
   (d) the length and width of the rectangle corresponding to the maximum area,
   (e) what shape of rectangle has the largest area.

2. A farmer has 60 m of wire fencing which he uses to make a rectangular pen for his sheep. He uses a stone wall as one side of the pen so the wire is used for only 3 sides of the pen.

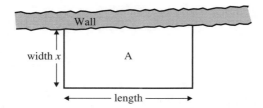

   If the width of the pen is x m, what is the length (in terms of x)?
   What is the area A of the pen?
   Draw a graph with area A on the vertical axis and the width x on the horizontal axis. Take values of x from 0 to 30.
   What dimensions should the pen have if the farmer wants to enclose the largest possible area?

3. A ball is thrown in the air so that t seconds after it is thrown, its height h metres above its starting point is given by the function $h = 25t - 5t^2$. Draw the graph of the function of $0 \leqslant t \leqslant 6$, plotting t on the horizontal axis with a scale of 2 cm to 1 second, and h on the vertical axis with a scale of 2 cm for 10 metres.
   Use the graph to find:
   (a) the time when the ball is at its greatest height,
   (b) the greatest height reached by the ball,
   (c) the interval of time during which the ball is at a height of more than 30 m.

4. The velocity v m/s of a missile t seconds after launching is given by the equation $v = 54t - 2t^3$.

Draw a graph, plotting t on the horizontal axis with a scale of 2 cm to 1 second, and v on the vertical axis with a scale of 1 cm for 10 m/s. Take values of t from 0 to 5.

Use the graph to find:

(a) the maximum velocity reached,

(b) the time taken to accelerate to a velocity of 70 m/s,

(c) the interval of time during which the missile is travelling at more than 100 m/s.

5. Consider the equation $y = \dfrac{1}{x}$.

When $x = \frac{1}{2}$, $y = \frac{1}{\frac{1}{2}} = 2$.   When $x = \frac{1}{100}$, $y = \frac{1}{\frac{1}{100}} = 100$.

As the denominator of $\dfrac{1}{x}$ gets smaller, the answer gets larger.

An 'infinitely small' denominator gives an 'infinitely large' answer.

We write $\dfrac{1}{0} \to \infty$ '$\dfrac{1}{0}$ tends to an infinitely large number.'

Draw the graph of $y = \dfrac{1}{x}$ for

$x = -4, -3, -2, -1, -0.5, -0.25, 0.25, 0.5, 1, 2, 3, 4$

(Scales: 2 cm for x and y)

6. Draw the graph of $y = x + \dfrac{1}{x}$ for

$x = -4, -3, -2, -1, -0.5, -0.25, 0.25, 0.5, 1, 2, 3, 4$

(Scales: 2 cm for x and y)

7. Draw the graph of $y = \dfrac{2^x}{x}$, for $-4 \leqslant x \leqslant 7$, including

$x = -0.5$, $x = 0.5$.

(Scales: 1 cm to 1 unit for x and y)

8. A firm makes a profit of P thousand pounds from producing x thousand tiles.

Corresponding values of P and x are given below

x	0	0·5	1·0	1·5	2·0	2·5	3·0
P	−1·0	0·75	2·0	2·75	3·0	2·75	2·0

Using a scale of 4 cm to one unit on each axis, draw the graph of P against x. [Plot x on the horizontal axis]. Use your graph to find:

(a) the number of tiles the firm should produce in order to make the maximum profit.

(b) the minimum number of tiles that should be produced to cover the cost of production.

(c) the range of values of x for which the profit is more than £2850.

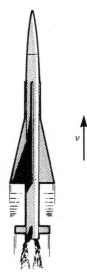

9. At time $t = 0$, one bacteria is placed in a culture in a laboratory. The number of bacteria doubles every 10 minutes.

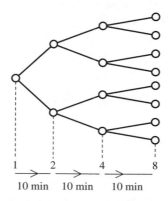

(a) Draw a graph to show the growth of the bacteria from $t = 0$ to $t = 120$ mins.
Use a scale of 1 cm to 10 minutes across the page and 5 cm to 1000 units up the page.

(b) Use your graph to estimate the time taken to reach 800 bacteria.

10. An economist estimates that the population of country A will be multiplied by 1·2 every 10 years and that the population of country B will be multiplied by 1·05 every 10 years. In 1980 the populations of A and B were 36 million and 100 million respectively.

(a) Draw a graph to show the projected populations of the two countries from 1980 to 2060.
Use a scale of 2 cm to 10 years across the page and 2 cm to 20 million up the page.

(b) Estimate when the population of A will exceed the population of B for the first time.

†**11.** Draw the graph of $\dfrac{x^4}{4^x}$, for $x = -1, -\frac{3}{4}, -\frac{1}{2}, -\frac{1}{4}, 0, \frac{1}{4}, \frac{1}{2}, \frac{3}{4},$

1, 1·5, 2, 2·5, 3, 4, 5, 6, 7.
(Scales: 2 cm to 1 unit for x, 5 cm to 1 unit for y)
(a) For what values of x is the gradient of the function zero?
(b) For what values of x is $y = 0·5$?

5.5 Graphical solution of equations

Accurately drawn graphs enable approximate solutions to be found for
a wide range of equations, many of which are impossible to solve
exactly by other methods.

Draw the graph of the function $y = 2x^2 - x - 3$ for $-2 \leqslant x \leqslant 3$.
Use the graph to find approximate solutions to the following
equations

(a) $2x^2 - x - 3 = 6$ (b) $2x^2 - x = x + 5$

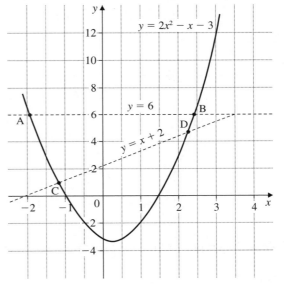

(a) To solve the equation
$2x^2 - x - 3 = 6$,
the line $y = 6$ is drawn.
At the points of intersection
(A and B), y simultaneously
equals both 6 and $(2x^2 - x - 3)$.
So we may write
$2x^2 - x - 3 = 6$

The solutions are the x-values of the
points A and B.
i.e. $x = -1.9$ and $x = 2.4$ approx.

(b) To solve the equation $2x^2 - x = x + 5$,
we rearrange the equation to obtain the
function $(2x^2 - x - 3)$ on the
left-hand side. In this case, subtract 3 from
both sides.
$$2x^2 - x - 3 = x + 5 - 3$$
$$2x^2 - x - 3 = x + 2$$
If we now draw the line $y = x + 2$, the solutions of the equation
are given by the x-values of C and D, the points of intersection.
i.e. $x = -1.2$ and $x = 2.2$ approx.

Assuming that the graph of $y = x^2 - 3x + 1$ has been drawn, find the
equation of the line which should be drawn to solve the equation
$x^2 - 4x + 3 = 0$

Rearrange $x^2 - 4x + 3 = 0$ in order to obtain $(x^2 - 3x + 1)$ on the
left-hand side.

$$x^2 - 4x + 3 = 0$$
add x $x^2 - 3x + 3 = x$
subtract 2 $x^2 - 3x + 1 = x - 2$

Therefore draw the line $y = x - 2$ to solve the equation.

Exercise 17

1. In the diagram shown, the graphs of $y = x^2 - 2x - 3$, $y = -2$ and $y = x$ have been drawn.

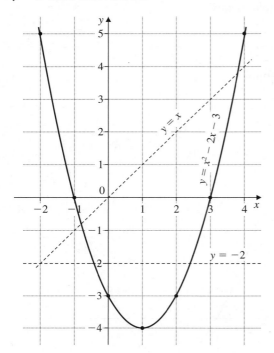

Use the graphs to find approximate solutions to the following equations.

(a) $x^2 - 2x - 3 = -2$ (b) $x^2 - 2x - 3 = x$

(c) $x^2 - 2x - 3 = 0$ (d) $x^2 - 2x - 1 = 0$

[Reminder: Only the x values are required.]

2. The graphs of $y = x^2 - 2$, $y = 2x$ and $y = 2 - x$ are shown.

Use the graphs to solve:

(a) $x^2 - 2 = 2 - x$

(b) $x^2 - 2 = 2x$

(c) $x^2 - 2 = 2$

(d) $2 - x = 2x$

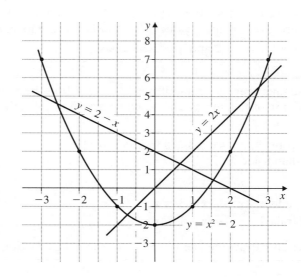

In Questions **3** to **5**, use a scale of 2 cm to 1 unit for x and 1 cm to 1 unit for y.

3. Draw the graphs of the functions $y = x^2 - 2x$ and $y = x + 1$ for $-1 \leqslant x \leqslant 4$. Hence find approximate solutions of the equation $x^2 - 2x = x + 1$.

4. Draw the graphs of the functions $y = x^2 - 3x + 5$ and $y = x + 3$ for $-1 \leqslant x \leqslant 5$. Hence find approximate solutions of the equation $x^2 - 3x + 5 = x + 3$.

5. Draw the graphs of the functions $y = 6x - x^2$ and $y = 2x + 1$ for $0 \leqslant x \leqslant 5$. Hence find approximate solutions of the equation $6x - x^2 = 2x + 1$.

6. (a) Complete the table and then draw the graph of $y = x^2 - 4x + 1$.

x	-1	0	1	2	3	4
y	6		-2		-2	1

   (b) On the same axes, draw the graph of $y = x - 3$.
   (c) Find solutions of the equations:
       (i) $x^2 - 4x + 1 = x - 3$
       (ii) $x^2 - 4x + 1 = 0$ [answers to 1 d.p.]

In Questions **7** to **9**, do *not* draw any graphs.

7. Assuming the graph of $y = x^2 - 5x$ has been drawn, find the equation of the line which should be drawn to solve the equations:
   (a) $x^2 - 5x = 3$ (b) $x^2 - 5x = -2$
   (c) $x^2 - 5x = x + 4$ (d) $x^2 - 6x = 0$
   (e) $x^2 - 5x - 6 = 0$

8. Assuming the graph of $y = x^2 + x + 1$ has been drawn, find the equation of the line which should be drawn to solve the equations:
   (a) $x^2 + x + 1 = 6$ (b) $x^2 + x + 1 = 0$
   (c) $x^2 + x - 3 = 0$ (d) $x^2 - x + 1 = 0$
   (e) $x^2 - x - 3 = 0$

9. Assuming the graph of $y = 6x - x^2$ has been drawn, find the equation of the line which should be drawn to solve the equations:
   (a) $4 + 6x - x^2 = 0$ (b) $4x - x^2 = 0$
   (c) $2 + 5x - x^2 = 0$ (d) $x^2 - 6x = 3$
   (e) $x^2 - 6x = -2$

For Questions **10** to **13**, use scales of 2 cm to 1 unit for x and 1 cm to 1 unit for y.

10. (a) Complete the table and then draw the graph of $y = x^2 + 3x - 1$.

x	-5	-4	-3	-2	-1	0	1	2
y	9	3		-3	-3			9

(b) By drawing other graphs, solve the equations:
    (i) $x^2 + 3x - 1 = 0$
    (ii) $x^2 + 3x = 7$
    (iii) $x^2 + 3x - 3 = x$

11. Draw the graph of $y = x^2 - 2x + 2$ for $-2 \leqslant x \leqslant 4$. By drawing other graphs, solve the equations:
    (a) $x^2 - 2x + 2 = 8$ (b) $x^2 - 2x + 2 = 5 - x$
    (c) $x^2 - 2x - 5 = 0$

12. Draw the graph of $y = x^2 - 7x$ for $0 \leqslant x \leqslant 7$. Draw suitable straight lines to solve the equations:
    (a) $x^2 - 7x + 9 = 0$ (b) $x^2 - 5x + 1 = 0$

13. Draw the graph of $y = 2x^2 + 3x - 9$ for $-3 \leqslant x \leqslant 2$. Draw suitable straight lines to find approximate solutions of the equations:
    (a) $2x^2 + 3x - 4 = 0$ (b) $2x^2 + 2x - 9 = 1$

14. Draw the graph of $y = \dfrac{18}{x}$ for $1 \leqslant x \leqslant 10$, using scales of 1 cm to one unit on both axes. Use the graph to solve approximately:
    (a) $\dfrac{18}{x} = x + 2$ (b) $\dfrac{18}{x} + x = 10$ (c) $x^2 = 18$

15. Draw the graph of $y = \frac{1}{2}x^2 - 6$ for $-4 \leqslant x \leqslant 4$, taking 2 cm to 1 unit on each axis.
    (a) Use your graph to solve approximately the equation
        $\frac{1}{2}x^2 - 6 = 1$.
    (b) Using tables or a calculator confirm that your solutions are approximately $\pm\sqrt{14}$ and explain why this is so.
    (c) Use your graph to find the square roots of 8.

16. Here are five sketch graphs.

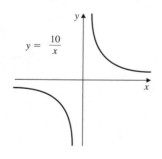

$y = \dfrac{10}{x}$

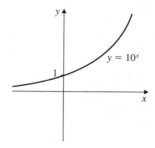

$y = 10^x$

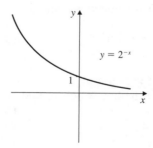

$y = 2^{-x}$

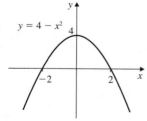

$y = 4 - x^2$

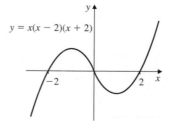

$y = x(x - 2)(x + 2)$

Make your own sketch graphs to find the *number of solutions* of the
following equations

(a) $\dfrac{10}{x} = 10^x$

(b) $4 - x^2 = 10^x$

(c) $x(x-2)(x+2) = \dfrac{10}{x}$

(d) $2^{-x} = 10^x$

(e) $4 - x^2 = 2^{-x}$

(f) $x(x-2)(x+2) = 0$

17. Draw the graph of $y = 3x^2 - x^3$ for $-2 \leqslant x \leqslant 3$. Use your graph to
find the range of values of k for which the equation $3x^2 - x^3 = k$
has three solutions.

†18. Draw the graph of $y = 6 - 2x - \frac{1}{2}x^3$ for $x = \pm 2, \pm 1\frac{1}{2}, \pm 1, \pm \frac{1}{2}, 0$.
Take 4 cm to 1 unit for x and 1 cm to 1 unit for y.
Use your graph to find approximate solutions of the equations:
(a) $\frac{1}{2}x^3 + 2x - 6 = 0$ (b) $x - \frac{1}{2}x^3 = 0$
Confirm that two of the solutions to the equation in part (b) are
$\pm\sqrt{2}$ and explain why this is so.

†19. Draw the graph of $y = 2^x$ for $-4 \leqslant x \leqslant 4$, taking 2 cm to one unit
for x and 1 cm to one unit for y. Find approximate solutions to the
equations:
(a) $2^x = 6$ (b) $2^x = 3x$ (c) $x2^x = 1$
Find also the approximate value of $2^{2.5}$.

5.6 Indices

Six basic rules

1. $a^n \times a^m = a^{n+m}$

$7^2 \times 7^4 = 7.7 \times 7.7.7.7 = 7^6$

so $7^2 \times 7^4 = 7^{2+4} = 7^6$

Remember

● $x^0 = 1$ *for any value of x which
is not zero*

2. $a^n \div a^m = a^{n-m}$

$3^5 \div 3^2 = \dfrac{3.3.3.\cancel{3}.\cancel{3}}{\cancel{3}.\cancel{3}} = 3^3$

so $3^5 \div 3^2 = 3^{5-2} = 3^3$

● $x^{-1} = \dfrac{1}{x}$

The reciprocal of x is $\dfrac{1}{x}$.

3. $(a^n)^m = a^{nm}$

$(2^2)^3 = (2.2)(2.2)(2.2) = 2^6$

so $(2^2)^3 = 2^{2\times 3} = 2^6$

Consider this sequence

2^3	2^2	2^1	2^0	2^{-1}	2^{-2}
↓	↓	↓	↓	↓	↓
8	4	2	1	$\frac{1}{2}$	$\frac{1}{4}$

4. $a^{-n} = \dfrac{1}{a^n}$

Check: $5^3 \div 5^5 = \dfrac{\cancel{5}.\cancel{5}.\cancel{5}}{\cancel{5}.\cancel{5}.\cancel{5}.5.5} = \dfrac{1}{5^2} = 5^{-2}$

5. $a^{\frac{1}{n}}$ means
'the nth root of a'

$3^{\frac{1}{2}}.3^{\frac{1}{2}} = \sqrt{3}.\sqrt{3}$
Using Rule 1:
$3^{\frac{1}{2}}.3^{\frac{1}{2}} = 3^1$

6. $a^{\frac{m}{n}}$ means
'the nth root of a
raised to the
power m'.

$$4^{\frac{3}{2}} = (\sqrt{4})^3 = 8$$

Simplify   (a) $x^7 . x^{13} = x^{7+13} = x^{20}$
   (b) $x^3 \div x^7 = x^{3-7} = x^{-4} = \dfrac{1}{x^4}$
   (c) $(x^4)^3 = x^{12}$
   (d) $(3x^2)^3 = 3^3 . (x^2)^3 = 27x^6$
   (e) $3y^2 \times 4y^3 = 12y^5$

Shown opposite is the graph of $y = 2^x$.

A calculator with a button marked $\boxed{x^y}$
can be used to work out difficult powers.

e.g. for $2^{1.7}$ press $\boxed{2}$ $\boxed{x^y}$ $\boxed{1.7}$ $\boxed{=}$

   $2^{1.7} = 3.249$ (4 s.f.)

Some calculators have a button
marked $\boxed{x^{\frac{1}{y}}}$ to work out roots
of numbers.
e.g. the fourth root of 100 is $100^{\frac{1}{4}}$

Press $\boxed{100}$ $\boxed{x^{\frac{1}{y}}}$ $\boxed{4}$ $\boxed{=}$

   $100^{\frac{1}{4}} = 3.162$ (4 s.f.)

You could press $\boxed{100}$ $\boxed{x^y}$ $\boxed{0.25}$ $\boxed{=}$ instead.

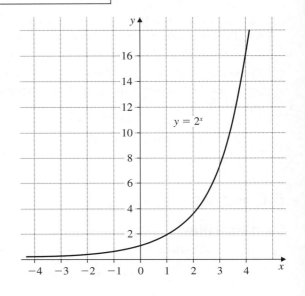

Exercise 18

Express in index form.

1. $3 \times 3 \times 3 \times 3$ **2.** $4 \times 4 \times 5 \times 5 \times 5$ **3.** $3 \times 7 \times 7 \times 7$

4. $2 \times 2 \times 2 \times 7$ **5.** $\dfrac{1}{10 \times 10 \times 10}$ **6.** $\dfrac{1}{2 \times 2 \times 3 \times 3 \times 3}$

7. $\sqrt{15}$ **8.** $\sqrt[3]{3}$ **9.** $a \times a \times a \times a \times a \times a \times a \times a$

10. $\sqrt[5]{10}$ **11.** $\frac{1}{11}$ **12.** $(\sqrt{5})^3$

In Questions **13** to **32** answer true or false.

13. $2^3 = 8$ **14.** $3^2 = 6$ **15.** $5^3 = 125$ **16.** $2^{-1} = \frac{1}{2}$
17. $10^{-2} = \frac{1}{20}$ **18.** $3^{-3} = \frac{1}{9}$ **19.** $2^2 > 2^3$ **20.** $2^3 < 3^2$
21. $2^{-2} > 2^{-3}$ **22.** $3^{-2} < 3$ **23.** $1^9 = 9$ **24.** $(-3)^2 = -9$
25. $5^{-2} = \frac{1}{10}$ **26.** $10^{-3} = \frac{1}{1000}$ **27.** $10^{-2} > 10^{-3}$ **28.** $5^{-1} = 0.2$
29. $10^{-1} = 0.1$ **30.** $2^{-2} = 0.25$ **31.** $5^0 = 1$ **32.** $16^0 = 0$

Write in a more simple form.

33. $5^2 \times 5^4$ **34.** $6^3 \times 6^2$ **35.** $2^3 \times 2^{10}$ **36.** $3^6 \times 3^{-2}$

37. $6^7 \div 6^2$ **38.** $8^5 \div 8^4$ **39.** $6^2 \div 6^{-2}$ **40.** Half of 2^6

41. Half of 2^{20} **42.** $10^{100} \div 10^0$ **43.** $5^{-2} \times 5^{-3}$ **44.** $3^{-2} \times 3^{-8}$

45. $\dfrac{3^4 \times 3^5}{3^2}$ **46.** $\dfrac{2^8 \times 2^4}{2^5}$ **47.** $\dfrac{7^3 \times 7^3}{7^4}$ **48.** $\dfrac{5^9 \times 5^{10}}{5^{20}}$

Exercise 19

Write in a more simple form.

1. $(3^3)^2$ **2.** $(5^4)^3$ **3.** $(7^2)^5$ **4.** $(x^2)^3$

5. $(2^{-1})^2$ **6.** $(-1)^{102}$ **7.** $(7^{-1})^{-2}$ **8.** $(y^4)^{\frac{1}{2}}$

9. Half of 2^{22} **10.** One-tenth of 10^{100}

Simplify:

11. $x^3 \times x^4$ **12.** $y^6 \times y^7$ **13.** $z^7 \div z^3$

14. $z^{50} \times z^{50}$ **15.** $m^3 \div m^2$ **16.** $e^{-3} \times e^{-2}$

17. $y^{-2} \times y^4$ **18.** $w^4 \div w^{-2}$ **19.** $y^{\frac{1}{2}} \times y^{\frac{1}{2}}$

20. $(x^2)^5$ **21.** $x^{-2} \div x^{-2}$ **22.** $w^{-3} \times w^{-2}$

23. $w^{-7} \times w^2$ **24.** $x^3 \div x^{-4}$ **25.** $(a^2)^4$

26. $(k^{\frac{1}{2}})^6$ **27.** $e^{-4} \times e^4$ **28.** $x^{-1} \times x^{30}$

29. $(y^4)^{\frac{1}{2}}$ **30.** $(x^{-3})^{-2}$ **31.** $z^2 \div z^{-2}$

32. $t^{-3} \div t$ **33.** $(2x^3)^2$ **34.** $(4y^5)^2$

35. $2x^2 \times 3x^2$ **36.** $5y^3 \times 2y^2$ **37.** $5a^3 \times 3a$

38. $(2a)^3$ **39.** $3x^3 \div x^3$ **40.** $8y^3 \div 2y$

41. $10y^2 \div 4y$ **42.** $8a \times 4a^3$ **43.** $(2x)^2 \times (3x)^3$

44. $4z^4 \times z^{-7}$ **45.** $6x^{-2} \div 3x^2$ **46.** $5y^3 \div 2y^{-2}$

47. $(x^2)^{\frac{1}{2}} \div (x^{\frac{1}{3}})^3$ **48.** $7w^{-2} \times 3w^{-1}$

49. $(2n)^4 \div 8n^0$ **50.** $4x^{\frac{3}{2}} \div 2x^{\frac{1}{2}}$

Evaluate the following.

(a) $9^{\frac{1}{2}} = \sqrt{9} = 3$ (b) $5^{-1} = \frac{1}{5}$

(c) $4^{-\frac{1}{2}} = \dfrac{1}{4^{\frac{1}{2}}} = \dfrac{1}{\sqrt{4}} = \dfrac{1}{2}$ (d) $25^{\frac{3}{2}} = (\sqrt{25})^3 = 5^3 = 125$

(e) $(5^{\frac{1}{2}})^3 . 5^{\frac{1}{2}} = 5^{\frac{3}{2}} . 5^{\frac{1}{2}} = 5^2$
$$= 25$$
     (f) $7^0 = 1$
         $[a^0 = 1$ for any non-zero value
         of $a]$

Exercise 20

Evaluate the following.

1. $3^2 \times 3$

2. 100^0

3. 3^{-2}

4. $(5^{-1})^{-2}$

5. $4^{\frac{1}{2}}$

6. $16^{\frac{1}{2}}$

7. $81^{\frac{1}{2}}$

8. $8^{\frac{1}{3}}$

9. $9^{\frac{3}{2}}$

10. $27^{\frac{1}{3}}$

11. $9^{-\frac{1}{2}}$

12. $8^{-\frac{1}{3}}$

13. $1^{\frac{5}{2}}$

14. $25^{-\frac{1}{2}}$

15. $1000^{\frac{1}{3}}$

16. $2^{-2} \times 2^5$

17. $2^4 \div 2^{-1}$

18. $8^{\frac{2}{3}}$

19. Copy and complete the crossnumber puzzle.

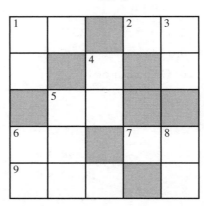

Across	**Down**
1. $\sqrt{2^8}$	1. $(1\,000\,000)^{\frac{1}{6}}$
2. $169^{\frac{1}{2}}$	3. $3^3 + 2^2 + 1^1$
5. $7^2 \times 5^0$	4. $10^2 - 7^0$
6. $2^4 \times 5$	5. $20^2 + 2^1$
7. 3×2^5	6. $3^7 \div 27$
9. 5^3	8. Half of 2^7

20. Find the missing index.
    (a) $\sqrt{80} = 80^{\square}$
    (b) $\dfrac{1}{\sqrt{3}} = 3^{\square}$
    (c) $(\sqrt{2})^3 = 2^{\square}$

21. Work out.

    (a) $27^{-\frac{2}{3}}$
    (b) $4^{-\frac{3}{2}}$
    (c) One per cent of 100^7

Evaluate the following.

22. $10\,000^{\frac{1}{4}}$

23. $100^{\frac{3}{2}}$

24. $(100^{\frac{1}{2}})^{-3}$

25. $(9^{\frac{1}{2}})^{-2}$

26. $(-16\cdot371)^0$

27. $81^{\frac{1}{4}} \div 16^{\frac{1}{4}}$

28. $(5^{-4})^{\frac{1}{4}}$

29. $1000^{-\frac{1}{3}}$

30. $(4^{-\frac{1}{2}})^2$

31. $8^{-\frac{2}{3}}$

32. $100^{\frac{5}{2}}$

33. $1^{\frac{4}{3}}$

34. 2^{-5}

35. $(0\cdot01)^{\frac{1}{2}}$

36. $(0\cdot04)^{\frac{1}{2}}$

37. $(2\cdot25)^{\frac{1}{2}}$

38. $(7\cdot63)^0$

39. $3^5 \times 3^{-3}$

40. $(3\frac{3}{8})^{\frac{1}{3}}$

41. $(11\frac{1}{9})^{-\frac{1}{2}}$

42. $(\frac{1}{8})^{-2}$

43. $(\frac{1}{1000})^{\frac{2}{3}}$

44. $(\frac{9}{25})^{-\frac{1}{2}}$

45. $(10^{-6})^{\frac{1}{3}}$

46. (a) Copy and complete the sentence.
    'The inverse operation of raising a positive number to power n
    is raising the result of this operation to power $\boxed{}$.'

    (b) Write down two numerical examples to illustrate this result.

Solve the equations   (a) $2^x = 16$ (b) $5^y = \frac{1}{25}$

(c) $6^z = (\sqrt{6})^3$ (d) $3^{x-1} = 81$

We use the principle that if

$\quad a^x = a^y$ then $x = y$

(a) $2^x = 16 = 2^4$ (b) $5^y = \frac{1}{25} = 5^{-2}$
$\quad \therefore \quad x = 4$ $\quad \therefore \quad y = -2$

(c) $6^z = (\sqrt{6})^3 = 6^{\frac{3}{2}}$ (d) $3^{x-1} = 81 = 3^4$
$\quad \therefore \quad z = \frac{3}{2}$ $\quad \therefore \quad x = 5$

Exercise 21

Solve the equations for x.

1. $2^x = 8$ **2.** $3^x = 81$ **3.** $5^x = \frac{1}{5}$

4. $10^x = \frac{1}{100}$ **5.** $3^{-x} = \frac{1}{27}$ **6.** $4^x = 64$

7. $6^{-x} = \frac{1}{6}$ **8.** $100\,000^x = 10$ **9.** $12^x = 1$

10. $10^x = 0{\cdot}0001$ **11.** $2^x + 3^x = 13$ **12.** $\left(\frac{1}{2}\right)^x = 32$

13. $5^{2x} = 25$ **14.** $1\,000\,000^{3x} = 10$

15. Here is a sketch of $y = 3^x$.

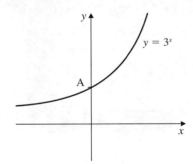

(a) Write down the coordinates of point A.
(b) Write down the solution of the equation $3^x = 5^x$.

16. Find a, b and c.
(a) $\frac{1}{8} = 2^a$ (b) $b^7 = 1$ (c) $2^{c-1} = 16$

17. Find p, q and r.
(a) $x^{\frac{1}{2}} \times x^2 = x^p$ (b) 10% of $100^2 = q$ (c) $x^{-\frac{1}{4}} \times x^2 = x^r$

18. Solve the equations.
(a) $10n^3 = 640$ (b) $10^n = 0{\cdot}1$ (c) $2n^3 = 0$

19. Find two solutions of the equation $x^2 = 2^x$

20. Use a calculator to find solutions correct to three significant figures.
(a) $x^x = 100$ (b) $x^x = 10\,000$

21. The last digit of 7^3 is 3 $[7^3 = 343]$.
(a) Copy and complete the table below, which gives the last digit of 7^n.

n	1	2	3	4	5	6	7	8	9	10
Last digit of 7^n	7	9	3		7			1		

(b) Write down the last digit of:
(i) 7^{48} (ii) 7^{101} (iii) 49^{35}

22. It is given that $10^x = 3$ and $10^y = 7$. What is the value of 10^{x+y}?

23. In a laboratory we start with 2 cells in a dish. The number of cells in the dish doubles every 30 minutes. This is an example of *exponential growth*.
(a) How many cells are in the dish after four hours?
(b) After what time are there 2^{13} cells in the dish?
(c) After $10\frac{1}{2}$ hours there are 2^{22} cells in the dish and an experimental fluid is added which eliminates half of the cells. How many cells are left?

24. Steve's bike is ill. Its computer-controlled ignition system has a virus. The doctor has advised Steve to keep the bike warm, in which case the number of germs in the bike will decay exponentially and will be $1\,000\,000 \times 2^{-n}$ after n hours.
(a) How many germs will there be after 10 hours?
(b) The bike will be cured when it contains less than one germ. After how many hours will it be cured?

25. A bank pays compound interest on money invested in an account. After n years a sum of £2000 will rise to £2000 $\times 1\cdot08^n$. This represents exponential growth.
(a) How much money is in the account after three years?
(b) After how many years will the original £2000 have nearly doubled?

†**26.** It is given that $x + \dfrac{1}{x} = 1$, where x is not zero. Show that
$x^2 = x - 1$ and $x^3 = x^2 - x$. Use these two expressions to show that $x^3 = -1$.
Using this value for x^3, find the value of x^6 and show that
$x^7 = x$. Hence show that $x^6 + \dfrac{1}{x^6} = 2$ and $x^7 + \dfrac{1}{x^7} = -2$.
Deduce the value of $x^{60} + \dfrac{1}{x^{60}}$ and of $x^{61} + \dfrac{1}{x^{61}}$.

6 SHAPE, SPACE AND MEASURES 2

6.1 Drawing 3D shapes, symmetry

Isometric drawing

When we draw a solid on paper we are making a 2-D representation of a 3-D object.
Here are two pictures of the same cuboid, measuring $4 \times 3 \times 2$ units.

(a) On ordinary squared paper

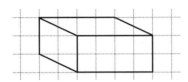

(b) On isometric paper
[a grid of equilateral triangles]

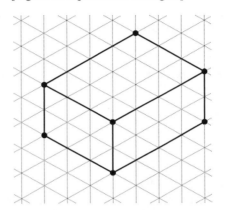

The dimensions of the cuboid cannot be taken from the first picture but
they can be taken from the picture drawn on isometric paper. Instead of
isometric paper you can also use 'triangular dotty' paper like this:
Be careful to use it the right way round (as shown here).

Exercise 1

In Questions **1** to **3** the objects consist of 1 cm cubes joined together.
Draw each object on isometric paper (or 'triangular dotty' paper).
Questions **1** and **2** are already drawn on isometric paper.

1.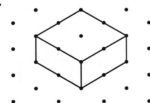

2.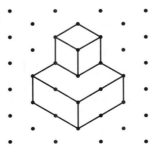

3.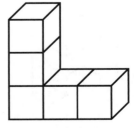

4. Here are two shapes made using four multilink cubes.

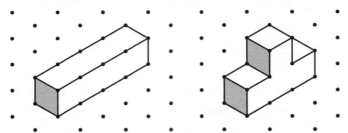

Make and then draw four more shapes, using four cubes, which are different from the two above.

5. The shape shown falls over onto its shaded face. Draw the shape after it has fallen over.

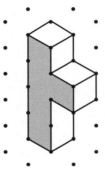

6. You need 16 small cubes.
Make the two shapes below and then arrange them into a $4 \times 4 \times 1$ cuboid by adding a third shape, which you have to find. Draw the third shape on isometric paper. (There are two possible shapes.)

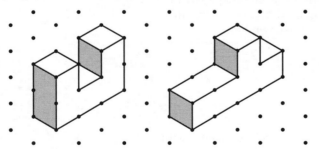

7. Three views of object A are shown.

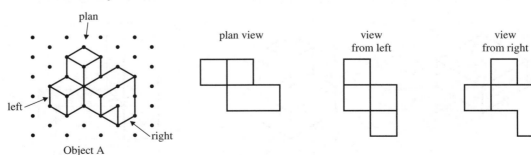

Make objects B and C below, using cubes. On squared paper, draw the plan view, the view from the left and the view from the right of each object.

8. Make each of the objects whose views are given below. Draw an isometric picture of each one.

(a)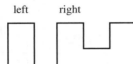

(b)

(b)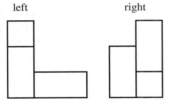

Nets

If the cube here was made of cardboard, and you cut along some of the edges and laid it out flat, you would have the *net* of the cube.

A cube has: 8 vertices;
6 faces;
12 edges

vertex

Here is the net for a square-based pyramid.

This pyramid has: 5 vertices;
5 faces;
8 edges

Exercise 2

1. Which of the nets below can be used to make a cube?

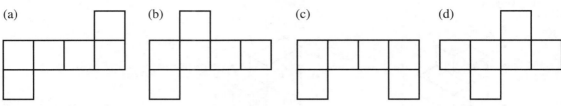

   (a)   (b)   (c)   (d)

2. The numbers on opposite faces of a dice add up to 7.
   Take one of the possible nets for a cube from **Question 1** and show
   the number of dots on each face.

3. Here we have started to draw the net of a cuboid
   (a closed rectangular box) measuring $4\,cm \times 3\,cm \times 1\,cm$.
   Copy and then complete the net.

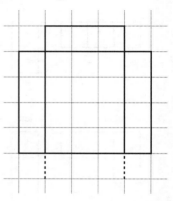

4. The diagram shown needs one more square to complete the net of a
   cube. Copy and cut out the shape and then draw the *four* possible
   nets which would make a cube.

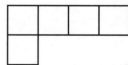

5. Describe the solid formed from each of these nets. State the number
   of vertices and faces for each object.

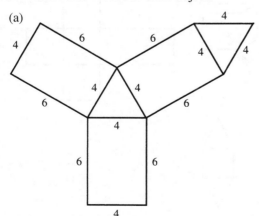

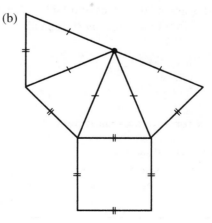

6. Sketch a possible net for each of the following:
   (a) a cuboid measuring 5 cm by 2 cm by 8 cm
   (b) a prism 10 cm long whose cross-section is a right-angled triangle
       with sides 3 cm, 4 cm and 5 cm.

7. The diagram shows the net of a pyramid.
   The base is shaded. The lengths are in cm.
   (a) How many edges will the pyramid have?
   (b) How many vertices will it have?
   (c) Find the lengths a, b, c, d.
   (d) Use the formula

   $$V = \frac{1}{3} \text{ base area} \times \text{height}$$

   to calculate the volume of the pyramid.

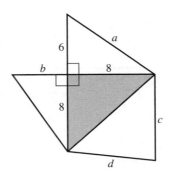

8. This is the net of a square-based pyramid.
   What are the lengths a, b, c, x, y?

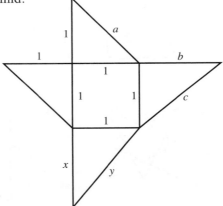

Symmetry

(a) Line symmetry

   The letter M has one
   line of symmetry, shown dotted.

(b) Rotational symmetry

   The shape may be turned about O into three
   identical positions. It has rotational
   symmetry of order three.

Exercise 3

For each shape state:
(a) the number of lines of symmetry
(b) the order of rotational symmetry.

1.

2.

3.

4.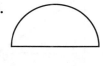

5.

6.

7.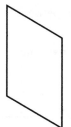

8.

9.

10.

11.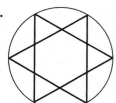

12.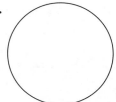

Planes of symmetry

- A plane of symmetry divides a 3-D shape into two congruent shapes. One shape must be a mirror image of the other shape.

The shaded plane is a plane of symmetry of the cube.

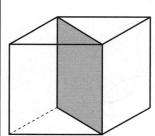

Exercise 4

1. How many planes of symmetry does this cuboid have?

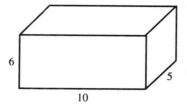

2. How many planes of symmetry do these shapes have?

(a) 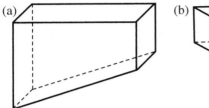   (b) 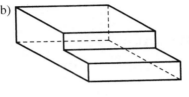   (c)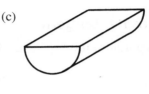

3. How many planes of symmetry does a cube have?

4. Draw a pyramid with a square base so that the vertex of the pyramid is vertically above the centre of the square base. Show any planes of symmetry by shading.

5. The diagrams show the plan view and the side view of an object.

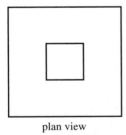   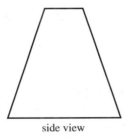

plan view   side view

How many planes of symmetry has this object?

6.2 Trigonometry

Trigonometry is used to calculate sides and angles in triangles.

Right-angled triangles

The side opposite the right angle is called the *hypotenuse* (use hyp.).
It is the longest side.

The side opposite the marked angle m is called the *opposite* (use opp.).
The other side is called the *adjacent* (use adj.).

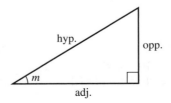

Consider two triangles, one of which is an enlargement of the other. It is clear that, for the angle 30°, the

$$\text{ratio} = \frac{\text{opposite}}{\text{hypotenuse}} = \frac{6}{12} = \frac{2}{4} = \frac{1}{2}$$

This is the same for both triangles.

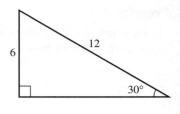

Sine, cosine, tangent

Three important ratios are defined for angle x.

$$\sin x = \frac{\text{opp.}}{\text{hyp.}} \quad \cos x = \frac{\text{adj.}}{\text{hyp.}} \quad \tan x = \frac{\text{opp.}}{\text{adj.}}$$

It is important to get the letters in the correct positions.

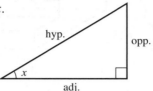

Some people find a simple sentence helpful where the first letters of each word describe sine, cosine or tangent, Hypotenuse, Opposite or Adjacent. An example is:

Silly Old Harry Caught A Herring Trawling Off Afghanistan

e.g.   S O H $\sin = \dfrac{\text{opp.}}{\text{hyp.}}$

Finding the length of an unknown side

Find the side marked x.

Label the sides of the triangle hyp., opp., adj.

In this example, we have only opp. and adj., so use the tangent ratio.

$$\tan 25\cdot4° = \frac{\text{opp.}}{\text{adj.}} = \frac{x}{10}$$

Find tan 25·4° on a calculator:

$$0\cdot4748 = \frac{x}{10}$$

Solve for x.

$$x = 10 \times 0\cdot4748 = 4\cdot748$$
$$x = 4\cdot75\,\text{cm (3 s.f.)}$$

Find the side marked z.

Label hyp., opp., adj.

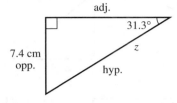

adj.

31.3°

z

7.4 cm
opp.

hyp.

we have opp. and hyp. so use the sine ratio.

$$\sin 31 \cdot 3° = \frac{\text{opp.}}{\text{hyp.}} = \frac{7 \cdot 4}{z}$$

Multiply by z. $z \times (\sin 31 \cdot 3°) = 7 \cdot 4$

$$z = \frac{7 \cdot 4}{\sin 31 \cdot 3°}$$

$$z = 14 \cdot 2 \text{ cm } (3 \text{ s.f.})$$

Exercise 5

Find the lengths of the sides marked with a letter. Give your answers to
three significant figures.

1.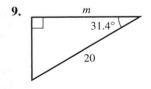
27°
10
a

2.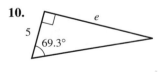
b
61°
4

3.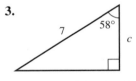
7
58°
c

4.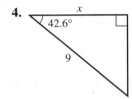
x
42.6°
9

5.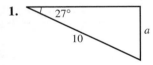
x
38.7°
10

6.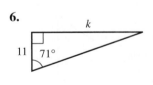
k
11
71°

7.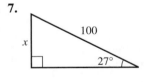
x
100
27°

8.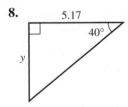
5.17
40°
y

9.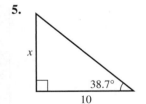
m
31.4°
20

10.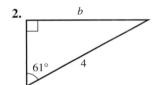
5
e
69.3°

11.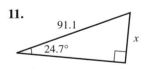
91.1
24.7°
x

12.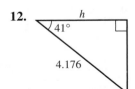
h
41°
4.176

13.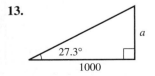
a
27.3°
1000

14.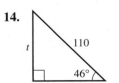
t
110
46°

15.
x
56.6°
7

16.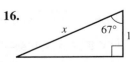
x
67°
1

17.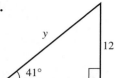

18.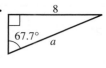

19.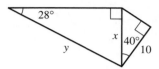

20.

21.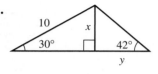

22.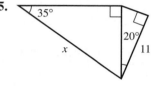

23.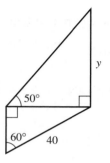

24.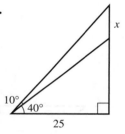

25.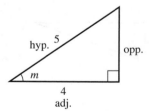

26.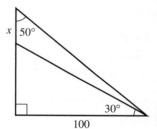

Finding an unknown angle

Find the angle marked m to one decimal place.

Label the sides of the triangle hyp., opp. and adj. in relation to angle m.

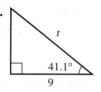

Here only adj. and hyp. are known, so use cosine.

$$\cos m = \frac{\text{adj.}}{\text{hyp.}} = \frac{4}{5}$$

Using a calculator, the angle m can be found as follows.

(a) Press ☐4☐ ☐÷☐ ☐5☐ ☐=☐

(b) Press ☐INV☐ and then ☐cos☐

This gives the angle as 36·869 898

$$m = 36\cdot9° \text{ (1 d.p.)}$$

Exercise 6

For Questions **1** to **15**, find the angle marked with a letter. All lengths
are in cm.

1.
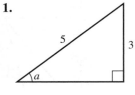

2.
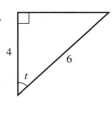

3.
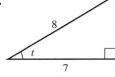

4.
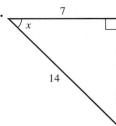

5.
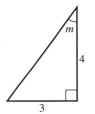

6.
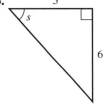

7.

8.
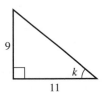

9.
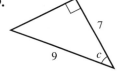

10.

11.
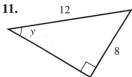

12.
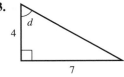

13.
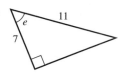

14.

15.
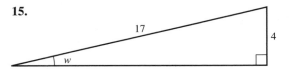

In Questions **16** to **20**, the triangle has a right angle at the middle
letter.

16. In $\triangle ABC$, $BC = 4$, $AC = 7$. Find $\widehat{A}$.

17. In $\triangle DEF$, $EF = 5$, $DF = 10$. Find $\widehat{F}$.

18. In $\triangle GHI$, $GH = 9$, $HI = 10$. Find $\widehat{I}$.

19. In $\triangle JKL$, $JL = 5$, $KL = 3$. Find $\widehat{J}$.

20. In $\triangle MNO$, $MN = 4$, $NO = 5$. Find $\widehat{M}$.

In Questions **21** to **24**, find the angle x.

21.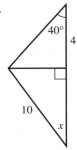

22.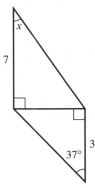

23.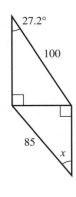

24.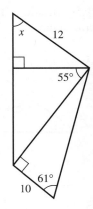

25. In $\triangle ABC$
$\widehat{BAC} = 42°$ and $AB = 12\,cm$.

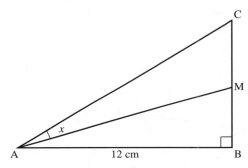

M is the mid-point of CB.
Find the angle x to 1 d.p.

Bearings

Bearings are measured *clockwise* from North.

(a)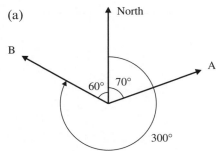

(b)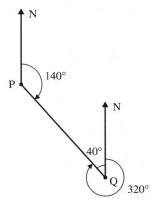

Ship A sails on a bearing 070°.
Ship B sails on a bearing 300°.

The bearing of Q from P is 140°.
The bearing of P from Q is 320°.

Angles of elevation and depression

(a)

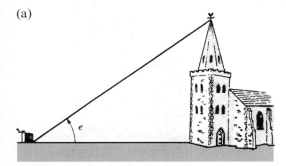

(b)

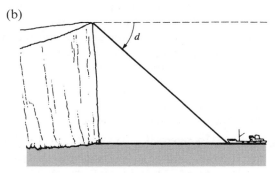

e is the angle of *elevation* of
the Steeple from the Gate.

d is the angle of *depression* of
the Boat from the Cliff top.

Exercise 7

In this exercise, start each question by drawing a clear diagram.

1. A ladder of length 6 m leans against a vertical wall so that the base
of the ladder is 2 m from the wall. Calculate the angle between the
ladder and the wall.

2. A ladder of length 8 m rests against a wall so that the angle between
the ladder and the wall is 31°. How far is the base of the ladder
from the wall?

3. A point P is 90 m away from a vertical flagpole, which is 11 m high.
What is the angle of elevation to the top of the flagpole from P?

4. A chord AB of length 10 cm is drawn
in a circle of radius 7 cm, centre O.
Work out the angle AOB.

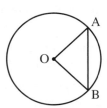

5. A ladder of length 5 m rests against a vertical wall. For safety
reasons the angle between a ladder and the ground must be between
65° and 75°. Work out the possible safe distances from the foot of a
ladder to the base of the wall.

6. An isosceles triangle has sides of length 8 cm, 8 cm and 5 cm. Find
the angle between the two equal sides.

7. The angles of an isosceles triangle are 66°, 66° and 48°. If the
shortest side of the triangle is 8·4 cm, find the length of one of the
two equal sides.

8. An arctic explorer reached the North Pole and decided
to erect his national flag and to sing his national anthem,
which lasted a rather chilly four minutes.
His eye was 1·6 m from the ground and the angle of
elevation to the top of the 3 m flagpole was 28°. How far
from the base of the flagpole did he stand?

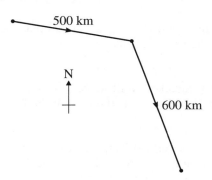

9. A ship sails 35 km on a bearing of 042°.
   (a) How far north has it travelled?
   (b) How far east has it travelled?

10. A ship sails 200 km on a bearing of 243·7°.
    (a) How far south has it travelled?
    (b) How far west has it travelled?

11. An aircraft flies 400 km from a point O on a bearing of 025° and
then 700 km on a bearing of 080° to arrive at B.
    (a) How far north of O is B?
    (b) How far east of O is B?
    (c) Find the distance and bearing of B from O.

12. An aircraft flies 500 km on a
bearing of 100° and then
600 km on a bearing of 160°.
Find the distance and bearing
of the finishing point from the
starting point.

13. Find the acute angle between the diagonals of a rectangle whose
sides are 5 cm and 7 cm.

14. A kite flying at a height of 55 m is attached to a string which makes
an angle of 55° with the horizontal. What is the length of the string?

15. A plane is flying at a constant height of 8000 m. It flies vertically
above me and 30 seconds later the angle of elevation is 74°.

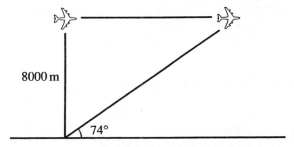

Find the speed of the plane in metres/second.

16. A rocket flies 10 km vertically, then 20 km at an angle of 15° to the vertical and finally 60 km at an angle of 26° to the vertical. Calculate the vertical height of the rocket at the end of the third stage.

17. Find x, given

AD = BC = 6 m.

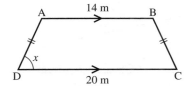

18. Ants can hear each other up to a range of 2 m. An ant A, 1 m from a wall sees her friend B about to be eaten by a spider. If the angle of elevation of B from A is 62°, will the spider have a meal or not? (Assume B escapes if he hears A calling.)

19. A regular pentagon is inscribed in a circle of radius 7 cm.
Find the angle a and then the length of a side of the pentagon.

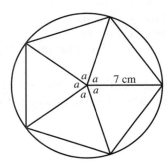

Exercise 8

1. Find the length d.

(a)

(b)

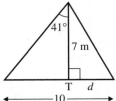

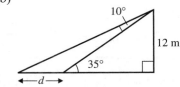

For Questions **2** to **5**, plot the points for each question on a sketch graph with x- and y-axes drawn to the same scale.

2. For the points A(5, 0) and B(7, 3), calculate the angle between AB and the x-axis.

3. For the points C(0, 2) and D(5, 9), calculate the angle between CD and the y-axis.

4. For the points A(3, 0), B(5, 2) and C(7, −2), calculate the angle BAC.

5. For the points P(2, 5), Q(5, 1) and R(0, −3), calculate the angle PQR.

6. From the top of a tower of height 75 m, a guard sees two prisoners, both due West of him.

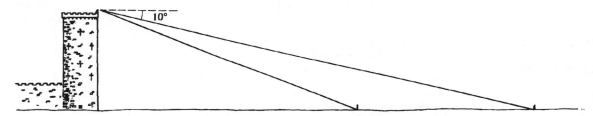

If the angles of depression of the two prisoners are 10° and 17°, calculate the distance between them.

7. A hedgehog wishes to cross a road without being run over. He observes the angle of elevation of a lamp post on the other side of the road to be 27° from the edge of the road and 15° from a point 10 m back from the road. How wide is the road? If he can run at 1 m/s, how long will he take to cross?
If cars are travelling at 20 m/s, how far apart must they be if he is to survive?

8. A symmetrical drawbridge is shown below. When lowered, the roads AX and BY just meet in the middle. Calculate the length XY.

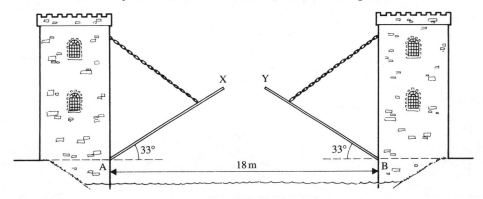

9. From a point 10 m from a vertical wall, the angles of elevation of the bottom and the top of a statue of Sir Isaac Newton, set in the wall, are 40° and 52°. Calculate the length of the statue.

10. Here are three triangles.

 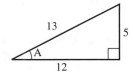 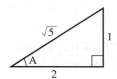

(a) For each triangle write down the values of sin A and cos A and work out $(\sin A)^2 + (\cos A)^2$.
(b) Write down what you notice.
(c) Does the same result hold for a general triangle with sides a, b, c?

11. The diagram shows part of a polygon with *n* sides and centre O.

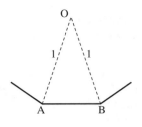
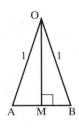

(a) What is angle AOB in terms of *n*?
(b) M is the mid-point of AB. What is angle MOB in terms of *n*?
(c) Find the length MB and hence the length AB in terms of *n*.
(d) Find an expression for the perimeter of the polygon.
(e) Work out the perimeter for *n* = 100 and *n* = 1000. What do you notice?

†**12.** A rectangular paving stone 3 m by 1 m rests against a vertical wall as shown. What is the height of the highest point of the stone above the ground?

†**13.** A rectangular piece of paper 30 cm by 21 cm is folded so that opposite corners coincide. How long is the crease?

†**14** The diagram shows the cross section of a rectangular fish tank. When AB is inclined at 40°, the water just comes up to A.

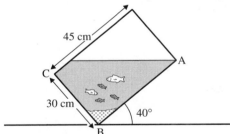

The tank is then lowered so that BC is horizontal. What is now the depth or water in the tank?

Three-dimensional problems

Projections and planes

A projection is like a shadow on a surface or plane.

● The angle between a line and a plane is the angle between the line and its *projection* in the plane.

PA is a vertical pole standing at the vertex A of a horizontal rectangular field ABCD.

The angle between the line PC and the plane ABCD is found as follows:

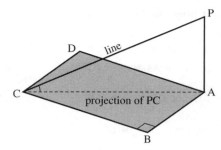

(a) The projection of PC in the plane ABCD is AC
(b) The angle between the line PC and the plane ABCD is the angle PCA.

Always draw a large, clear diagram. It is often helpful to redraw the particular triangle which contains the length or angle to be found.

A rectangular box with top WXYZ and base ABCD has AB = 6 cm, BC = 8 cm and WA = 3 cm.
Calculate:

(a) the length of AC
(b) the angle between the line WC and the plane ABCD.

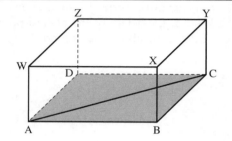

(a) Redraw triangle ABC.
$AC^2 = 6^2 + 8^2 = 100$
$AC = 10$ cm.

(b) The projection of WC on the plane ABCD is AC. The angle required is $W\widehat{C}A$.

Redraw triangle WAC.
let $W\widehat{C}A = \theta$

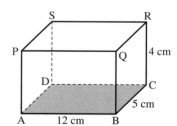

$\tan \theta = \frac{3}{10}$
$\theta = 16.7°$.

The angle between WC and the plane ABCD is 16.7°.

Exercise 9

1. In the rectangular box shown, find:
   (a) AC
   (b) AR
   (c) the angle between AC and AR.

2. A vertical pole BP stands at one corner of a horizontal rectangular field as shown.

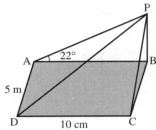

If AB = 10 m, AD = 5 m and the angle of elevation of P from A is 22°, calculate:
(a) the height of the pole
(b) the angle of elevation of P from C
(c) the length of a diagonal of the rectangle ABCD
(d) the angle of elevation of P from D.

3. In the cube shown, find:
   (a) BD
   (b) AS
   (c) BS
   (d) the angle SBD.

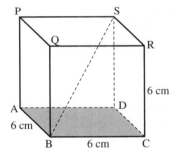

4. In the square-based pyramid shown, V is vertically above the middle of the base, AB = 10 cm and VC = 20 cm. Find:
   (a) AC
   (b) the height of the pyramid
   (c) the angle between VC and the base ABCD.

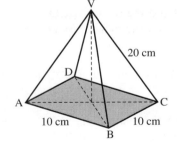

5. The figure shows a cuboid. Calculate:
   (a) the lengths of AC and AY
   (b) the angle between AY and the plane ABCD.

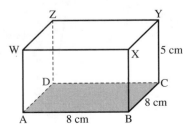

6. The figure shows a cuboid.
Calculate:
(a) the lengths ZX and KX
(b) the angle between NX and
    the plane WXYZ
(c) the angle between KY and
    the plane KLWX.

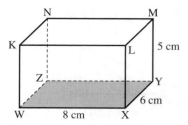

7. In the wedge shown, PQRS is perpendicular to ABRQ; PQRS and
ABRQ are rectangles with AB = QR = 6 m, BR = 4 m, RS = 2 m.
Find:
(a) BS   (b) AS
(c) the angle between AS and the plane ABRQ.

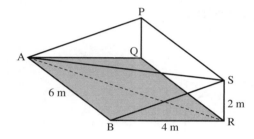

8. The pyramid VPQRS has a square base PQRS.
VP = VQ = VR = VS = 12 cm and PQ = 9 cm.
Calculate the angle between VP and the plane PQRS.

9. The pyramid VABCD has a rectangular base ABCD.
VA = VB = VC = VD = 15 cm, AB = 14 cm and BC = 8 cm.
Calculate:
(a) the angle between VB and the plane ABCD
(b) the angle between VX and the plane ABCD where X is the mid-
    point of BC.

10. In the diagram A, B and O are points in a horizontal plane and P is
vertically above O, where OP = h m.
A is due West of O, B is due South of O
and AB = 60 m. The angle of elevation of
P from A is 25° and the angle of elevation
of P from B is 33°.
(a) Find the length AO in terms of h.
(b) Find the length BO in terms of h.
(c) Find the value of h.

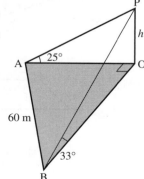

†**11.** The angle of elevation of the top of a tower is 38° from a point A due South of it. The angle of elevation of the top of the tower from another point B, due East of the tower is 29°. Find the height of the tower if the distance AB is 50 m.

†**12.** An observer at the top of a tower of height 15 m sees a man due West of him at an angle of depression 31°. He sees another man due South at an angle of depression 17°. Find the distance between the men.

6.3 Dimensions of formulas

Here are some formulas for finding volumes, areas and lengths met earlier in this book.

sphere: volume $= \dfrac{4}{3}\pi \boldsymbol{r}^3$

surface area $= 4\pi \boldsymbol{r}^2$

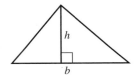

triangle: area $= \dfrac{1}{2}\boldsymbol{bh}$

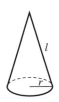

cylinder: volume $= \pi \boldsymbol{r}^2\boldsymbol{h}$

curved surface area $= 2\pi \boldsymbol{rh}$

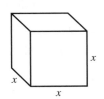

cube: volume $= \boldsymbol{x}^3$

surface area $= 6\boldsymbol{x}^2$

cone: volume $= \dfrac{1}{3}\pi \boldsymbol{r}^2\boldsymbol{h}$

curved surface area $= \pi \boldsymbol{rl}$

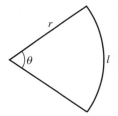

sector area $= \dfrac{\theta}{360} \times \pi \boldsymbol{r}^2$

arc length $= \dfrac{\theta}{360} \times 2\pi \ \boldsymbol{r}$

All the symbols in bold type are lengths. They have the *dimension* of length and are measured in cm, metres, km, etc. The other symbols are numbers (including π) or angles and have no dimensions.

(a) It is not hard to see that all the formulas for volume have *three* lengths multiplied together. They have three dimensions.

(b) All the formulas for area have *two* lengths multiplied together. They have two dimensions.

(c) Any formula for the length of an object will involve just *one* length (or one dimension).

> **Remember:**
> Numbers, like 3, $\frac{1}{2}$, or π, have **no dimensions**.

It is quite possible that a formula can have more than one term. The formula $A = \pi r^2 + 3rd$ has two terms and each term has two dimensions.

It is *not* possible to have a mixture of terms some with, say, two
dimensions and some with three dimensions.
So the formula $A = \pi r^2 + 3r^2 d$ could not possibly represent area. Nor
could it represent volume.

The formula $z = \dfrac{2\pi r^2 h}{L}$ has three dimensions on the top line and one

dimension on the bottom. The dimensions can be 'cancelled' so the
expression for z has only two dimensions and can only represent an
area.
We can use these facts to check that any formula we may be using has
the correct number of dimensions.

Here are four formulas where the letters c, d, r represent lengths:

(a) $t = 3c^2$ (b) $k = \dfrac{\pi}{3}r^3 + 4r^2 d$

(c) $m = \pi(c + d)$ (d) $f = 4c + 3cd$

State whether the formula gives:
>   (i) a length   (ii) an area
>   (iii) a volume   (iv) an impossible expression.

(a) $t = 3c^2 = 3c \times c$
    This has *two* dimensions so t is an *area*. [Notice that the number '3' has no dimensions.]

(b) $k = \dfrac{\pi}{3}r^3 + 4r^2 d$.

    Both $\dfrac{\pi}{3}r^3$ and $4r^2 d$ have *three* dimensions so k is a *volume*. $\left[\dfrac{'\pi'}{3}\text{ and '4' have no dimensions}\right]$

(c) $m = \pi(c + d) = \pi c + \pi d$
    πc is a length and πd is a length.
    So m is a length plus a length.
    $\therefore$ m is a length.

(d) $f = 4c + 3cd$
    $4c$ is a length.
    $3cd$ is a length multiplied by a length and is an area.
    So f is a length plus an area which is an *impossible* expression.

Exercise 10

The symbols a, b, d, h, l, r represent lengths.

1. For each expression, decide if it represents a length, an area or a
   volume:
   (a) $a + 3b$ (b) $a^2 b$ (c) ab
   (d) $5(b - a)$ (e) $b^3 + a^3$ (f) $7ab + a^2$

2. State the number of dimensions for each of the following:

(a) πl^2 (b) $3\pi lr$ (c) $\dfrac{\pi}{2}b^2h$

(d) $\pi(a+b)$ (e) $\dfrac{ab+h^2}{6}$ (f) $abd \sin 30°$

3. Give the number of dimensions that a formula for each of the following should have.
(a) Total area of windows in a room.
(b) Volume of sand in a lorry.
(c) Area of a sports field.
(d) The diagonal of a rectangle.
(e) The capacity of an oil can.
(f) The perimeter of a trapezium.
(g) The number of people in a cinema.
(h) The surface area of the roof of a house.

4. One of these expressions gives the area of this shape and another gives the perimeter.

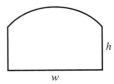

$\boxed{2h + w + \frac{5}{4}w}$ $\boxed{w^2h + h^2}$

h

w

$\boxed{w + h + h + wh}$ $\boxed{wh + \dfrac{\pi}{6}wh}$

Write down the correct expression for:
(a) the area   (b) the perimeter.

5. From this list of expressions, choose the *two* that represent volume, the four that represent area and the *one* that represents a length. The other expression is impossible.

(a) $\pi rh + \pi r^2$ (b) $5a + 6c$

(c) $3·5\,abd$ (d) $4hl + \pi rh$

(e) $3r^2hl$ (f) $2\pi r(r+h)$

(g) $2(rb^2 + h^3)$ (h) $\dfrac{\pi}{2}(l+d)^2$

6. State whether each expression represents length (L), area (A), volume (V) or is impossible (I).

(a) $2a - b$ (b) $a^2 + c$ (c) $2a^3 + abc$
(d) $\pi(ab + c^2)$ (e) $a^2b + b^2a$ (f) $(a-c)(b+a)$
(g) $a(b^2 + c)$ (h) $\pi r^2(a+b)$ (i) $\pi^2(b+a)$

(j) $4\pi a + b$ (k) $\sqrt{\dfrac{a^2+b^2}{c}}$ (l) $\dfrac{\pi r^3}{4} + a^2$

7. In Sam's notes, the formula for the volume of a container was written with Tippex over the index for r. The formula was
$V = \dfrac{\pi}{3}r$ h. What was the missing index?

8. Work out the missing index numbers in these formulas:

(a) Area $= 3(a^{\square} + bd)$

(b) Volume $= \dfrac{\pi L^{\square}}{3}$

(c) Length $= \dfrac{\pi r^{\square}}{3}$

(d) Area $= 3(a^{\square} + b^{\square}d)$

(e) Volume $= 2\pi(r^{\square} + b^{\square})$

(f) Area $= \dfrac{3\pi}{4}(a+b)^{\square}$

9. A physicist worked out a formula for the surface area of a complicated object and got

$$S = 3\pi(a+b)^2 \sin 20° + \frac{\pi}{2}a$$

Explain why the formula could not be correct.

6.4 Transformations

Reflection and rotation

- A reflection is specified by the choice of a mirror line.

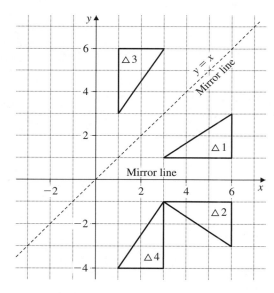

In the diagram above:
(a) △2 is the image of △1 after reflection in the x-axis (the mirror line).
(b) △3 is the image of △1 after reflection in the mirror line $y = x$.
- A rotation requires specification of *three* things: angle, direction and centre of rotation.

(c) △4 is the image of △2 after rotation through 90° clockwise about $(3, -1)$.
(d) △4 is the image of △3 after rotation through 180° in either direction about $(2,1)$.
- By convention, an anticlockwise rotation is positive and a clockwise rotation is negative. So the rotation of △2 onto △4 is $-90°$, centre $(3, -1)$.
  It is helpful to use tracing paper to obtain the result of a rotation.
- Under reflection and rotation (and translation) the object and image are always **congruent**.
  [The object and image are the same size.]

Exercise 11

In Questions **1** and **2**, draw the object and its image after reflection in the broken line.

1.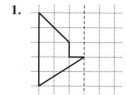

2.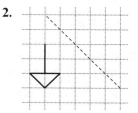

In Questions **3** to **8**, draw x- and y-axes with values from -8 to $+8$.

3. (a) Draw the triangle DEF at D$(-6, 8)$, E$(-2, 8)$, F$(-2, 6)$. Draw the lines $x = 1$, $y = x$, $y = -x$.
   (b) Draw the image of $\triangle$DEF after reflection in:
       (i) the line $x = 1$. Label it $\triangle 1$.
       (ii) the line $y = x$. Label it $\triangle 2$.
       (iii) the line $y = -x$. Label it $\triangle 3$.
   (c) Write down the coordinates of the image of point D in each case.

4. (a) Draw $\triangle 1$ at $(3, 1)$, $(7, 1)$, $(7, 3)$.
   (b) Reflect $\triangle 1$ in the line $y = x$ onto $\triangle 2$.
   (c) Reflect $\triangle 2$ in the x-axis onto $\triangle 3$.
   (d) Reflect $\triangle 3$ in the line $y = -x$ onto $\triangle 4$.
   (e) Reflect $\triangle 4$ in the line $x = 2$ onto $\triangle 5$.
   (f) Write down the coordinates of $\triangle 5$.

5. (a) Draw $\triangle 1$ at $(2, 6)$, $(2, 8)$, $(6, 6)$.
   (b) Reflect $\triangle 1$ in the line $x + y = 6$ onto $\triangle 2$.
   (c) Reflect $\triangle 2$ in the line $x = 3$ onto $\triangle 3$.
   (d) Reflect $\triangle 3$ in the line $x + y = 6$ onto $\triangle 4$.
   (e) Reflect $\triangle 4$ in the line $y = x - 8$ onto $\triangle 5$.
   (f) Write down the coordinates of $\triangle 5$.

6. (a) Draw a triangle PQR at P$(1, 2)$, Q$(3, 5)$, R$(6, 2)$.
   (b) Find the image of PQR under the following rotations:
       (i) 90° anticlockwise, centre $(0, 0)$; label the image P′Q′R′
       (ii) 90° clockwise, centre $(-2, 2)$; label the image P″Q″R″
       (iii) 180°, centre, $(1, 0)$; label the image P*Q*R*.
   (c) Write down the coordinates of P′, P″, P*.

7. (a) Draw $\triangle 1$ at $(1, 2)$, $(1, 6)$, $(3, 5)$.
   (b) Rotate $\triangle 1$ 90° clockwise, centre $(1, 2)$ onto $\triangle 2$.
   (c) Rotate $\triangle 2$ 180°, centre $(2, -1)$ onto $\triangle 3$.
   (d) Rotate $\triangle 3$ 90° clockwise, centre $(2, 3)$ onto $\triangle 4$.
   (e) Write down the coordinates of $\triangle 4$.

8. (a) Draw and label the following triangles:
       $\triangle 1 : (3, 1)$, $(6, 1)$, $(6, 3)$
       $\triangle 2 : (-1, 3)$, $(-1, 6)$, $(-3, 6)$
       $\triangle 3 : (1, 1)$, $(-2, 1)$, $(-2, -1)$
       $\triangle 4 : (3, -1)$, $(3, -4)$, $(5, -4)$
       $\triangle 5 : (4, 4)$, $(1, 4)$, $(1, 2)$
   (b) Describe fully the following rotations:
       (i) $\triangle 1$ onto $\triangle 2$ (ii) $\triangle 1$ onto $\triangle 3$
       (iii) $\triangle 1$ onto $\triangle 4$ (iv) $\triangle 1$ onto $\triangle 5$
       (v) $\triangle 5$ onto $\triangle 4$ (vi) $\triangle 3$ onto $\triangle 2$

Translation and enlargement

- A translation can be specified by the choice of a *column vector*.

  In the diagram at the top of the next column
  (a) $\triangle 5$ is the image of $\triangle 4$ after translation with the column vector $\begin{pmatrix} 4 \\ 2 \end{pmatrix}$.

  (b) $\triangle 4$ is the image of $\triangle 1$ after translation with the column vector $\begin{pmatrix} 2 \\ -3 \end{pmatrix}$.

- An enlargement requires specification of two things: *scale factor* and *centre of enlargement*.
  (c) $\triangle 2$ is the image of $\triangle 1$ after enlargement by a scale factor 3 with centre of enlargement $(0, 0)$.
  (d) $\triangle 1$ is the image of $\triangle 2$ after enlargement by a scale factor $\frac{1}{3}$ with centre of enlargement $(0, 0)$. Note the construction lines.

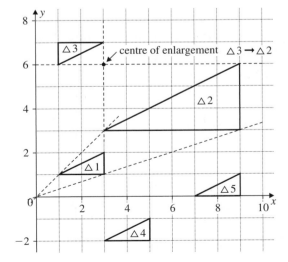

(e) △2 is the image of △3 after enlargement by a scale factor −3 with centre of enlargement (3, 6). Note that the negative scale factor causes the object to be inverted to form the image.

Exercise 12

1. For the diagram below, write down the column vector for each of the following translations.

(a) D onto A   (b) B onto F
(c) E onto A   (d) A onto C
(e) E onto C   (f) C onto B
(g) F onto E   (h) B onto C.

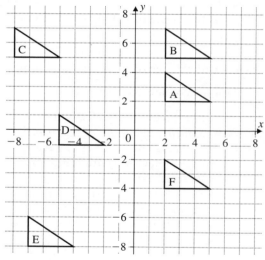

For Questions **2** to **5**, copy the diagram and draw an enlargement using the centre O and the scale factor given.

2. Scale factor 2   **3.** Scale factor 3

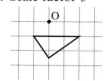

4. Scale factor 3   **5.** Scale factor −2

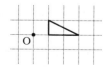   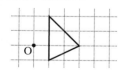

For Questions **6** to **11**, draw x- and y-axes with values from 0 to 15. Enlarge the object using the centre of enlargement and scale factor given.

	object			centre	scale factor
6.	(2, 4)	(4, 2)	(5, 5)	(1, 2)	+2
7.	(1, 1)	(4, 2)	(2, 3)	(1, 1)	+3
8.	(1, 2)	(13, 2)	(1, 10)	(0, 0)	$+\frac{1}{2}$
9.	(5, 10)	(5, 7)	(11, 7)	(2, 1)	$+\frac{1}{3}$
10.	(1, 1)	(3, 1)	(3, 2)	(4, 3)	−2
11.	(9, 2)	(14, 2)	(14, 6)	(7, 4)	$-\frac{1}{2}$

12. Copy the diagram below, leaving space for construction lines.

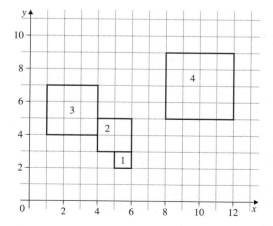

(a) Mark the centres of enlargement for the following:
   (i) △1 → △2
   (ii) △1 → △3
   (iii) △2 → △3
(b) Write down the scale factor for the enlargement △2 → △3.

13. Copy the diagram below.

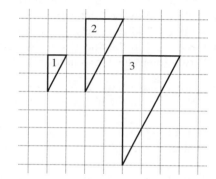

Describe fully each of the following enlargements
(a) square 1 → square 2
(b) square 1 → square 3
(c) square 1 → square 4
(d) square 2 → square 4
(e) square 3 → square 2
(f) square 4 → square 1
(g) Draw the image of square 4 under enlargement with centre (10, 7) and scale factor $\frac{1}{4}$. Label the image square 5.

Questions **14** to **16** involve rotation, reflection, enlargement and translation.

14. Copy the diagram below.

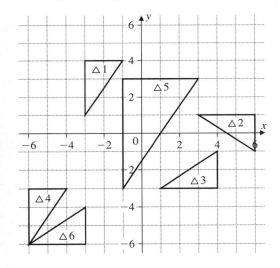

Describe fully the following transformations:
(a) △1 → △2   (b) △1 → △3
(c) △4 → △1   (d) △1 → △5
(e) △3 → △6   (f) △6 → △4

15. Draw x- and y-axes from −8 to +8. Plot and label the following triangles:
△1 : (−5, −5), (−1, −5), (−1, −3)
△2 : (1, 7), (1, 3), (3, 3)
△3 : (3, −3), (7, −3), (7, −1)
△4 : (−5, −5), (−5, −1), (−3, −1)
△5 : (1, −6), (3, −6), (3, −5)
△6 : (−3, 3), (−3, 7), (−5, 7)

Describe fully the following transformations:
(a) △1 → △2   (b) △1 → △3
(c) △1 → △4   (d) △1 → △5
(e) △1 → △6   (f) △5 → △3
(g) △2 → △3

16. Draw x- and y-axes from −8 to +8. Plot and label the following triangles:
△1 : (−3, −6), (−3, −2), (−5, −2)
△2 : (−5, −1), (−5, −7), (−8, −1)
△3 : (−2, −1), (2, −1), (2, 1)
△4 : (6, 3), (2, 3), (2, 5)
△5 : (8, 4), (8, 8), (6, 8)
△6 : (−3, 1), (−3, 3), (−4, 3)

Describe fully the following transformations:
(a) △1 → △2   (b) △1 → △3
(c) △1 → △4   (d) △1 → △5
(e) △1 → △6   (f) △3 → △5
(g) △6 → △2

Successive transformations

In the diagram △2 is the image of △1 after rotation 90° clockwise about (1, 3).

△3 is the image of △2 after translation $\begin{pmatrix} 3 \\ 3 \end{pmatrix}$.

So △1 has been moved onto △3 by successive transformations.
The equivalent single transformation is rotation 90° clockwise about (3, 3). [Check this.]

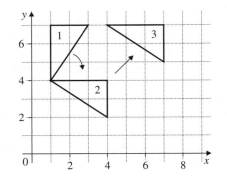

Exercise 13

1. Copy the diagram shown.
   (a) Rotate △1 180° about (4, 2).
       Label the image △2.
   (b) Reflect △2 in the line $y = 2$.
       Label the image △3.
   (c) Describe the *single* transformation
       which maps △1 onto △3.

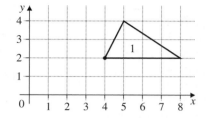

2. Copy the diagram.
   (a) Reflect △A in the line $y = x$.
       Label the image △B.
   (b) Reflect △B in the x-axis.
       Label the image △C.
   (c) Describe fully the single transformations
       which maps △A onto △C.

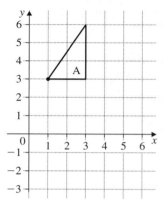

In Questions 3 to 8 transformations A, B, ... H, are as follows:

A   denotes reflection in $x = 2$
B   denotes 180° rotation, centre (1, 1)
C   denotes translation $\begin{pmatrix} -6 \\ 2 \end{pmatrix}$
D   denotes reflection in $y = x$
E   denotes reflection in $y = 0$
F   denotes translation $\begin{pmatrix} 4 \\ 3 \end{pmatrix}$
G   denotes 90° rotation clockwise, centre (0, 0)
H   denotes enlargement, scale factor $+\frac{1}{2}$, centre (0, 0).

Draw x- and y-axes with values from -8 to $+8$.

3. Draw triangle LMN at L(2, 2), M(6, 2), N(6, 4). Find the image of
   LMN under the following transformations:
   (a) **A** followed by **C**   (b) **D** followed by **E**
   (c) **B** followed by **D**   (d) **E** followed by **B**.
   Write down the coordinates of the image of point L in each case.

4. Draw triangle PQR at P(2, 2), Q(6, 2), R(6, 4). Find the image of
   PQR under the following transformations:
   (a) **F** followed by **A**   (b) **G** followed by **C**
   (c) **G** followed by **A**   (d) **E** followed by **H**.
   Write down the coordinates of the image of point P in each case.

5. Draw triangle XYZ at X(−2, 4), Y(−2, 1), Z(−4, 1). Find the image of XYZ under the following combinations of transformations and state the equivalent single transformation in each case.
   (a) **E**, then **G**, then **G**   (b) **B**, then **C**
   (c) **A**, then **D**.

6. Draw triangle OPQ at O(0, 0), P(0, 2), Q(3, 2). Find the image of OPQ under the following combinations of transformations and state the equivalent single transformation in each case.
   (a) **E**, then **D**   (b) **C**, then **F**
   (c) **C**, then **E**, then **D**   (d) **E**, then **F**, then **D**.

7. Draw triangle RST at R(−4, −1), S(−2½, −2), T(−4, −4). Find the image of RST under the following combinations of transformations and state the equivalent single transformation in each case.
   (a) **G**, then **A**, then **E**   (b) **H**, then **F**
   (c) **F**, then **G**.

8. This inverse of a transformation is the transformation which takes the image back to the object.
   Write down the inverses of the transformations **A**, **B**, ... **H**.

6.5 Vectors

A vector quantity has both magnitude and direction. A translation is described by a vector and a vector can also represent physical quantities such as velocity, force, acceleration, etc. The symbol for a vector is a bold letter, e.g. **a**, **x**. On a coordinate grid, the magnitude and direction of the vector can be shown by a column vector, e.g. $\begin{pmatrix} 2 \\ 1 \end{pmatrix}$, $\begin{pmatrix} 1 \\ -3 \end{pmatrix}$

where the upper number shows the distance across the page and the lower number shows the distance up the page.

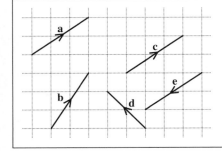

$$\mathbf{a} = \begin{pmatrix} 3 \\ 2 \end{pmatrix} \qquad \mathbf{d} = \begin{pmatrix} -2 \\ 2 \end{pmatrix}$$

$$\mathbf{b} = \begin{pmatrix} 2 \\ 3 \end{pmatrix} \qquad \mathbf{e} = \begin{pmatrix} -3 \\ -2 \end{pmatrix}$$

$$\mathbf{c} = \begin{pmatrix} 3 \\ 2 \end{pmatrix}$$

Equal vectors

Two vectors are equal if they have the same length *and* the same direction. The actual position of the vector on the diagram or in space is of no consequence.

Thus in the example, vectors **a** and **c** are equal because they have the same magnitude and direction. Even though vector **b** also has the same length (magnitude) as **a** and **c**, it is not equal to **a** or **c** because it acts in a different direction. Likewise, vector **e** has the same length as vector **c** but acts in the reverse direction and so cannot equal **c**.
Equal vectors have identical column vectors.

Addition of vectors

Vectors **a** and **b** are represented by the line segments shown below, they can be added by using the 'nose-to-tail' method to give a single equivalent vector.

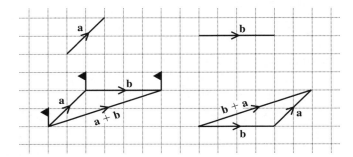

The 'tail' of vector **b** is joined to the 'nose' of vector **a**.

Alternatively the tail of **a** can be joined to the 'nose' of vector **b**.

In both cases the vector $\overrightarrow{XY}$ has the same length and direction and therefore **a** + **b** = **b** + **a**.

In the first diagram the flag is moved by translation **a** and then translation **b**. The translation **a** + **b** is the equivalent or *resultant* translation.

Multiplication by a scalar

A scalar quantity has magnitude but no direction (e.g. mass, volume, temperature). Ordinary numbers are scalars.

When vector **x** is multiplied by 2, the result is 2**x**.

When **x** is multiplied by −3 the result is −3**x**.

Note
(1) The negative sign reverses the direction of the vector.
(2) The result **a** − **b** is **a** + −**b**.
    So, subtracting **b** is equivalent to adding the negative of **b**.

The diagram on the right shows vectors
a and **b**. Draw a diagram to show $\overrightarrow{OP}$
and $\overrightarrow{OQ}$ such that

$\overrightarrow{OP} = 3\mathbf{a} + \mathbf{b}$ $\overrightarrow{OQ} = -2\mathbf{a} - 3\mathbf{b}$

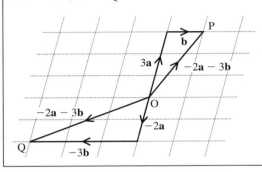

Exercise 14

In Questions **1** to **15**, use the diagram below to describe the vectors
given in terms of **c** and **d** where $\mathbf{c} = \overrightarrow{QN}$ and $\mathbf{d} = \overrightarrow{QR}$, e.g. $\overrightarrow{QS} = 2\mathbf{d}$,
$\overrightarrow{TD} = \mathbf{c} + \mathbf{d}$.

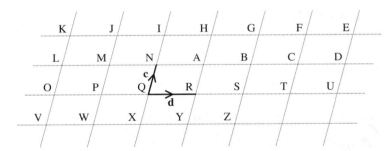

1. $\overrightarrow{AB}$	2. $\overrightarrow{SG}$	3. $\overrightarrow{VK}$	4. $\overrightarrow{KH}$	5. $\overrightarrow{OT}$
6. $\overrightarrow{WJ}$	7. $\overrightarrow{FH}$	8. $\overrightarrow{FT}$	9. $\overrightarrow{KV}$	10. $\overrightarrow{NQ}$
11. $\overrightarrow{OM}$	12. $\overrightarrow{SD}$	13. $\overrightarrow{PI}$	14. $\overrightarrow{YG}$	15. $\overrightarrow{OI}$

In Questions **16** to **21**, use the same diagram above to find vectors for
the following in terms of the capital letters, starting from Q each time.

e.g. $3\mathbf{d} = \overrightarrow{QT}$, $\mathbf{c} + \mathbf{d} = \overrightarrow{QA}$.

16. $2\mathbf{c}$ **17.** $4\mathbf{d}$ **18.** $2\mathbf{c} + \mathbf{d}$

19. $2\mathbf{d} + \mathbf{c}$ **20.** $3\mathbf{d} + 2\mathbf{c}$ **21.** $2\mathbf{c} - \mathbf{d}$

In Questions **22** and **23**, use the diagram below. $\overrightarrow{LM} = \mathbf{a}$, $\overrightarrow{LQ} = \mathbf{b}$.

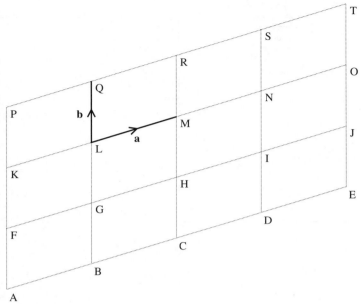

22. Write these vectors in terms of **a** and **b**.

(a) $\overrightarrow{GN}$ (b) $\overrightarrow{CO}$ (c) $\overrightarrow{TN}$

(d) $\overrightarrow{FT}$ (e) $\overrightarrow{KC}$ (f) $\overrightarrow{CJ}$

23. From your answers to Question **22**, find the vector which is:

(a) parallel to $\overrightarrow{LR}$

(b) 'opposite' to $\overrightarrow{LR}$

(c) parallel to $\overrightarrow{CJ}$ with twice the magnitude

(d) parallel to the vector $(\mathbf{a} - \mathbf{b})$.

In Questions **24** to **27**, write each vector in terms of **a**, **b**, or **a** and **b**.

24. (a) $\overrightarrow{BA}$

(b) $\overrightarrow{AC}$

(c) $\overrightarrow{DB}$

(d) $\overrightarrow{AD}$

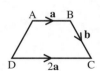

25. (a) $\overrightarrow{ZX}$

(b) $\overrightarrow{YW}$

(c) $\overrightarrow{XY}$

(d) $\overrightarrow{XZ}$

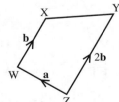

26. (a) $\overrightarrow{MK}$

(b) $\overrightarrow{NL}$

(c) $\overrightarrow{NK}$

(d) $\overrightarrow{KN}$

27. (a) $\overrightarrow{FE}$

(b) $\overrightarrow{BC}$

(c) $\overrightarrow{FC}$

(d) $\overrightarrow{DA}$

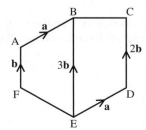

28. The points A, B and C lie on a straight line and the vector $\overrightarrow{AB}$ is
a + 2**b**. Which of the following vectors is possible for $\overrightarrow{AC}$:
(a) 3**a** + 6**b**   (b) 4**a** + 4**b**   (c) **a** − 2**b**   (d) 5**a** + 10**b**?

29. Find three pairs of parallel vectors from those below.

a + 3**b**	**a** − **b**	6**a** − 3**b**	2**a** + 6**b**	3**a** − 3**b**	2**a** − **b**	**a** + **b**
A	B	C	D	E	F	G

Vector geometry

In the diagram, OA = AP and BQ = 3OB.
N is the mid-point of PQ;
$\overrightarrow{OA}$ = **a** and $\overrightarrow{OB}$ = **b**.

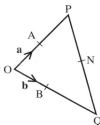

Express each of the following vectors
in terms of **a**, **b**, or **a** and **b**.

(a) $\overrightarrow{AP}$ (b) $\overrightarrow{AB}$ (c) $\overrightarrow{OQ}$ (d) $\overrightarrow{PO}$ (e) $\overrightarrow{PQ}$ (f) $\overrightarrow{PN}$ (g) $\overrightarrow{ON}$ (h) $\overrightarrow{AN}$

(a) $\overrightarrow{AP}$ = **a**   (b) $\overrightarrow{AB}$ = − **a** + **b**   (c) $\overrightarrow{OQ}$ = 4 **b**   (d) $\overrightarrow{PO}$ = −2 **a**

(e) $\overrightarrow{PQ}$ = $\overrightarrow{PO}$ + $\overrightarrow{OQ}$ (f) $\overrightarrow{PN}$ = $\frac{1}{2}\overrightarrow{PQ}$ (g) $\overrightarrow{ON}$ = $\overrightarrow{OP}$ + $\overrightarrow{PN}$
= 2 **a** + 4 **b**   = − **a** + 2 **b**   = 2 **a** + (− **a** + 2 **b**)
= **a** + 2 **b**

(h) $\overrightarrow{AN}$ = $\overrightarrow{AP}$ + $\overrightarrow{PN}$
= **a** + (− **a** + 2 **b**)
= 2 **b**

Exercise 15

In Questions **1** to **4**, $\overrightarrow{OA}$ = **a** and $\overrightarrow{OB}$ = **b**. Copy each diagram and use
the information given to express the following vectors in terms of **a**, **b**
or **a** and **b**.

(a) $\overrightarrow{AP}$ (b) $\overrightarrow{AB}$ (c) $\overrightarrow{OQ}$ (d) $\overrightarrow{PO}$ (e) $\overrightarrow{PQ}$
(f) $\overrightarrow{PN}$ (g) $\overrightarrow{ON}$ (h) $\overrightarrow{AN}$ (i) $\overrightarrow{BP}$ (j) $\overrightarrow{QA}$

1. A, B and N are mid-points of OP, OB and   **2.** A and N are mid-points of OP and PQ;
PQ respectively.   BQ = 2OB.

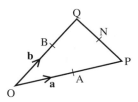   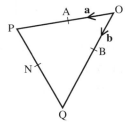

3. AP = 2OA, BQ = OB, PN = NQ.

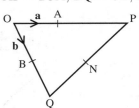

4. OA = 2AP, BQ = 3OB, PN = 2NQ.

5. In △XYZ, the mid-point of YZ is M. If $\overrightarrow{XY}$ = **s** and $\overrightarrow{ZX}$ = **t**, find $\overrightarrow{XM}$ in terms of **s** and **t**.

6. In △AOB, AM : MB = 2 : 1. If $\overrightarrow{OA}$ = **a**, and $\overrightarrow{OB}$ = **b** find $\overrightarrow{OM}$ in terms of **a** and **b**.

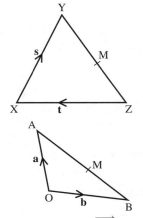

7. O is any point in the plane of the square ABCD. The vectors $\overrightarrow{OA}$, $\overrightarrow{OB}$, and $\overrightarrow{OC}$, are **a, b** and **c** respectively. Find the vector $\overrightarrow{OD}$, in terms of **a, b** and **c**.

8. ABCDEF is a regular hexagon with $\overrightarrow{AB}$, representing the vector **m** and $\overrightarrow{AF}$, representing the vector **n**. Find the vector representing $\overrightarrow{AD}$.

9. ABCDEF is a regular hexagon with centre O. $\overrightarrow{FA}$ = **a** and $\overrightarrow{FB}$ = **b**.

Express the following vectors in terms of **a** and/or **b**.

(a) $\overrightarrow{AB}$ (b) $\overrightarrow{FO}$ (c) $\overrightarrow{FC}$

(d) $\overrightarrow{BC}$ (e) $\overrightarrow{AO}$ (f) $\overrightarrow{FD}$

10. In the diagram, M is the mid-point of CD, BP : PM = 2 : 1, $\overrightarrow{AB}$ = **x**, and $\overrightarrow{AC}$ = **y** and $\overrightarrow{AD}$ = **z**.

Express the following vectors in terms of **x, y** and **z**.

(a) $\overrightarrow{DC}$ (b) $\overrightarrow{DM}$ (c) $\overrightarrow{AM}$

(d) $\overrightarrow{BM}$ (e) $\overrightarrow{BP}$ (f) $\overrightarrow{AP}$

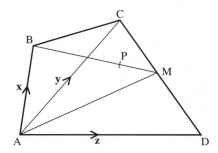

11. In the quadrilateral shown $\overrightarrow{OA} = 2\mathbf{a}$, $\overrightarrow{OB} = 2\mathbf{b}$, $\overrightarrow{OC} = 2\mathbf{c}$.
Points P, Q, R and S are the mid-points of the sides shown.

(a) Express in terms of $\mathbf{a}$, $\mathbf{b}$ and $\mathbf{c}$:
   (i) $\overrightarrow{AB}$
   (ii) $\overrightarrow{BC}$
   (iii) $\overrightarrow{PQ}$
   (iv) $\overrightarrow{QR}$
   (v) $\overrightarrow{PS}$.

(b) Describe the relationship between QR and PS.

(c) What sort of quadrilateral is PQRS?

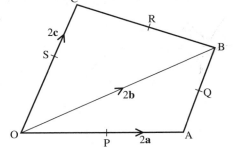

12. In the diagram, $\overrightarrow{OA} = \mathbf{a}$, $\overrightarrow{OB} = \mathbf{b}$, OC = CA,
OB = BE and BD : DA = 1 : 2.

(a) Express in terms of $\mathbf{a}$ and $\mathbf{b}$:
   (i) $\overrightarrow{BA}$
   (ii) $\overrightarrow{BD}$
   (iii) $\overrightarrow{CD}$
   (iv) $\overrightarrow{CE}$.

(b) Explain why points C, D and E lie
on a straight line.

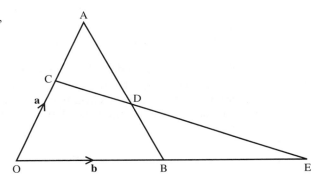

6.6 Sine, cosine, tangent for any angle

So far we have used sine, cosine and tangent
only in right-angled triangles. For angles
greater than 90°, we will see that there is a close
connection between trigonometric ratios and
circles.

The circle on the right is of radius 1 unit with
centre (0, 0). A point P with coordinates (x, y)
moves round the circumference of the circle.
The angle that OP makes with the positive
x-axis as it turns in an anticlockwise direction is θ.

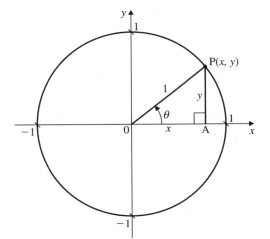

In triangle OAP, $\cos \theta = \dfrac{x}{1}$ and $\sin \theta = \dfrac{y}{1}$

The x-coordinate of P is $\cos \theta$
The y-coordinate of P is $\sin \theta$

This idea is used to define the cosine and the sine of any angle,
including angles greater than 90°.

Here are two angles that are greater than 90°.

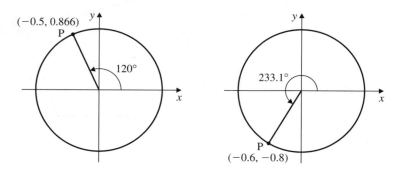

$\cos 120° = -0.5$
$\sin 120° = 0.866$

$\cos 233.1° = -0.6$
$\sin 233.1° = -0.8$

A graphics calculator can be used to show the graph of $y = \sin x$ for any range of angles. The graph below shows $y = \sin x$ for x from 0° to 360°. The curve above the x-axis has symmetry about $x = 90°$ and that below the x-axis has symmetry about $x = 270°$.

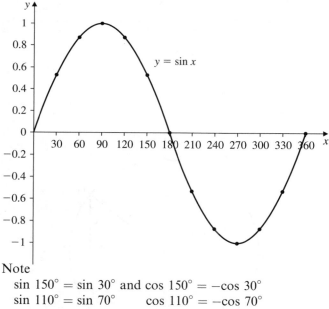

Note

$\sin 150° = \sin 30°$ and $\cos 150° = -\cos 30°$
$\sin 110° = \sin 70°$ $\cos 110° = -\cos 70°$
$\sin 163° = \sin 17°$ $\cos 163° = -\cos 17°$

or $\sin x = \sin (180° - x)$
and $\cos x = -\cos (180° - x)$

These two results are particularly important for use with obtuse angles ($90° < x < 180°$) in Section 6.7 when applying the sine formula or the cosine formula.

Exercise 16

1. (a) Use a calculator to find the cosine of all the angles $0°$, $30°$, $60°$, $90°$, $120°$, ... $330°$, $360°$.
   (b) Draw a graph of $y = \cos x$ for $0 \leqslant x \leqslant 360°$. Use a scale of 1 cm to $30°$ on the x-axis and 5 cm to 1 unit on the y-axis.

2. Draw the graph of $y = \sin x$, using the same angles and scales as in Question **1**.

3. Find the tangent of the angles $0°$, $20°$, $40°$, $60°$, ... $320°$, $340°$, $360°$.
   (a) Notice that we have deliberately omitted $90°$ and $270°$. Why has this been done?
   (b) Draw a graph of $y = \tan x$. Use a scale of 1 cm to $20°$ on the x-axis and 1 cm to 1 unit on the y-axis.
   (c) Draw a vertical dotted line at $x = 90°$ and $x = 270°$. These lines are *asymptotes* to the curve. As the value of x approaches $90°$ from either side, the curve gets nearer and nearer to the asymptote but it *never* quite reaches it.
   [Use a calculator to find $\tan 89°$, $\tan 89 \cdot 9°$, $\tan 89 \cdot 9°$, $\tan 89 \cdot 99°$, $\tan 89 \cdot 999°$ etc.]

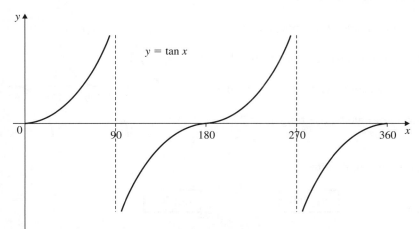

4. Sort the following into pairs of equal value.

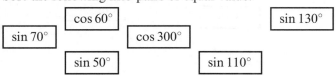

$\cos 60°$ $\sin 130°$ $\sin 70°$ $\cos 300°$ $\sin 50°$ $\sin 110°$

5. Sort the following into pairs of equal value.

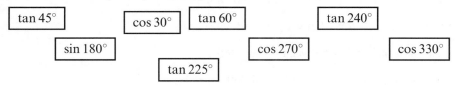

$\tan 45°$ $\cos 30°$ $\tan 60°$ $\tan 240°$ $\sin 180°$ $\cos 270°$ $\cos 330°$ $\tan 225°$

In Questions **6** to **17** do not use a calculator. Use the symmetry of the graphs $y = \sin x$, $y = \cos x$, and $y = \tan x$.
Angles are from 0° to 360° and answers should be given to the nearest degree.

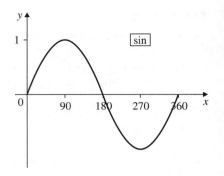

6. If $\sin 18° = 0\cdot309$, give another angle whose sine is $0\cdot309$.

7. If $\sin 27° = 0\cdot454$, give another angle whose sine is $0\cdot454$.

8. Give another angle which has the same sine as
   (a) 40°   (b) 70°   (c) 130°

9. If $\cos 70° = 0\cdot342$, give another angle whose cosine is $0\cdot342$.

10. If $\cos 45° = 0\cdot707$, give another angle whose cosine is $0\cdot707$.

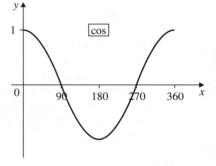

11. Give another angle which has the same cosine as
   (a) 10°   (b) 56°   (c) 300°

12. If $\tan 40° = 0\cdot839$, give another angle whose tangent is $0\cdot839$.

13. If $\sin 20° = 0\cdot342$, what other angle has a sine of $0\cdot342$?

14. If $\sin 98° = 0\cdot990$, give another angle whose sine is $0\cdot990$.

15. If $\tan 135° = -1$, give another angle whose tangent is -1.

16. If $\cos 120° = -0\cdot5$, give another angle whose cosine is $-0\cdot5$.

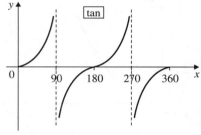

17. If $\tan 70° = 2\cdot75$, give another angle whose tangent is $2\cdot75$.

18. Sort the following into pairs of equal value.

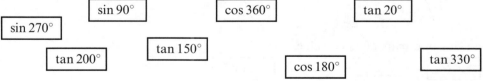

Exercise 17

In Question **1** to **8** give angles correct to one decimal place.

1. Find two values for x, between 0° and 360°, if $\sin x = 0\cdot848$.

2. If $\sin x = 0\cdot35$, find two solutions for x between 0° and 360°.

3. If $\cos x = 0\cdot6$, find two solutions for x between 0° and 360°.

4. If $\tan x = 1$, find two solutions for x between 0° and 360°.

5. Find two solutions between $0°$ and $360°$.

(a) $\sin x = 0.339$

(b) $\sin x = 0.951$

(c) $\sin x = \frac{1}{2}$

(d) $\sin x = \frac{\sqrt{3}}{2}$

6. Find two solutions between $0°$ and $360°$.

(a) $\sin x = 0.72$

(b) $\cos x = 0.3$

(c) $\tan x = 5$

(d) $\sin x = -0.65$

7. Find *four* solutions of the equation $(\sin x)^2 = \frac{1}{4}$, for x between $0°$ and $360°$.

8. Find four solutions of the equation $(\tan x)^2 = 1$, for x between $0°$ and $360°$.

9. The triangle is drawn to find sin, cos and tan of $30°$ and $60°$. For example, $\sin 30° = \frac{1}{2}$ and $\tan 60° = \sqrt{3}$.

Copy and complete this table.

	30°	60°	120°	150°
sin	$\frac{1}{2}$			
cos				
tan		$\sqrt{3}$		

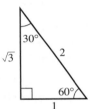

10. (a) Between $0°$ and $360°$, for what values of x is $\sin x > 0$?

(b) Between $0°$ and $360°$, for what values of x is $\cos x < 0$?

11.

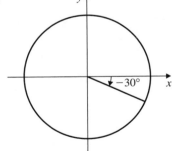

A negative angle is measured *clockwise* from the x-axis.

$\sin(-30°) = -\frac{1}{2}$

$\cos(-30°) = -\frac{\sqrt{3}}{2}$

An angle greater than $360°$ is shown.

$\sin 390° = \sin 30°$

$\cos 400° = \cos 40°$ etc.

Sketch the graphs of $\sin\theta$, $\cos\theta$ and $\tan\theta$ for θ between $-360°$ and $360°$.

12. (a) Find three values of θ for which $\sin\theta = 1$.

(b) Find four values of θ for which $\cos\theta = 0$.

13. (a) Write down the value of $\sin(360n°)$, where n is an integer.

(b) Write down the value of $\cos(360n°)$, where n is an integer.

14. (a) Copy and complete the following.

$$\sin\theta = \frac{a}{c}, \quad \cos\theta = \boxed{-}, \quad \tan\theta = \boxed{-}$$

$$\sin\theta \div \cos\theta = \frac{a}{c} \div \boxed{-} = \boxed{-}$$

$$\therefore \quad \frac{\sin\theta}{\cos\theta} = \tan\theta$$

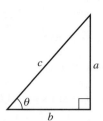

(b) Check if this relationship holds for angles greater than 90°.
  Work out:
  (i) $\dfrac{\sin 135°}{\cos 135°}$, $\tan 135°$ (ii) $\dfrac{\sin 240°}{\cos 240°}$, $\tan 240°$ (iii) $\dfrac{\sin 320°}{\cos 320°}$, $\tan 320°$.

15. Suppose the $\boxed{\cos}$ button on your calculator was broken, but the
$\boxed{\sin}$ button was working. Explain how you could work out $\cos 40°$.

16. Find all possible solutions between 0° and 360°.
  (a) $\sin x = \cos x$ (b) $\sin 2x = \sin 60°$
  (c) $\tan 3x = 1$ [Hint: There are six solutions.]

17. Draw the graph of $y = 2\sin x + 1$ for $0 \leqslant x \leqslant 180°$, taking 1 cm
  to 10° for x and 5 cm to 1 unit for y. Find approximate solutions to
  the equations:
  (a) $2\sin x + 1 = 2\cdot3$ (b) $\dfrac{1}{(2\sin x + 1)} = 0\cdot5$

18. Draw the graph of $y = 2\sin x + \cos x$ for $0 \leqslant x \leqslant 180°$, taking
  1 cm to 10° for x and 5 cm to 1 unit for y.
  (a) Solve approximately the equations:
      (i) $2\sin x + \cos x = 1\cdot5$
      (ii) $2\sin x + \cos x = 0$
  (b) Estimate the maximum value of y.
  (c) Find the value of x at which the maximum occurs.

19. Draw the graph of $y = 3\cos x - 4\sin x$ for $0° \leqslant x \leqslant 220°$, taking
  1 cm to 10° for x and 2 cm to 1 unit for y.
  Solve approximately the equations:
  (a) $3\cos x - 4\sin x + 1 = 0$
  (b) $3\cos x = 4\sin x$

6.7 Sine and cosine rules

The sine rule enables us to calculate sides and angles in
some triangles where there is not a right angle.

In triangle ABC, we use the convention that

a is the side opposite $\widehat{A}$
b is the side opposite $\widehat{B}$

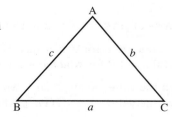

Sine rule

either $\dfrac{a}{\sin A} = \dfrac{b}{\sin B} = \dfrac{c}{\sin C}$ ... [1]

or $\dfrac{\sin A}{a} = \dfrac{\sin B}{b} = \dfrac{\sin C}{c}$ ... [2]

Use [1] when finding a *side*,
and [2] when finding an *angle*.

> **Remember**
> To find a *side*, have the *sides* on *top*
> To find an *angle*, have the *angles* on *top*

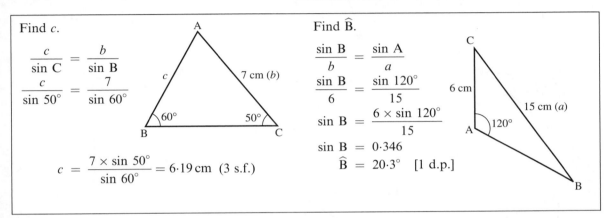

Find c.

$\dfrac{c}{\sin C} = \dfrac{b}{\sin B}$

$\dfrac{c}{\sin 50°} = \dfrac{7}{\sin 60°}$

$c = \dfrac{7 \times \sin 50°}{\sin 60°} = 6.19 \text{ cm}$ (3 s.f.)

Find $\widehat{B}$.

$\dfrac{\sin B}{b} = \dfrac{\sin A}{a}$

$\dfrac{\sin B}{6} = \dfrac{\sin 120°}{15}$

$\sin B = \dfrac{6 \times \sin 120°}{15}$

$\sin B = 0.346$

$\widehat{B} = 20.3°$ [1 d.p.]

Exercise 18

For Questions **1** to **6**, find each side marked with a letter.

1.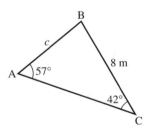

2.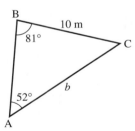

3.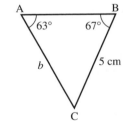

4.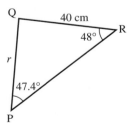

5.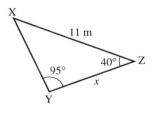

6.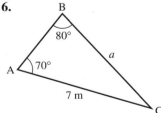

7. In $\triangle$ABC, $\widehat{A} = 61°$, $\widehat{B} = 47°$,
AC = 7.2 cm. Find BC.

8. In $\triangle$XYZ, $\widehat{Z} = 32°$, $\widehat{Y} = 78°$,
XY = 5.4 cm. Find XZ.

9. In $\triangle$PQR, $\widehat{Q} = 100°$, $\widehat{R} = 21°$,
PQ = 3.1 cm. Find PR.

10. In $\triangle$LMN, $\widehat{L} = 21°$, $\widehat{N} = 30°$,
MN = 7 cm. Find LN.

In Questions **11** to **18**, find each angle marked *. All lengths are in centimetres.

11.
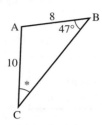

12.
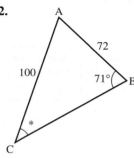

13.
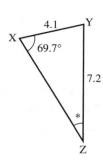

14.
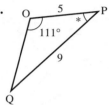

15.
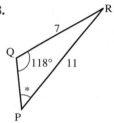

16.
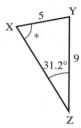

17.
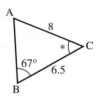

18.
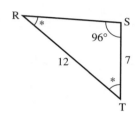

19. In $\triangle ABC$, $\widehat{A} = 62°$, $BC = 8$, $AB = 7$. Find $\widehat{C}$.

20. In $\triangle XYZ$, $\widehat{Y} = 97·3°$, $XZ = 22$, $XY = 14$. Find $\widehat{Z}$.

21. In $\triangle DEF$, $\widehat{D} = 58°$, $EF = 7·2$, $DE = 5·4$. Find $\widehat{F}$.

22. In $\triangle LMN$, $\widehat{M} = 127·1°$, $LN = 11·2$, $LM = 7·3$. Find $\widehat{L}$.

23. The sine rule can be ambiguous.

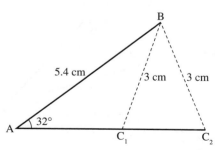

The diagram shows two possible triangles with $AB = 5·4$, $\widehat{A} = 32°$ and $BC = 3$.

Find the two possible values of angle C.

Cosine rule

We use the cosine rule when we know either
(a) two sides and the included angle or
(b) all three sides.

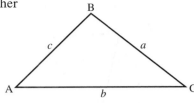

There are two forms.

1. To find the length of a side.

$$a^2 = b^2 + c^2 - (2bc\,\cos A)$$

or $\quad b^2 = c^2 + a^2 - (2ac\,\cos B)$

or $\quad c^2 = a^2 + b^2 - (2ab\,\cos C)$

2. To find an angle when given
all three sides.

$$\cos A = \frac{b^2 + c^2 - a^2}{2bc}$$

or $\quad \cos B = \dfrac{a^2 + c^2 - b^2}{2ac}$

or $\quad \cos C = \dfrac{a^2 + b^2 - c^2}{2ab}$

For an obtuse angle x we have $\cos x = -\cos(180° - x)$

Examples $\qquad \cos 120° = -\cos 60°$

$\qquad\qquad\quad \cos 142° = -\cos 38°$

(a) Find b.

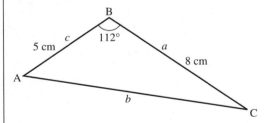

$b^2 = a^2 + c^2 - (2ac\,\cos B)$

$b^2 = 8^2 + 5^2 - (2 \times 8 \times 5 \times \cos 112°)$

$b^2 = 64 + 25 - [80 \times (-0{\cdot}3746)]$

$b^2 = 64 + 25 + 29{\cdot}968$

(Notice the change of sign for the obtuse
angle)

$b = \sqrt{(118{\cdot}968)} = 10{\cdot}9\,\text{cm}$ (to 3 s.f.)

(b) Find angle C.

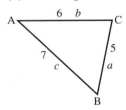

$$\cos C = \frac{a^2 + b^2 - c^2}{2ab}$$

$$\cos C = \frac{5^2 + 6^2 - 7^2}{2 \times 5 \times 6} = \frac{12}{60} = 0{\cdot}200$$

$\hat{C} = 78{\cdot}5°$ (to 1 d.p.)

Exercise 19

Find the sides marked *.

1.

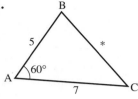

2.

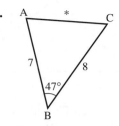

3.

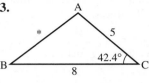

4.

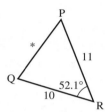

5.

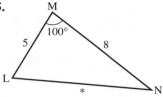

6.

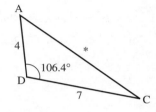

7. In △ABC, AB = 4 cm, AC = 7 cm,
$\hat{A}$ = 57°. Find BC.

8. In △XYZ, XY = 3 cm, YZ = 3 cm,
$\hat{Y}$ = 90°. Find XZ

9. In △LMN, LM = 5·3 cm, MN = 7·9 cm,
$\hat{M}$ = 127°. Find LN.

10. In △PQR, $\hat{Q}$ = 117°, PQ = 80 cm,
QR = 100 cm. Find PR.

In Questions **11** to **16**, find the angles marked *.

11.

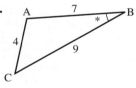

12.

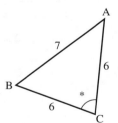

13.

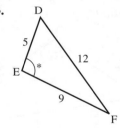

14.

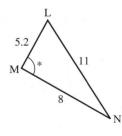

15.

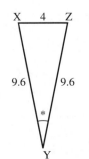

16.

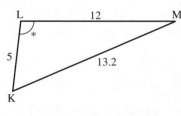

17. In △ABC, a = 4·3, b = 7·2, c = 9. Find $\hat{C}$.

18. In △DEF, d = 30, e = 50, f = 70. Find $\hat{E}$.

19. In △PQR, p = 8, q = 14, r = 7. Find $\hat{Q}$.

20. In △LMN, l = 7, m = 5, n = 4. Find $\hat{N}$.

A ship sails from a port P a distance of 7 km on a bearing of 306°
and then a further 11 km on a bearing of 070° to arrive at X.
Calculate the distance from P to X.

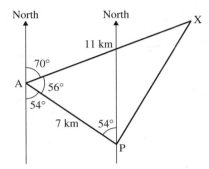

We know two sides and the included angle. Using the cosine rule
gives

$$PX^2 = 7^2 + 11^2 - (2 \times 7 \times 11 \times \cos 56°)$$
$$= 49 + 121 - (86·12)$$

$$PX^2 = 83·88$$
$$PX = 9·16 \text{ km} \quad \text{(to 3 S.F.)}$$

The distance from P to X is 9·16 km.

Exercise 20

1. Ship B is 58 km south-east of Ship A.
   Ship C is 70 km due south of Ship A.

   (a) How far is Ship B from Ship C?
   (b) What is the bearing of Ship B from
       Ship C?

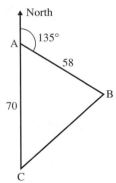

2. A destroyer D and a cruiser C leave port P at the same time. The
   destroyer sails 25 km on a bearing 040° and the cruiser sails 30 km
   on a bearing of 320°. How far apart are the ships?

3. Two honeybees A and B leave the hive H at the same time; A flies
   27 m due south and B flies 9 m on a bearing of 111°. How far apart
   are they?

4. Find the sides and angles marked with letters. All lengths are in cm.

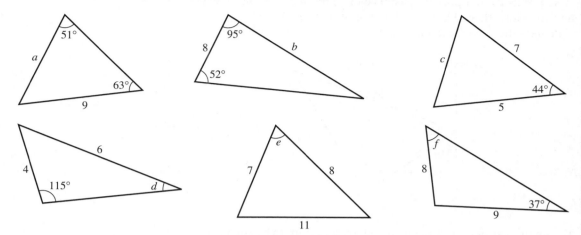

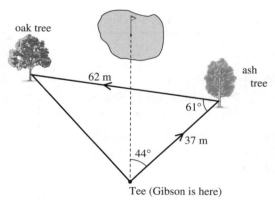

5. Find the largest angle in a triangle in which the sides are in the ratio
$5:6:8$.

6. A golfer hits his ball a distance of 170 m on a hole which measures
195 m from the tee T to the hole. If his shot is directed 10° away
from the direct line to the hole, find the distance between his ball
and the hole.

7. Mr Gibson has taken up golf with
alarming consequences for the local
tree population. The diagram shows
a typical effort, even before he has
reached the 19th hole. His shot travels
from the tee, hits an ash tree then an
oak tree and then hits him on the head.
How far is our hero from the oak tree?

8. A rhombus has sides of length 8 cm and angles of 50° and 130°.
Find the length of the longer diagonal of the rhombus.

9. From A, B lies 11 km away on a bearing of 041° and C lies 8 km
away on a bearing of 341°. Find:
(a) the distance between B and C
(b) the bearing of B from C.

10. From a lighthouse L an aircraft carrier A is 15 km away on a
bearing of 112° and a submarine S is 26 km away on a bearing of
200°. Find:
(a) the distance between A and S
(b) the bearing of A from S.

11. •A

The diagram show three towns A, B, C.
The bearing of B from A is 110°.
The bearing of C from A is 160°.
The bearing of C from B is 240°.
The distance from C to B is 110 km.
Find the distance from A to B.

B
•

•C

12. Find: (a) AE
          (b) EÂC.
If the line BCD is horizontal, find the
angle of elevation of E from A.

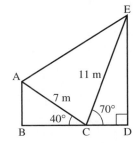

13. An aircraft flies from its base 200 km on a bearing 162°, then
350 km on a bearing 260°, and then returns directly to base.
Calculate the length and bearing of the return journey.

14. Town Y is 9 km due north of town Z. Town X is 8 km from Y,
5 km from Z and somewhere to the west of the line YZ.
(a) Draw triangle XYZ and find angle YZX
(b) During an earthquake, town X moves due south until it is due
     west of Z. Find how far it has moved.

15.

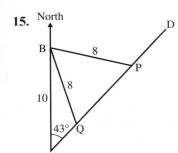

The diagram shows a point A which lies
10 km due south of a point B. A straight
road AD is such that the bearing of D
from A is 043°.
P and Q are two points on this road
which are 8 km from B.
Calculate the bearing of P from B.
(Hint: Find angle BPA first)

16. The diagram shows wires attached
to a communications antenna.
Find the length *h*, correct to the
nearest metre.

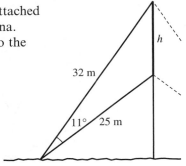

17. In a triangle ABC, $a = \sqrt{28}$, $b = 6$ and $A = 60°$.
  (a) Use the cosine rule to write down a quadratic equation involving c.
  (b) Solve the equation to find two values of c.
  (c) Use the Sine rule to find two values of c independently.
      [Find angle B first.] Compare your answers with those obtained in (b).

18. Proof of the cosine rule.
  (a) Using $\triangle ADC$ show that $x = b \cos C$.
  (b) Using Pythagoras' theorem in $\triangle ADC$,
      find h^2 in terms of b and x.
  (c) Using $\triangle ABD$ find c^2 in terms of h, a and x.
  (d) Use your results to prove that $c^2 = a^2 + b^2 - 2ab \cos C$.

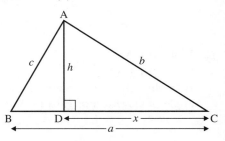

19. Calculate WX, given $YZ = 15$ m.

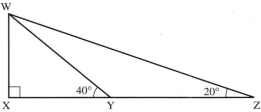

†**20.** Find angle DCB. Check your answer by making a scale drawing.

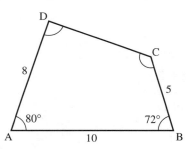

†**21.** Find   (a) WX   (b) WZ   (c) WY.

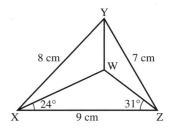

†**22.** The diagram shows a cube of side 10 cm from which one corner has been cut.
Calculate the angle PQR.

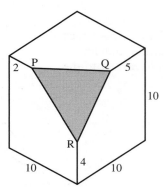

6.8 Circle theorems

Theorem 1

The angle subtended at the centre of a circle is twice the angle subtended at the circumference.

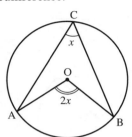

$$A\widehat{O}B = 2 \times A\widehat{C}B$$

Theorem 2

Angles subtended by an arc in the same segment of a circle are equal.

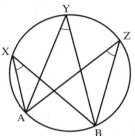

$$A\widehat{X}B = A\widehat{Y}B = A\widehat{Z}B$$

The proof of this theorem is given in the section on proof.

(a) Given $A\widehat{B}O = 50°$, find $B\widehat{C}A$.

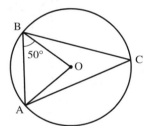

Triangle OBA is isosceles (OA = OB).

$\therefore$ $O\widehat{A}B = 50°$

$\therefore$ $B\widehat{O}A = 80°$ (angle sum of a triangle)

$\therefore$ $B\widehat{C}A = 40°$ (angle at the centre)

(b) Given $B\widehat{D}C = 62°$ and $D\widehat{C}A = 44°$ find $B\widehat{A}C$ and $A\widehat{B}D$.

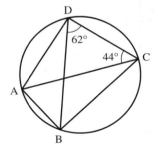

$B\widehat{D}C = B\widehat{A}C$
(both subtended by arc BC)

$\therefore$ $B\widehat{A}C = 62°$

$D\widehat{C}A = A\widehat{B}D$
(both subtended by arc DA)

$\therefore$ $A\widehat{B}D = 44°$

Exercise 21

Find the angles marked with letters. A line passes through the centre only when point O is shown.

1.

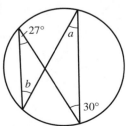

2.

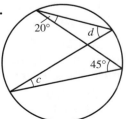

3.

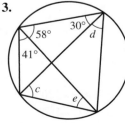

4.

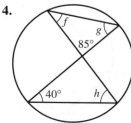

5.

6.

7.

8.

9.

10.

11.

12.

13.

14.

15.

16.

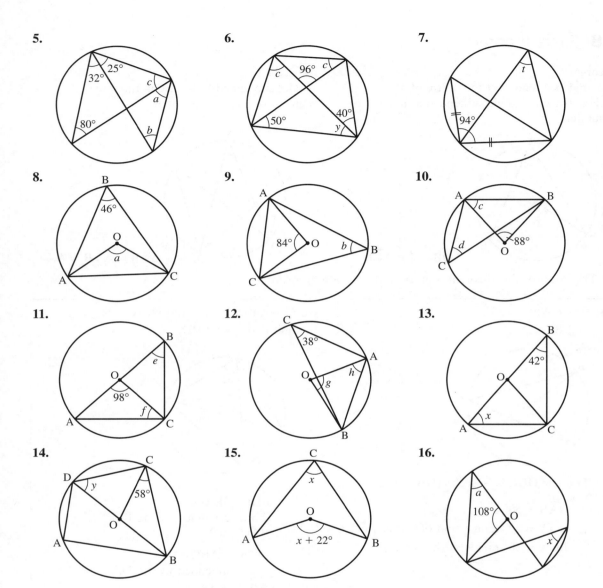

Theorem 3

The opposite angles in a cyclic quadrilateral add up to 180° (the angles are supplementary).

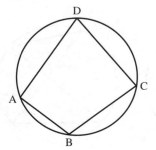

$$\widehat{A} + \widehat{C} = 180°$$
$$\widehat{B} + \widehat{D} = 180°$$

Find a and x.

$a = 180° - 81°$

(opposite angles of a cyclic quadrilateral)

$\therefore \quad a = 99°$

$x + 2x = 180°$

(opposite angles of a cyclic quadrilateral)

$3x = 180°$

$\therefore \quad x = 60°$

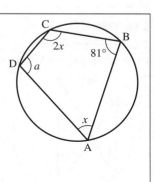

Theorem 4

The angle in a semicircle is a right angle.

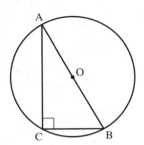

In the diagram,
AB is a diameter.
$A\widehat{C}B = 90°$.

Find b given that AOB is a diameter.

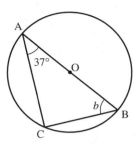

$A\widehat{C}B = 90°$ (angle in a semicircle)

$\therefore \quad b = 180° - (90 + 37)°$

$\quad\quad = 53°$

Exercise 22

Find the angles marked with a letter.

1.

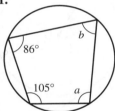

2.

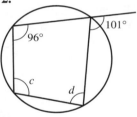

3.

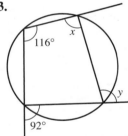

4.

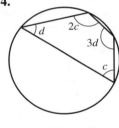

5.

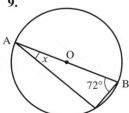

6.

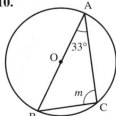

7.

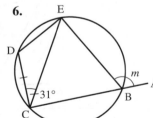

8.

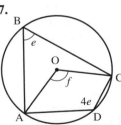

9.

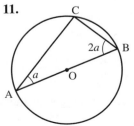

10.

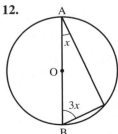

11.

12.

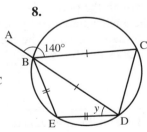

13.

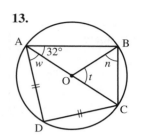

14.

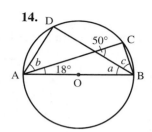

15.

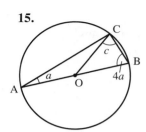

16.

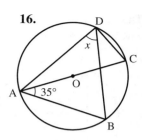

17.

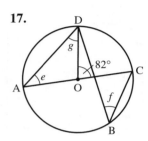

18.

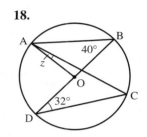

19.

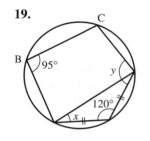

20.

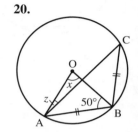

Tangents to circles

Theorem 1

The angle between a tangent and the radius
drawn to the point of contact is $90°$.

$$T\widehat{A}O = 90°$$
$$TA = TB$$

Theorem 2

From any point outside a circle just two
tangents to the circle may be drawn and
they are of equal length.

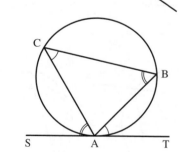

Theorem 3 – Alternate segment theorem

The angle between a tangent and a chord
through the point of contact is equal to
the angle subtended by the chord in the
alternate segment. The theorem is proved
in Question 16 below.

$$T\widehat{A}B = B\widehat{C}A$$
$$S\widehat{A}C = C\widehat{B}A$$

TA and TB are tangents to the circle, centre O.

Given $A\widehat{T}B = 50°$, find: (a) $A\widehat{B}T$ (b) $O\widehat{B}A$ (c) $A\widehat{C}B$.

(a) △TBA is isosceles (TA = TB)

  ∴ $A\widehat{B}T = \frac{1}{2}(180 - 50) = 65°$

(b) $O\widehat{B}T = 90°$ (tangent and radius)

  ∴ $O\widehat{B}A = 90 - 65 = 25°$

(c) $A\widehat{C}B = A\widehat{B}T$ (alternate segment theorem)

      $A\widehat{C}B = 65°$

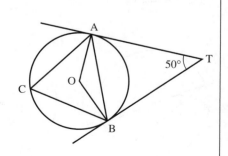

Exercise 23

Find the angles marked with letters.

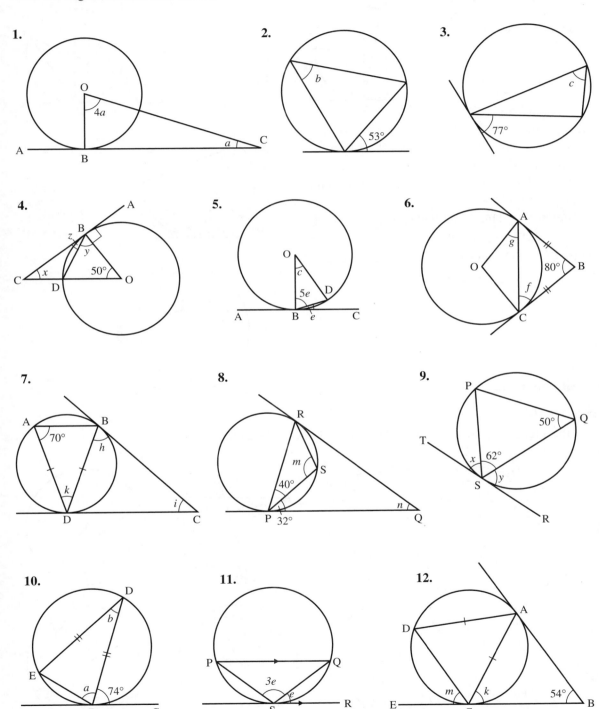

13.

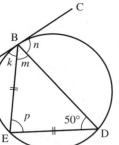

14. Find, in terms of p
(a) $\widehat{BAC}$ (b) $\widehat{XCA}$ (c) $\widehat{ACO}$

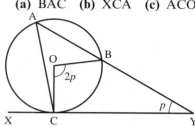

15. Find x, y and z.

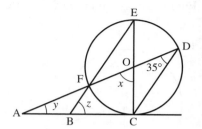

16. Copy and complete the following proof of the alternate segment theorem [ie to prove $a = b$].

Draw a diameter RP.
$\widehat{RPA} = 90°$ [angle between tangent and radius]
$\therefore$ $\widehat{RPQ} = 90 - a$
$\widehat{RQP} = \square$ [angle in a semicircle]
$\therefore$ $\widehat{PRQ} = \square$ [angles in a triangle]
$\widehat{PRQ} = \widehat{PTQ}$ [angles in the same segment]
$\therefore$ $\widehat{QPA} = \widehat{QTP}$ as required.

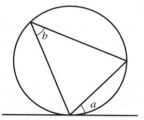

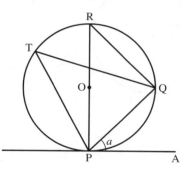

Mixed questions

Exercise 24

Find the angles marked with letters. The centre of the circle is O.

1.

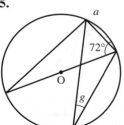

2.

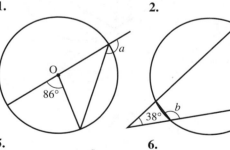

3.

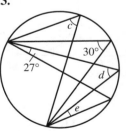

4.

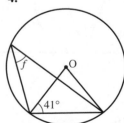

5.

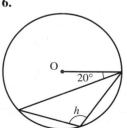

6.

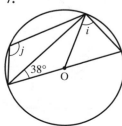

7.

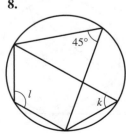

8.

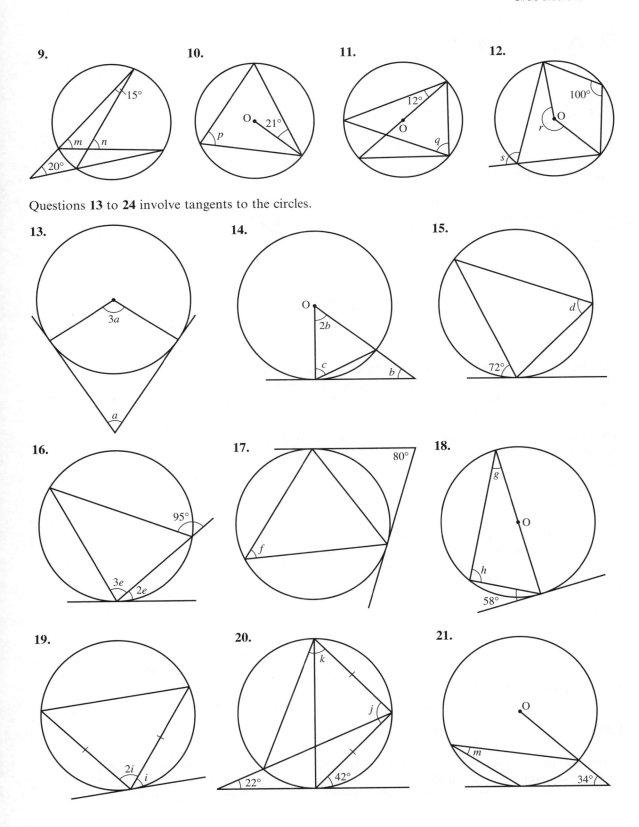

Questions **13** to **24** involve tangents to the circles.

7 ALGEBRA 3

7.1 Gradient of a curve

Mathematicians have been interested in finding the gradient of a curve for hundreds of years. When we say 'gradient of the curve', we really mean the 'gradient of the tangent to the curve' at that point.

In the sketch, the gradient of the curve at B is greater than the gradient of the curve at A.

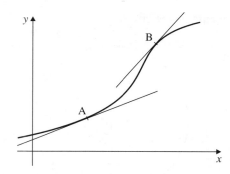

Newton and Leibnitz independently discovered a method for *calculating* the gradient of a curve in about 1680. This branch of mathematics is beyond the scope of this book. We will find the gradient of a curve by drawing a tangent to the curve.

The gradient of a curve is used to find speed or acceleration and such questions appear later in this chapter.

The graph of $y = \dfrac{12}{x} + x - 6$ is drawn on the right.

Find the gradient of the tangent to the curve drawn at the point where $x = 5$.

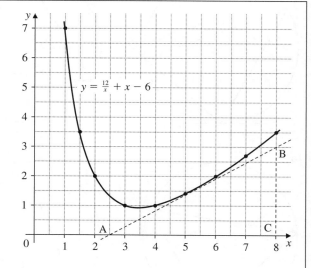

The tangent AB is drawn to touch the curve at $x = 5$.

The gradient of AB $= \dfrac{BC}{AC}$.

gradient $= \dfrac{3}{8 - 2 \cdot 4} = \dfrac{3}{5 \cdot 6} \approx 0 \cdot 54$

It is difficult to obtain an accurate value for the gradient of a tangent so the above result is more realistically 'approximately 0·5'.

Exercise 1

1. The graph shows $y = x^2 - 4x + 3$ with two tangents drawn.
   (a) Find the gradient of the tangent to the curve at $x = 3$.
   (b) Find the gradient of the tangent to the curve at $x = 0$.
   [Remember to count *units*, not just squares!]

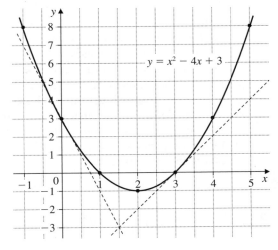

2. Draw the graph of $y = x^2 - 3x$, for $-2 \leqslant x \leqslant 5$.
   (Scales: $2\,\text{cm} = 1$ unit for x; $1\,\text{cm} = 1$ unit for y). Find
   (a) the gradient of the tangent to the curve at $x = 3$,
   (b) the gradient of the tangent to the curve at $x = -1$,
   (c) the value of x where the gradient of the curve is zero.

3. Draw the graph of $y = 3^x$, for $-3 \leqslant x \leqslant 3$.
   (Scales: $2\,\text{cm} = 1$ unit for x; $1\,\text{cm} = 2$ units for y).
   Find the gradient of the curve at $x = 1$.

4. Two tangents are drawn to the curve $y = \dfrac{10}{x}$.

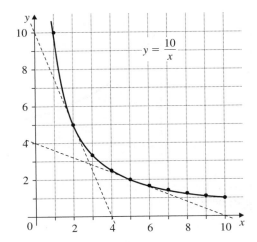

   (a) Find the gradient of the curve at $x = 2$.
   (b) Find the gradient of the curve at $x = 5$.
   (c) Copy and complete this sentence:
   'As $x \to \infty$, the gradient of the curve $y = \dfrac{10}{x} \to \boxed{}$.'

5. Draw the graph of $y = x^2$, for $0 \leqslant x \leqslant 6$.
   (Scales: $2\,\text{cm} = 1$ unit for x; $1\,\text{cm} = 2$ units for y).
   (a) Find
       (i) the gradient of the curve at $x = 2$,
       (ii) the gradient of the curve at $x = 4$.
       Give both answers to the nearest whole number.
   (b) Finding the gradient of a tangent to a curve is notoriously
       inaccurate. You will get more reliable results if you take the
       'class average' of your answers to (i) and (ii) above.
   (c) Use your answers to *predict* the gradient of the curve at $x = 5$.
   (d) Draw a tangent at $x = 5$ and find its gradient to check if your
       prediction was correct.

7.2 Area under a curve

It is sometimes useful to know the area under a curve. In the next
section we look at the area under a speed-time graph to calculate
distance travelled.
The method below, using trapeziums, gives only an approximate value
for the area. When a computer is used, the answer can be found to a
high degree of accuracy.

Find an approximate value for the area under
the curve $y = \dfrac{12}{x}$ between $x = 2$ and $x = 5$.

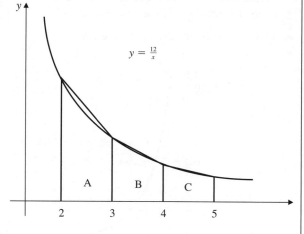

(a) Draw lines parallel to the y-axis at $x = 2, 3,$
   4, 5 to form three trapeziums. The area
   under the curve is approximately equal to
   the area of the three trapeziums A, B and C.

(b) Calculate the values of y at $x = 2, 3, 4, 5$
   at $x = 2$, $y = \frac{12}{2} = 6$
   at $x = 3$, $y = \frac{12}{3} = 4$
   at $x = 4$, $y = \frac{12}{4} = 3$
   at $x = 5$, $y = \frac{12}{5} = 2\!\cdot\!4$

(c) Calculate the area of each of the trapeziums using the formula area $= \frac{1}{2}(a + b)h$, where a and b
   are the parallel sides and h is the distance in between.

   Area A $= \frac{1}{2}(6 + 4) \times 1 \quad = 5$ sq. units
   Area B $= \frac{1}{2}(4 + 3) \times 1 \quad = 3\!\cdot\!5$ sq. units
   Area C $= \frac{1}{2}(3 + 2\!\cdot\!4) \times 1 = 2\!\cdot\!7$ sq. units

   Total area under the curve $\approx (5 + 3\!\cdot\!5 + 2\!\cdot\!7)$
   $\approx 11\!\cdot\!2$ sq. units

Exercise 2

1. Find an approximate value for the area under the curve
$y = 3 + 4x - x^2$ between $x = 0$ and $x = 4$. Divide the area
into four trapeziums as shown.

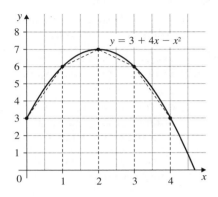

2. Estimate the area under each curve below.

(a)

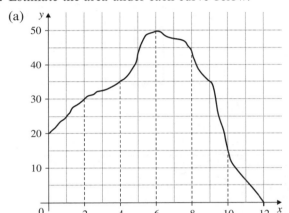

(b)

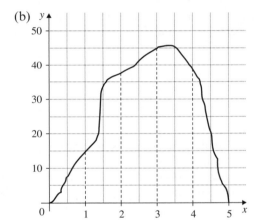

3. Find an approximate value for the area
under the curve $y = \dfrac{8}{x}$ between
$x = 2$ and $x = 5$.

Divide the area into three trapeziums as shown.

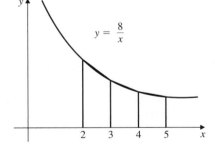

4. Find an approximate value for the area
under the curve $y = \dfrac{12}{(x + 2)}$ between
$x = 0$ and $x = 4$.

Divide the area into four trapeziums as shown.

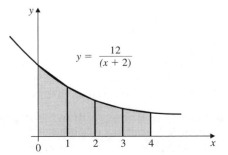

5. Find an approximate value for the area under the curve
$y = 6x - x^2$ between $x = 1$ and $x = 4$. Divide the area into three
trapeziums of equal width.

6. (a) Sketch the curve $y = 16 - x^2$ for $-4 \leqslant x \leqslant 4$.
   (b) Find an approximate value for the area under the curve
   $y = 16 - x^2$ between $x = 0$ and $x = 3$.
   (Divide the area into three trapeziums of equal width.)
   (c) State whether your approximate value is greater than or less
   than the actual value for the area.

7. (a) Find an approximate value for the area
   under the curve $y = x^2 + 2x + 5$
   between $x = 0$ and $x = 4$.
   Divide the area into four trapeziums.
   (b) State whether your approximate value is
   greater than or less than the actual value
   for the area.

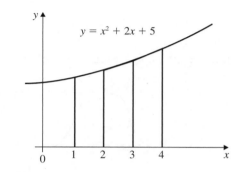

† 8. The sketch shows the curve $y = x^2 - 5x + 8$
   and the line $y = 4$.
   (a) Calculate the x-values at the points
   A and B.
   (b) Find an approximate value for the
   area shaded.

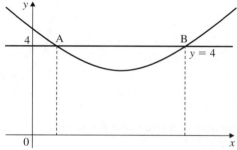

7.3 Distance, velocity, acceleration

When a *distance–time* graph is drawn, the gradient of
the graph gives the speed of the object.

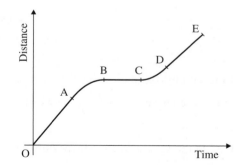

From O to A : constant speed
    A to B : speed goes down to zero
    B to C : at rest
    C to D : speed increases
    D to E : constant speed (not as fast as O to A)

- When a *velocity–time* graph is drawn two quantities can be found.
  (i) acceleration = gradient of graph.
  (ii) distance travelled = area under graph.

The diagram is the speed–time graph of the first 30 seconds of a car journey.

(a) The gradient of line OA $= \frac{20}{10} = 2$.

   $\therefore$ The acceleration in the first 10 seconds is $2\,\text{m/s}^2$.

(b) The distance travelled in the first 30 seconds is given by the area of OAD plus the area of ABCD.

   Distance $= (\frac{1}{2} \times 10 \times 20) + (20 \times 20) = 500\,\text{m}$

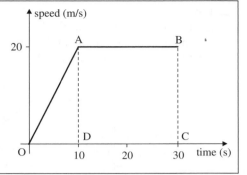

The sketch shows the velocity–time graph of a rocket as it takes off. The equation of the curve is $v = 11t^2$

Find an approximate value for the distance travelled by the rocket in the first 3 seconds.

We use 'distance travelled = area under graph'.

Divide the area into 3 trapeziums (one of which is a triangle).

$AB = 11$, $CD = 44$, $EF = 99$
(from $v = 11t^2$)

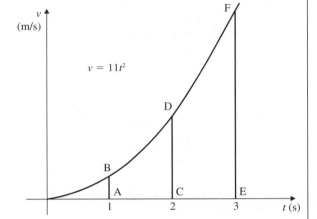

Area $\approx \frac{1}{2}(0 + 11) \times 1 + \frac{1}{2}(11 + 44) \times 1 + \frac{1}{2}(44 + 99) \times 1$
$\approx 104{\cdot}75$

Distance travelled in the first 3 seconds is about 105 m.

Note: To find the acceleration of the rocket at $t = 2$, we could draw a tangent to the curve at this point and so find its gradient.

Exercise 3

The graphs show speed v in m/s and time t in seconds.

1. Find:
   (a) the acceleration when $t = 4$
   (b) the total distance travelled.

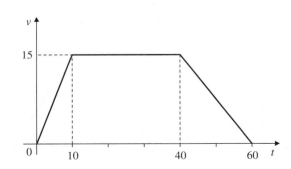

2. Find:
   (a) the total distance travelled,
   (b) the distance travelled in the first 10 seconds
   (c) the acceleration when $t = 20$.

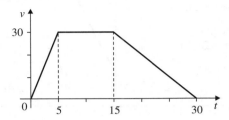

3. Find:
   (a) the total distance travelled
   (b) the distance travelled in the first 40 seconds
   (c) the acceleration when $t = 15$.

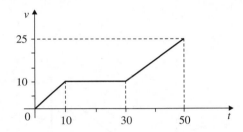

4. Find:
   (a) V if the total distance travelled is 900 m
   (b) the distance travelled in the first 60 seconds.

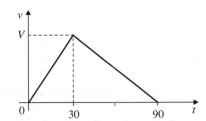

5. Find:
   (a) T if the initial acceleration is $2\,\text{m/s}^2$
   (b) the total distance travelled
   (c) the average speed for the whole journey.

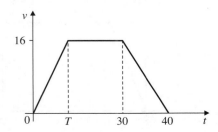

6. Each graph represents one of the following situations.
A A car moving in congested traffic
B A ball rolling along a lane at a bowling alley.
C An apple thrown vertically in the air
D A parachutist after jumping from a stationary hot-air balloon
E A lift moving from level 2 to level 3
F A table tennis ball during a game.

1.
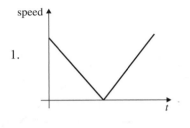

2.
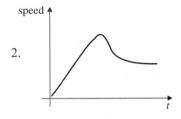

3.
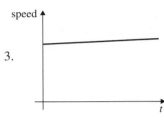

4.
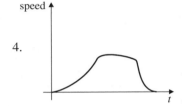

5.
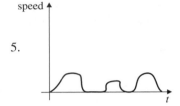

6.
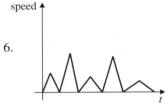

Decide which graph matches each situation.

7. The speed of a train is measured at regular intervals of time from
$t = 0$ to $t = 60$ s, as shown below.

t s	0	10	20	30	40	50	60
v m/s	0	10	16	19·7	22·2	23·8	24·7

Draw a speed–time graph to illustrate the motion. Plot t on the
horizontal axis with a scale of 1 cm to 5 s and plot v on the vertical
axis with a scale of 2 cm to 5 m/s.
Use the graph to estimate:
(a) the acceleration at $t = 10$
(b) the distance travelled by the train from $t = 30$ to $t = 60$.

8. The speed of a car is measured at regular intervals of time from
$t = 0$ to $t = 60$ s, as shown below.

t s	0	10	20	30	40	50	60
v m/s	0	1·3	3·2	6	10·1	16·5	30

Draw a speed–time graph using the same scales as in Question **6**.
Use the graph to estimate:
(a) the acceleration at $t = 30$
(b) the distance travelled by the car from $t = 20$ to $t = 50$.

9. The sketch graph shows the distance–time graph of an object in an experiment. [N.B. Not a velocity-time graph.]

The graph passes through $(0, 0)$, $(0 \cdot 2, 1 \cdot 2)$, $(0 \cdot 5, 1 \cdot 6)$, $(1, 2)$, $(2, 2 \cdot 5)$, $(3, 2 \cdot 9)$.

Draw your own accurate graph and then draw a suitable tangent to find the speed of the object at $t = 0 \cdot 5$ seconds.
Use a scale of 5 cm to 1 second across the page and 5 cm to 1 unit up the page.

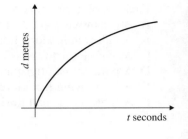

†**10.** Given that the average speed for the whole journey is $37 \cdot 5 \, \text{m/s}$ and that the deceleration between T and $2T$ is $2 \cdot 5 \, \text{m/s}^2$, find:
(a) the value of V (b) the value of T.

†**11.** Given that the total distance travelled is $4 \, \text{km}$ and that the initial deceleration is $4 \, \text{m/s}^2$, find:
(a) the value of V (b) the value of T.

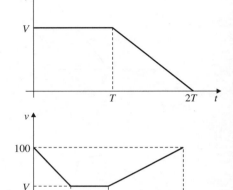

7.4 Quadratic equations

Factorising quadratic expressions

Earlier in the book a pair of brackets like $(x + 4)(x - 3)$ were multiplied to give $x^2 + x - 12$.

The reverse of this process is called factorising.

Factorise $x^2 + 6x + 8$

(a) Find two numbers which multiply to give 8 and add up to 6.

(b) Put these numbers into brackets.
So $x^2 + 6x + 8 = (x + 4)(x + 2)$

Factorise (a) $x^2 + 2x - 15$
           (b) $x^2 - 6x + 8$

(a) Two numbers which multiply to give -15 and add up to $+2$ are -3 and 5.
$\therefore \quad x^2 + 2x - 15 = (x - 3)(x + 5)$

(b) Two numbers which multiply to give $+8$ and add up to -6 are -2 and -4.
$\therefore \quad x^2 - 6x + 8 = (x - 2)(x - 4)$

Exercise 4

Factorise the following:

1. $x^2 + 7x + 10$	**2.** $x^2 + 7x + 12$	**3.** $x^2 + 8x + 15$
4. $x^2 + 10x + 21$	**5.** $x^2 + 8x + 12$	**6.** $y^2 + 12y + 35$
7. $y^2 + 11y + 24$	**8.** $y^2 + 10y + 25$	**9.** $y^2 + 15y + 36$
10. $a^2 - 3a - 10$	**11.** $a^2 - a - 12$	**12.** $z^2 + z - 6$
13. $x^2 - 2x - 35$	**14.** $x^2 - 5x - 24$	**15.** $x^2 - 6x + 8$
16. $y^2 - 5y + 6$	**17.** $x^2 - 8x + 15$	**18.** $a^2 - a - 6$
19. $a^2 + 14a + 45$	**20.** $b^2 - 4b - 21$	**21.** $x^2 - 8x + 16$
22. $y^2 + 2y + 1$	**23.** $y^2 - 3y - 28$	**24.** $x^2 - x - 20$
25. $x^2 - 8x - 240$	**26.** $x^2 - 26x + 165$	**27.** $y^2 + 3y - 108$
28. $x^2 - 49$	**29.** $x^2 - 9$	**30.** $x^2 - 16$

31. The terms in the expression $2x^2 + 12x + 16$ have a common factor of 2. So $2x^2 + 12x + 16 = 2(x^2 + 6x + 8)$. Complete the factorisation.

32. Factorise:
(a) $2x^2 + 4x - 30$ (b) $3x^2 + 21x + 30$ (c) $3x^2 + 24x + 45$
(d) $2n^2 - 6n - 20$ (e) $5a^2 + 5a - 30$ (f) $4x^2 - 64$

Factorise $3x^2 + 13x + 4$

(a) Find two numbers which multiply to give (3×4), i.e. 12, and add up to 13. In this case the numbers are 1 and 12.

(b) Split the '13x' term, $3x^2 + x + 12x + 4$

(c) Factorise in pairs, $x(3x + 1) + 4(3x + 1)$

(d) $(3x + 1)$ is common, $(3x + 1)(x + 4)$

Exercise 5

Factorise the following:

1. $2x^2 + 5x + 3$	**2.** $2x^2 + 7x + 3$	**3.** $3x^2 + 7x + 2$	**4.** $2x^2 + 11x + 12$
5. $3x^2 + 8x + 4$	**6.** $2x^2 + 7x + 5$	**7.** $3x^2 - 5x - 2$	**8.** $2x^2 - x - 15$
9. $2x^2 + x - 21$	**10.** $3x^2 - 17x - 28$	**11.** $6x^2 + 7x + 2$	**12.** $3x^2 - 11x + 6$
13. $3y^2 - 11y + 10$	**14.** $6y^2 + 7y - 3$	**15.** $10x^2 + 9x + 2$	**16.** $6x^2 - 19x + 3$
17. $8x^2 - 10x - 3$	**18.** $12x^2 + 23x + 10$	**19.** $4y^2 - 23y + 15$	**20.** $6x^2 - 27x + 30$

The difference of two squares

$x^2 - y^2 = (x - y)(x + y)$
Remember this result.

Factorise (a) $y^2 - 16$ (b) $4a^2 - b^2$

(a) $y^2 - 16 = (y - 4)(y + 4)$ (b) $4a^2 - b^2 = (2a - b)(2a + b)$

Exercise 6

Factorise the following:

1. $y^2 - a^2$ **2.** $m^2 - n^2$ **3.** $x^2 - t^2$ **4.** $y^2 - 1$

5. $x^2 - 9$ **6.** $a^2 - 25$ **7.** $x^2 - \dfrac{1}{4}$ **8.** $x^2 - \dfrac{1}{9}$

9. $4x^2 - y^2$ **10.** $a^2 - 4b^2$ **11.** $25x^2 - 4y^2$ **12.** $9x^2 - 16y^2$

13. $4x^2 - \dfrac{z^2}{100}$ **14.** $x^3 - x$ **15.** $a^3 - ab^2$ **16.** $4x^3 - x$

17. $8x^3 - 2xy^2$ **18.** $y^3 - 9y$

19. Find the exact value of $100\,003^2 - 99\,997^2$.

20. Find the exact value of $1\,500\,002^2 - 1\,499\,998^2$.

21. Rewrite 9991 as the difference of two squares. Use your answer to find the prime factors of 9991.

Methods of solving quadratic equations

The last three exercises have been about factorising expressions. These contain no 'equals' sign, but equations do.

Quadratic equations always have an x^2 term, and often an x term as well as a number term. They generally have two different solutions. Here three different methods of solution are considered.

(a) Solution by factors

Consider the equation $a \times b = 0$, where a and b are numbers. The product $a \times b$ can only be zero if either a or b (or both) is equal to zero. Can you think of other possible pairs of numbers which multiply together to give zero?

(a) Solve the equation $x^2 + x - 12 = 0$

Factorising, $(x - 3)(x + 4) = 0$

either $x - 3 = 0$ or $x + 4 = 0$

          $x = 3$ $x = -4$

(b) Solve the equation $6x^2 + x - 2 = 0$

Factorising, $(2x - 1)(3x + 2) = 0$

either $2x - 1 = 0$ or $3x + 2 = 0$

          $2x = 1$ $3x = -2$

          $x = \frac{1}{2}$ $x = -\frac{2}{3}$

Exercise 7

Solve the following equations.

1. $x^2 + 7x + 12 = 0$ **2.** $x^2 + 7x + 10 = 0$ **3.** $x^2 + 2x - 15 = 0$

4. $x^2 + x - 6 = 0$ **5.** $x^2 - 8x + 12 = 0$ **6.** $x^2 + 10x + 21 = 0$

7. $x^2 - 5x + 6 = 0$ **8.** $x^2 - 4x - 5 = 0$ **9.** $x^2 + 5x - 14 = 0$

10. $2x^2 - 3x - 2 = 0$ **11.** $3x^2 + 10x - 8 = 0$ **12.** $2x^2 + 7x - 15 = 0$

13. $6x^2 - 13x + 6 = 0$ **14.** $4x^2 - 29x + 7 = 0$ **15.** $10x^2 - x - 3 = 0$

16. $y^2 - 15y + 56 = 0$ **17.** $12y^2 - 16y + 5 = 0$ **18.** $y^2 + 2y - 63 = 0$

19. $x^2 + 2x + 1 = 0$ **20.** $x^2 - 6x + 9 = 0$ **21.** $x^2 + 10x + 25 = 0$

22. $x^2 - 14x + 49 = 0$

23. $6a^2 - a - 1 = 0$

24. $4a^2 - 3a - 10 = 0$

25. $z^2 - 8z - 65 = 0$

26. $6x^2 + 17x - 3 = 0$

27. $10k^2 + 19k - 2 = 0$

28. $y^2 - 2y + 1 = 0$

29. $36x^2 + x - 2 = 0$

30. $20x^2 - 7x - 3 = 0$

†**31.** $x^4 - 5x^2 + 4 = 0$
  [Hint: Put $y = x^2$]

†**32.** $x^4 - 13x^2 + 36 = 0$

†**33.** $4x^4 - 17x^2 + 4 = 0$

†**34.** $x^6 - 9x^3 + 8 = 0$

Quadratics with only two terms

(a) Solve the equation $x^2 - 7x = 0$

  Factorising, $x(x - 7) = 0$

  either $x = 0$ or $x - 7 = 0$
                             $x = 7$

  The solutions are $x = 0$ and $x = 7$.

(b) Solve the equation $x^2 - 100 = 0$
  Rearranging, $x^2 = 100$
  Take square root, $x = 10$ or -10
  [Alternatively use the difference of squares.]

Exercise 8

Solve the following equations.

1. $x^2 - 3x = 0$

2. $x^2 + 7x = 0$

3. $2x^2 - 2x = 0$

4. $3x^2 - x = 0$

5. $x^2 - 16 = 0$

6. $x^2 - 49 = 0$

7. $4x^2 - 1 = 0$

8. $9x^2 - 4 = 0$

9. $6y^2 + 9y = 0$

10. $6a^2 - 9a = 0$

11. $10x^2 - 55x = 0$

12. $16x^2 - 1 = 0$

13. $y^2 - \frac{1}{4} = 0$

14. $56x^2 - 35x = 0$

15. $36x^2 - 3x = 0$

16. $x^2 = 6x$

17. $x^2 = 11x$

18. $2x^2 = 3x$

19. $x^2 = x$

20. $4x = x^2$

(b) Solution by formula

• The solutions of the quadratic equation $ax^2 + bx + c = 0$ are given
  by the formula $x = \dfrac{-b \pm \sqrt{(b^2 - 4ac)}}{2a}$.

Use this formula only after trying (and failing) to factorise.

Solve the equation $2x^2 - 3x - 4 = 0$.

In this case $a = 2$, $b = -3$, $c = -4$.

$$x = \frac{-(-3) \pm \sqrt{[(-3)^2 - (4 \times 2 \times -4)]}}{2 \times 2}$$

$$x = \frac{3 \pm \sqrt{[9 + 32]}}{4} = \frac{3 \pm \sqrt{41}}{4}$$

$$x = \frac{3 \pm 6\cdot403}{4}$$

either $x = \dfrac{3 + 6\cdot403}{4} = 2\cdot35$ (2 decimal places)

or $x = \dfrac{3 - 6\cdot403}{4} = \dfrac{-3\cdot403}{4} = -0\cdot85$ (2 decimal places).

Exercise 9

Solve the following, giving answers to two decimal places where necessary.

1. $2x^2 + 11x + 5 = 0$ 2. $3x^2 + 11x + 6 = 0$ 3. $6x^2 + 7x + 2 = 0$

4. $3x^2 - 10x + 3 = 0$ 5. $5x^2 - 7x + 2 = 0$ 6. $6x^2 - 11x + 3 = 0$

7. $2x^2 + 6x + 3 = 0$ 8. $x^2 + 4x + 1 = 0$ 9. $5x^2 - 5x + 1 = 0$

10. $x^2 - 7x + 2 = 0$ 11. $2x^2 + 5x - 1 = 0$ 12. $3x^2 + x - 3 = 0$

13. $3x^2 + 8x - 6 = 0$ 14. $3x^2 - 7x - 20 = 0$ 15. $2x^2 - 7x - 15 = 0$

16. $x^2 - 3x - 2 = 0$ 17. $2x^2 + 6x - 1 = 0$ 18. $6x^2 - 11x - 7 = 0$

19. $3x^2 + 25x + 8 = 0$ 20. $3y^2 - 2y - 5 = 0$ 21. $2y^2 - 5y + 1 = 0$

22. $\frac{1}{2}y^2 + 3y + 1 = 0$ 23. $2 - x - 6x^2 = 0$ 24. $3 + 4x - 2x^2 = 0$

25. (a) To find the x-coordinate of A and B, solve the equation

$$x^2 - 5x + 1 = 0$$

(b) Explain why the equation $x^2 - 5x + 10 = 0$ has no solutions.

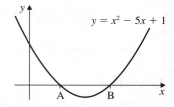

†26. Copy and complete the sentence: 'The equation $ax^2 + bx + c = 0$ has solutions if $(b^2 - 4ac)$ is _____.'

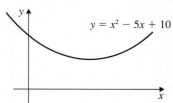

The solution to a problem can involve an equation which does not at first appear to be quadratic. The terms in the equation may need to be rearranged as shown below.

$$
\begin{aligned}
\text{Solve:} \qquad 2x(x - 1) &= (x + 1)^2 - 5 \\
2x^2 - 2x &= x^2 + 2x + 1 - 5 \\
2x^2 - 2x - x^2 - 2x - 1 + 5 &= 0 \\
x^2 - 4x + 4 &= 0 \\
(x - 2)(x - 2) &= 0 \\
x &= 2
\end{aligned}
$$

In this example the quadratic has a repeated root of $x = 2$.

Exercise 10

Solve the following, giving answers to two decimal places where necessary.

1. $x^2 = 6 - x$ 2. $x(x + 10) = -21$

3. $3x + 2 = 2x^2$ 4. $x^2 + 4 = 5x$

5. $6x(x + 1) = 5 - x$ 6. $(2x)^2 = x(x - 14) - 5$

7. $(x - 3)^2 = 10$ 8. $(x + 1)^2 - 10 = 2x(x - 2)$

9. $(2x - 1)^2 = (x - 1)^2 + 8$ 10. $3x(x + 2) - x(x - 2) + 6 = 0$

11. $x = \dfrac{15}{x} - 22$

12. $x + 5 = \dfrac{14}{x}$

13. $4x + \dfrac{7}{x} = 29$

14. $10x = 1 + \dfrac{3}{x}$

15. $2x^2 = 7x$

16. $16 = \dfrac{1}{x^2}$

17. $2x + 2 = \dfrac{7}{x} - 1$

18. $\dfrac{2}{x} + \dfrac{2}{x+1} = 3$

19. $\dfrac{3}{x-1} + \dfrac{3}{x+1} = 4$

20. $\dfrac{2}{x-2} + \dfrac{4}{x+1} = 3$

21. Hassan says you can make it easier to solve some equations by dividing through by a common factor.

So, by dividing through by 2, the equation $2x^2 + 10x - 48 = 0$
can be written as $x^2 + 5x - 24 = 0$

Is this OK?
Check by solving the two equations above.
Are the solutions the same?

22. Solve the equations:

(a) $4x^2 - 10x - 6 = 0$

(b) $3x^2 - 6x - 72 = 0$

†**23.** One of the solutions published by Cardan in 1545 for the solution of cubic equations is given below. For an equation in the form $x^3 + px = q$

$$x = \sqrt[3]{\left[\sqrt{\left(\dfrac{p}{3}\right)^3 + \left(\dfrac{q}{2}\right)^2} + \dfrac{q}{2}\right]} - \sqrt[3]{\left[\sqrt{\left(\dfrac{p}{3}\right)^3 + \left(\dfrac{q}{2}\right)^2} - \dfrac{q}{2}\right]}$$

Use the formula to solve the following equations, giving answers to 4 sig. fig. where necessary.

(a) $x^3 + 7x = -8$

(b) $x^3 + 6x = 4$

(c) $x^3 + 3x = 2$

(d) $x^3 + 9x - 2 = 0$

Girolamo Cardan (1501–1576) was a colourful character who became Professor of Mathematics at Milan. As well as being a distinguished academic, he was an astrologer, a physician, a gambler and a heretic, yet he received a pension from the Pope. His mathematical genius enabled him to open up the general theory of cubic and quartic equations, although a method for solving cubic equations which he claimed as his own was pirated from Niccolo Tartaglia.

(c) Solution by completing the square

Look at the function $f(x) = x^2 + 6x$

Completing the square, this becomes $f(x) = (x + 3)^2 - 9$

This is done as follows.
1. 3 is half of 6 and gives $6x$.
2. Having added 3 to the square term, 9 needs to be subtracted from the expression to cancel the $+9$ obtained.

Here are some more examples.

(a) $x^2 - 12x = (x - 6)^2 - 36$

(b) $x^2 + 3x = (x + \frac{3}{2})^2 - \frac{9}{4}$

(c) $x^2 + 6x + 1 = (x + 3)^2 - 9 + 1$
$= (x + 3)^2 - 8$

d) $x^2 - 10x - 17 = (x - 5)^2 - 25 - 17$
$= (x - 5)^2 - 42$

(e) $2x^2 - 12x + 7 = 2[x^2 - 6x + \frac{7}{2}]$
$= 2[(x - 3)^2 - 9 + \frac{7}{2}]$
$= 2[(x - 3)^2 - \frac{11}{2}]$

Solve the quadratic equation $x^2 - 6x + 7 = 0$ by completing the square.

$(x - 3)^2 - 9 + 7 = 0$
$(x - 3)^2 \qquad = 2$
$\therefore \quad x - 3 = +\sqrt{2} \ \text{ or } \ -\sqrt{2}$
$\qquad = 3 + \sqrt{2} \ \text{ or } \ 3 - \sqrt{2}$
So, $\qquad x = 4{\cdot}41 \ \text{ or } \ 1{\cdot}59$ to 2 d.p.

Given $f(x) = x^2 - 8x + 18$, show that $f(x) \geqslant 2$ for all values of x.

Completing the square, $f(x) = (x - 4)^2 - 16 + 18$
$f(x) = (x - 4)^2 + 2.$

Now $(x - 4)^2$ is always greater than or equal to zero because it is 'something squared.'

$$\therefore \quad f(x) \geqslant 2$$

Exercise 11

In Questions **1** to **10**, complete the square for each expression by writing each one in the form $(x + a)^2 + b$ where a and b can be positive or negative.

1. $x^2 + 8x$ **2.** $x^2 - 12x$ **3.** $x^2 + x$

4. $x^2 + 4x + 1$ **5.** $x^2 - 6x + 9$ **6.** $x^2 + 2x - 15$

7. $x^2 + 16x + 5$ **8.** $x^2 - 10x$ **9.** $x^2 + 3x$

10. Solve the equations by completing the square.
(a) $x^2 - 8x + 12 = 0$ (b) $x^2 + 10x + 21 = 0$
(c) $x^2 - 4x - 5 = 0$

11. Solve these equations by completing the square.
(a) $x^2 + 4x - 3 = 0$ (b) $x^2 - 3x - 2 = 0$
(c) $x^2 + 12x = 1$

12. Try to solve the equation $x^2 + 6x + 10 = 0$, by completing the square. Explain why you can find no solutions.

13. Given $f(x) = x^2 + 6x + 12$, show that $f(x) \geqslant 3$ for all values of x.

14. Given $g(x) = x^2 - 7x + \frac{1}{4}$, show that the least possible value of $g(x)$ is -12.

15. If $f(x) = x^2 + 4x + 7$ find:
   (a) the smallest possible value of $f(x)$,
   (b) the value of x for which this smallest value occurs,
   (c) the greatest possible value of $\dfrac{1}{(x^2 + 4x + 7)}$.

†**16.** Given $y = x^2 - x + 1$, find:
   (a) the lowest value of y,
   (b) the value of x for which this lowest value occurs,
   (c) the greatest possible value of $\dfrac{1}{(x^2 - x + 1)}$.

†**17.** Simplify $(x + 4)(x + 2) - 2(x + 2)$, and explain why the expression can never be negative, whatever the value of x.

Using quadratic equations to solve problems

The perimeter of a rectangle is 42 cm. If the diagonal is 15 cm, find the width of the rectangle.

Let the width of the rectangle be x cm.
Since the perimeter is 42 cm, the sum of the length and the width is 21 cm.
∴   length of rectangle $= (21 - x)$ cm

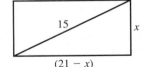

$(21 - x)$

By Pythagoras' theorem

$$x^2 + (21 - x)^2 = 15^2$$
$$x^2 + (21 - x)(21 - x) = 15^2$$
$$x^2 + 441 - 42x + x^2 = 225$$
$$2x^2 - 42x + 216 = 0$$
$$x^2 - 21x + 108 = 0$$
$$(x - 12)(x - 9) = 0$$
$$x = 12$$
$$\text{or} \quad x = 9$$

Note that the dimensions of the rectangle are 9 cm and 12 cm, whichever value of x is taken.
∴   The width of the rectangle is 9 cm.

Exercise 12

Solve by forming a quadratic equation.

1. Two numbers, which differ by 3, have a product of 88. Find them.
   [Call the numbers x and $x + 3$.]

2. The product of two consecutive odd numbers is 143. Find the
   numbers. [Hint: If the first odd number is x, what is the next odd
   number?]

3. The height of a photo exceeds the width by 7 cm. If the area is
   60 cm^2, find the height of the photo.

4. The length of a rectangle exceeds the width by 2 cm. If the diagonal
   is 10 cm long, find the width of the rectangle.

5. The area of the rectangle exceeds the area of
   the square by 24 m^2. Find x.

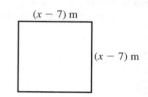

6. Three consecutive integers are written as x, $x + 1$, $x + 2$.
   The square of the largest number is 45 less than the sum of the
   squares of the other numbers. Find the three numbers.

7. $(x - 1)$, x and $(x + 1)$ represent three positive integers.
   The product of the three numbers is five times their sum.
   (a) Write an equation in x.
   (b) Show that your equation simplifies to $x^3 - 16x = 0$
   (c) Factorise $x^3 - 16x$ completely.
   (d) Hence find the three positive integers.

8. An aircraft flies a certain distance on a bearing of 045° and then twice
   the distance on a bearing of 135°. Its distance from the starting point
   is then 350 km. Find the length of the first part of the journey.

9. The area of rectangle A is twice the area of rectangle B.
   Find x.

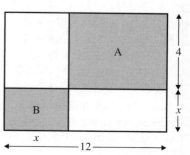

10. The perimeter of a rectangle is 68 cm. If the diagonal is 26 cm, find the dimensions of the rectangle.

11. A stone is thrown in the air. After t seconds its height, h, above sea level is given by the formula $h = 80 + 3t - 5t^2$. Find the value of t when the stone falls into the sea.

12. The total surface area of a cylinder, A, is given by the formula
$A = 2\pi r^2 + 2\pi rh$.
Given that $A = 200 \, \text{cm}^2$ and $h = 10 \, \text{cm}$, find the value of r, correct to 1 d.p.

13. A rectangular pond, 6 m × 4 m, is surrounded by a uniform path of width x. The area of the path is equal to the area of the pond. Find x.

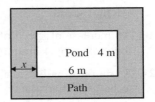

14. The perimeters of a square and a rectangle are equal. The length of the rectangle is 11 cm and the area of the square is 4 cm² more than the area of the rectangle. Find the side of the square.

15. The sequence 3, 6, 15, 24, ... can be written (1×3), (2×4), (3×5), (4×6)... Write down an expression for the nth term of the sequence. One term in the sequence is 255. Form an equation and hence find what number term it is.

16. In figure 1, ABCD is a rectangle with AB = 12 cm and BC = 7 cm. AK = BL = CM = DN = x cm. If the area of KLMN is 54 cm² find x.

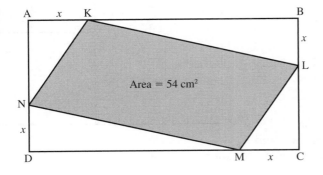

Fig. 1

17. In figure 1, AB = 14 cm BC = 11 cm and AK = BL = CM = DN = x cm. If the area of KLMN is now 97 cm², find x.

18. When each edge of a cube is decreased by 1 cm, its volume is decreased by 91 cm^3. Find the length of a side of the original cube.

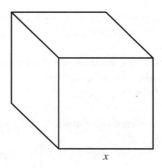 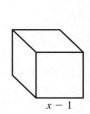

x $x - 1$

19. A cyclist travels 40 km at a speed x km/h. Find the time taken in terms of x. Find the time taken when his speed is reduced by 2 km/h. If the difference between the times is 1 hour, find the original speed.

20. An increase of speed of 4 km/h on a journey of 32 km reduces the time taken by 4 hours. Find the original speed.

21. A train normally travels 60 miles at a certain speed. One day, due to bad weather, the train's speed is reduced by 10 mph so that the journey takes 3 hours longer. Find the normal speed.

22. A number exceeds four times its reciprocal by 3. Find the number.

23. Two numbers differ by 3. The sum of their reciprocals is $\frac{7}{10}$; find the numbers.

24. The numerator of a fraction is 1 less than the denominator. When both numerator and denominator are increased by 2, the fraction is increased by $\frac{1}{12}$. Find the original fraction.

25. A lot of paper comes in standard sizes A0, A1, A2, A3 etc. The large sheet shown oposite is size A0; A0 can be cut in half to give two sheets of A1; A1 can be cut in half to give two sheets of A2 and so on.

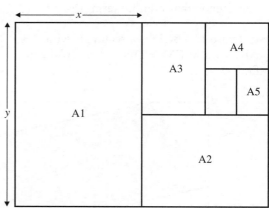

Call the sides of A1 x and y, as shown.
(a) What are the sides of A0 in terms of x and y?
(b) All the sizes of paper are similar. Form an equation involving x and y.
(c) Size A0 has an area of 1 m^2. Calculate the values of x and y correct to the nearest mm.
(d) Hence calculate the long side of a sheet of A4. Check your answer by measuring a sheet of A4.

7.5 Algebraic fractions

(a) (i) $\dfrac{3a}{5a^2} = \dfrac{3 \times \cancel{a}}{5 \times a \times \cancel{a}} = \dfrac{3}{5a}$ (ii) $\dfrac{3y + y^2}{6y} = \dfrac{y(3 + y)}{6y}$

$$= \dfrac{3 + y}{6}$$

(b) Write as a single fraction: (i) $\dfrac{4\cancel{x}}{3} \times \dfrac{2}{x^{\cancel{2}}}$ (ii) $\dfrac{5}{n} \div \dfrac{3}{n}$

$$= \dfrac{8}{3x} \qquad\qquad = \dfrac{5}{\cancel{n}} \times \dfrac{\cancel{n}}{3} = \dfrac{5}{3}$$

Exercise 13

1. Simplify as far as possible.

(a) $\dfrac{25}{35}$ (b) $\dfrac{5y^2}{y}$ (c) $\dfrac{y}{2y}$ (d) $\dfrac{8x^2}{2x^2}$

(e) $\dfrac{2x}{4y}$ (f) $\dfrac{6y}{3y}$ (g) $\dfrac{5ab}{10b}$ (h) $\dfrac{8ab^2}{12ab}$

2. Sort these into four pairs of equivalent expressions.

A $\dfrac{x^2}{2x}$ B $\dfrac{x(x+1)}{x^2}$ C $\dfrac{x^2 + x}{x^2 - x}$ D $\dfrac{3x + 6}{3x}$

E $\dfrac{x+1}{x}$ F $\dfrac{x+2}{x}$ G $\dfrac{x}{2}$ H $\dfrac{x+1}{x-1}$

3. Write as a single fraction.

(a) $\dfrac{3x}{2} \times \dfrac{2a}{3x}$ (b) $\dfrac{5mn}{3} \times \dfrac{2}{n}$ (c) $\dfrac{3y^2}{3} \times \dfrac{2x}{9y}$ (d) $\dfrac{2}{q} \div \dfrac{a}{2}$

(e) $\dfrac{4x}{3} \div \dfrac{x}{2}$ (f) $\dfrac{x}{5} \times \dfrac{y^2}{x^2}$ (g) $\dfrac{a^2}{5} \div \dfrac{a}{10}$ (h) $\dfrac{x^2}{x^2 + 2x} \div \dfrac{x}{x + 2}$

4. Simplify as far as possible.

(a) $\dfrac{7a^2b}{35ab^2}$ (b) $\dfrac{(2a)^2}{4a}$ (c) $\dfrac{7yx}{8xy}$ (d) $\dfrac{3x}{4x - x^2}$

(e) $\dfrac{5x + 2x^2}{3x}$ (f) $\dfrac{9x + 3}{3x}$ (g) $\dfrac{4a + 5a^2}{5a}$ (h) $\dfrac{5ab}{15a + 10a^2}$

5. Copy and complete.

(a) $\dfrac{x^2}{3} \times \dfrac{\square}{x} = 2x$ (b) $\dfrac{8}{x} \div \dfrac{2}{x} = \square$ (c) $\dfrac{a}{3} + \dfrac{a}{3} = \dfrac{\square}{3}$

6. Sort these into four pairs of equivalent expressions.

A $\dfrac{x^2}{3x}$ B $\dfrac{x}{2} \times \dfrac{x}{2}$ C $\dfrac{12x + 6}{6}$ D $\dfrac{x}{5} - \dfrac{2}{5}$

E $\dfrac{2x^2 + x}{x}$ F $\dfrac{x(x+1)}{3x + 3}$ G $\dfrac{x - 2}{5}$ H $\dfrac{ax^2}{4a}$

7. Write in a more simple form.

(a) $\dfrac{5x + 10y}{15xy}$ (b) $\dfrac{18a - 3ab}{6a^2}$ (c) $\dfrac{4ab + 8a^2}{2ab}$ (d) $\dfrac{(2x)^2 - 8x}{4x}$

8. Simplify as far as possible.

(a) $\dfrac{x^2 + 2x}{x^2 - 3x}$ (b) $\dfrac{x^2 - 3x}{x^2 - 2x - 3}$ (c) $\dfrac{x^2 + 4x}{2x^2 - 10x}$

(d) $\dfrac{x^2 + 6x + 5}{x^2 - x - 2}$ (e) $\dfrac{x^2 - 4x - 21}{x^2 - 5x - 14}$ (f) $\dfrac{x^2 + 7x + 10}{x^2 - 4}$

Addition and subtraction of algebraic fractions

(a) Write as a single fraction $\dfrac{2}{x} + \dfrac{3}{y}$

The L.C.M. of x and y is xy.

$$\therefore \quad \frac{2}{x} + \frac{3}{y} = \frac{2y}{xy} + \frac{3x}{xy} = \frac{2y + 3x}{xy}$$

(b) Write as a single fraction $\dfrac{4}{x} + \dfrac{5}{x - 1}$

The L.C.M. of x and $(x - 1)$ is $x(x - 1)$

$$\therefore \quad \frac{4}{x} + \frac{5}{x - 1} = \frac{4(x - 1) + 5x}{x(x - 1)}$$

$$= \frac{9x - 4}{x(x - 1)}$$

Exercise 14

1. Write as a single fraction.

(a) $\dfrac{2x}{5} + \dfrac{x}{5}$ (b) $\dfrac{2}{x} + \dfrac{1}{x}$ (c) $\dfrac{x}{7} + \dfrac{3x}{7}$ (d) $\dfrac{1}{7x} + \dfrac{3}{7x}$

(e) $\dfrac{5x}{8} + \dfrac{x}{4}$ (f) $\dfrac{5}{8x} + \dfrac{1}{4x}$ (g) $\dfrac{2x}{3} + \dfrac{x}{6}$ (h) $\dfrac{2}{3x} + \dfrac{1}{6x}$

2. Sort into four pairs of equivalent fractions.

A $\dfrac{x}{2} + \dfrac{x}{4}$ B $\dfrac{2}{x} + \dfrac{2}{x}$ C $\dfrac{9x}{8} - \dfrac{x}{4}$ D $\dfrac{7x}{8}$

E $\dfrac{5x}{x^2} - \dfrac{2}{x}$ F $\dfrac{3x}{4}$ G $\dfrac{3}{x}$ H $\dfrac{4}{x}$

3. Simplify.

(a) $\dfrac{3x}{4} + \dfrac{2x}{5}$ (b) $\dfrac{3}{4x} + \dfrac{2}{5x}$ (c) $\dfrac{3x}{4} - \dfrac{2x}{3}$ (d) $\dfrac{3}{4x} - \dfrac{2}{3x}$

(e) $\dfrac{x}{2} + \dfrac{x + 1}{3}$ (f) $\dfrac{x - 1}{3} + \dfrac{x + 2}{4}$

4. Work out these subtractions.

(a) $\dfrac{x + 1}{3} - \dfrac{(2x + 1)}{4}$ (b) $\dfrac{x - 3}{3} - \dfrac{(x - 2)}{5}$ (c) $\dfrac{(x + 2)}{2} - \dfrac{(2x + 1)}{7}$

5. Copy and complete.

(a) $\dfrac{x}{5} - \dfrac{\square}{\square} = \dfrac{x}{10}$ (b) $\dfrac{\square}{2} + \dfrac{x}{4} = \dfrac{7x}{4}$ (c) $\dfrac{3}{2x} - \dfrac{1}{8x} = \dfrac{\square}{8x}$

6. Write as a single fraction. [Hint: Use brackets freely.]

(a) $\dfrac{1}{x} + \dfrac{2}{x+1}$ (b) $\dfrac{3}{x-2} + \dfrac{4}{x}$ (c) $\dfrac{5}{x-2} + \dfrac{3}{x+3}$

(d) $\dfrac{7}{x+1} - \dfrac{3}{x+2}$ (e) $\dfrac{2}{x+3} - \dfrac{5}{x-1}$ (f) $\dfrac{3}{x-2} - \dfrac{4}{x+1}$

7. The perimeter of the rectangle shown is 24 units. Form an equation and solve it to find x.

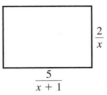

$\dfrac{2}{x}$

$\dfrac{5}{x+1}$

8. A rectangle measuring $\dfrac{3}{x}$ by $\dfrac{6}{x+1}$ has an area of 75 square units. Find x.

7.6 Transformation of curves

The notation $f(x)$ means 'function of x'. A function of x is an expression which (usually) varies, depending on the value of x. Examples of functions are:

$$f(x) = x^2 + 3; \quad f(x) = \dfrac{1}{x} + 7; \quad f(x) = \sin x.$$

We can imagine a box which performs the function f on any input.

If $f(x) = x^2 + 7x + 2, \quad f(3) = 3^2 + 7 \times 3 + 2 = 32$

Here are *four* ways in which any function $f(x)$ can be transformed.

$$f(x) + a; \qquad f(x-a); \qquad af(x); \qquad f(ax).$$

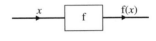

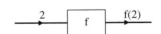

(a) $y = f(x) + a$

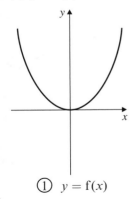

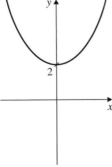

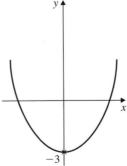

① $y = f(x)$ ② $y = f(x) + 2$ ③ $y = f(x) - 3$

- Can you describe the transformation from ① to ②?
- Can you describe the transformation from ① to ③?

(b) $y = f(x - a)$

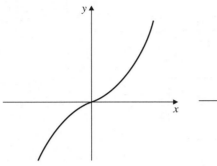

① $y = f(x)$ ② $y = f(x - 2)$ ③ $y = f(x + 1)$

- Can you describe the transformation from ① to ②?
- Can you describe the transformation from ① to ③?

(c) $y = af(x)$

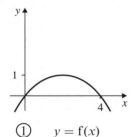 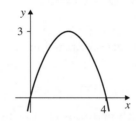 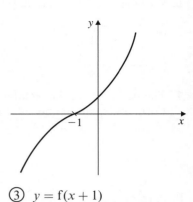 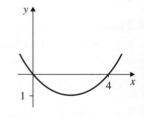

① $y = f(x)$ ② $y = 2f(x)$ ③ $y = 3f(x)$ ④ $y = -f(x)$

- Describe the transformations: from ① to ②
  from ① to ③
  from ① to ④

(d) $y = f(ax)$

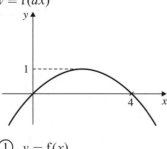 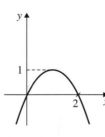

① $y = f(x)$ ② $y = f(2x)$ ③ $y = f(\frac{1}{2}x)$

- Can you describe the transformation from ① to ②?
- Can you describe the transformation from ① to ③?

This is the most difficult one.

From ① to ② the transformation is a stretch parallel to the x-axis by a scale factor $\frac{1}{2}$.

From ②to ①the transformation is a stretch parallel to the x-axis by a scale factor 2.

Similarly $f(3x)$ would be a stretch with scale factor $\frac{1}{3}$ and $f(\frac{1}{5}x)$ would be a stretch with scale factor 5.

Summary of rules for curve transformations

(a) $y = f(x) + a$: Translation by a units parallel to the y-axis.

(b) $y = f(x - a)$: Translation by a units parallel to the x-axis.
      (note the negative sign).

(c) $y = af(x)$: Stretch parallel to the y-axis by a scale factor a.

(d) $y = f(ax)$: Stretch parallel to the x-axis by a scale factor $\dfrac{1}{a}$.

      (note the inverse of a).

From the sketch of $y = f(x)$, draw

(a) $y = f(x + 3)$,
(b) $y = f(3x)$

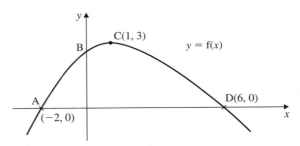

(a) This is a translation of -3 units parallel to the x-axis.

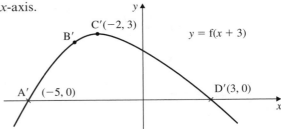

(b) This is a stretch parallel to the x-axis by a scale factor $\frac{1}{3}$.
      Notice that B remains in the same place and that the
      y-coordinate of C remains unchanged.

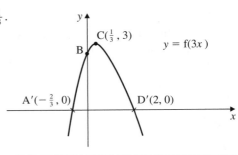

Exercise 15

A computer or calculator which sketches curves can be used effectively in this exercise, although it is not essential.

1. (a) Draw an accurate graph of $y = x^2$ for values of x from -3 to 3. Label the graph $y = f(x)$
   (b) On the same axes draw an accurate graph of $y = x^2 + 4$ and label the graph $y = f(x) + 4$.
   (c) On the same axes draw the graph of $y = (x - 1)^2$ and label the graph $y = f(x - 1)$.
   Scales: x from -3 to $+3$, 2 cm $= 1$ unit
   y from   0 to 14, 1 cm $= 1$ unit.

2. This is the sketch graph of $y = f(x)$.
   (a) Sketch the graph of $y = f(x) + 3$
   (b) Sketch the graph of $y = f(x + 1)$
   (c) Sketch the graph of $y = -f(x)$.

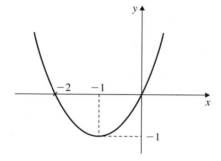

3. This is the sketch graph of $y = f(x)$

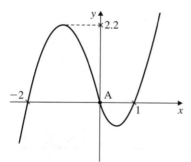

   (a) Sketch $y = f(x) - 2$
   (b) Sketch $y = f(x - 7)$
   Give the new coordinates of the point A on the two sketches.

4. This is the sketch of $y = f(x)$ which passes through A,B,C.

   Sketch the following curves, giving the new coordinates of A,B,C in each case.

   (a) $y = -f(x)$
   (b) $y = f(x - 2)$
   (c) $y = f(2x)$

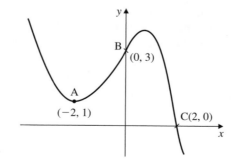

In Questions **5** to **16** each graph shows a different function $f(x)$. On squared paper draw a sketch to show the given transformation. The scales are 1 square = 1 unit on both axes.

5.

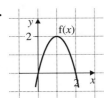

Sketch $f(x) + 3$.

6.

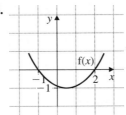

Sketch $f(x - 1)$.

7.

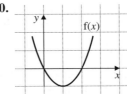

Sketch $2f(x)$.

8.

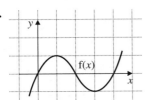

Sketch $f(x + 2)$.

9.

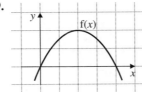

Sketch $f(2x)$.

10.

Sketch $f\left(\dfrac{x}{2}\right)$.

11.

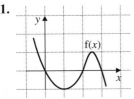

Sketch $3f(x)$.

12.

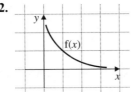

Sketch $f(x) - 2$.

13.

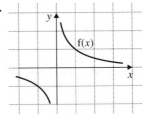

Sketch $f(x - 2)$.

14.

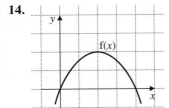

Sketch $f(4x)$.

15.

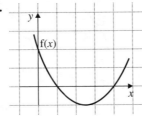

Sketch $2f(x)$.

16.

Sketch $-f(x)$.

Exercise 16

1. On the same axes, sketch and label the graphs of:
   (a) $y = x^2$ (b) $y = x^2 - 4$ (c) $y = x^2 + 2$

2. On the same axes, sketch and label:
   (a) $y = x^2$ (b) $y = (x - 3)^2$ (c) $y = (x + 2)^2$

3. On the same axes, sketch and label:
   (a) $y = x^2$ (b) $y = 4x^2$ (c) $y = \frac{1}{2}x^2$

4. On the same axes, sketch and label:
   (a) $y = x^3$ (b) $y = (x-1)^3$ (c) $y = (x-1)^3 + 4$

5. On the same axes, sketch and label:
   (a) $y = x(x-1)$ (b) $y = 2x(2x-1)$

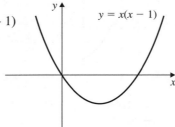

6. On the same axes, sketch and label:
   (a) $y = \dfrac{1}{x}$ (b) $y = \dfrac{1}{x-2}$

7. On squared paper copy the sketch of $y = \cos x$.
   Using different colours, sketch:
   (a) $y = \cos(x + 90°)$
   (b) $y = 2\cos x$
   (c) $y = \cos 2x$

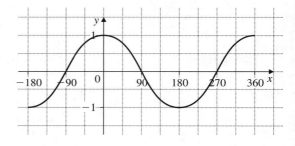

8. On squared paper copy the sketch graphs of $y = \sin x$ and $y = \tan x$.

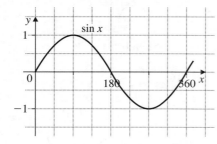 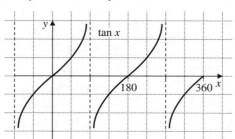

Using different colours, sketch the graphs of:

(a) $y = \sin\dfrac{x}{2}$ (b) $y = \tan(x - 90°)$ (c) $y = 2\sin x$ (d) $y = \tan 2x$

9. (a) Draw an accurate graph of $y = (x + 2)^2$ for values of x from -5 to 1. Label the graph $y = f(x)$.
   (b) Draw an accurate graph of $y = (\frac{1}{2}x + 2)^2$ for values of x from -9 to 1. Label the graph $y = f(\frac{1}{2}x)$.
   Scales: x from -10 to 1, 1 cm $= 1$ unit
            y from   0 to 10, 1 cm $= 1$ unit.
   Describe the transformation from $y = (x + 2)^2$ onto $y = (\frac{1}{2}x + 2)^2$.

10. It is possible to perform two or more successive transformations on the same curve. This is a sketch of $y = f(x)$.

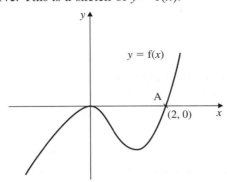

(a) Sketch $y = f(x + 1) + 5$.
(b) Sketch $y = f(x - 3) - 4$.

Show the new coordinates of the point A on each sketch.

†**11.** Find the equation of the curve obtained when the graph of $y = x^2 + 3x$ is:
(a) translated 5 units in the direction ↑
(b) translated 2 units in the direction →
(c) reflected in the x-axis.

†**12.** $f(x) = x^2$ and $g(x) = x^2 - 4x + 7$.
(a) If $g(x) = f(x - a) + b$, find the values of a and b.
(b) Hence sketch the graphs of $y = f(x)$ and $y = g(x)$ showing the transformation from f to g.

†**13.** Here are the sketches of $y = f(x)$ and $y = g(x)$.

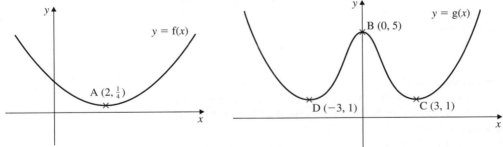

(a) Sketch $y = \dfrac{1}{f(x)}$, showing the new coordinates of A.

(b) Sketch $y = \dfrac{1}{g(x)}$, showing the new coordinates of B, C, D.

8 HANDLING DATA

8.1 Averages and range

A single number called an average can be used to represent the whole range of a set of data, whether the data is discrete (e.g. exam marks) or continuous (e.g. heights).

(a) The median

The data is arranged in order from the smallest to the largest; the middle number is then selected. This is really the central number of the range and is called the median.
If there are two 'middle' numbers, the median is in the middle of these two numbers.

(b) The mean

All the data is added up and the total divided by the number of items. This is called the mean and is equivalent to sharing out all the data evenly.

(c) The mode

The item which occurs most frequently in a frequency table is selected. This is the most popular value and is called the mode (from the French 'a la mode' meaning 'fashionable').

Each 'average' has its purpose and sometimes one is preferable to the other.

The median is fairly easy to find and has an advantage in being hardly affected by untypical values such as very large or very small values that occur at the ends of a distribution.
Consider these examination marks.

20, 21, 21, 22, 23, 23, 25, 27, 27, 27, 29, 98, 98
                                  ↑

Clearly the median value is 25.
The mean of the above data is 35.5. It is easier to use in further work such as standard deviation (page 350), but clearly it does not give a true picture of the centre of distribution of the data.
The mode of this data is 27. It is easy to calculate and it eliminates some of the effects of extreme values. However it does have disadvantages, particularly in data which has two 'most popular' values, and it is not widely used.

Range

In addition to knowing the centre of a distribution, it is useful to know the range or spread of the data.

range = (largest value) − (smallest value)

For the examination marks, range = $98 - 20 = 78$.

Find the median, the mean, the mode and the range of this set of 10 numbers: 5, 4, 10, 3, 3, 4, 7, 4, 6, 5.

(a) Arrange the numbers in order of size to find the median.

3,   3,   4,   4,   4,   5,   5,   6,   7,   10

            ↑

the median is the 'average' of 4 and 5.
∴   median = 4·5

(b) mean = $\dfrac{(5 + 4 + 10 + 3 + 3 + 4 + 7 + 4 + 6 + 3)}{10} = \dfrac{51}{10} = 5·1$

(c) mode = 4 because there are more 4's then any other number
(d) range = 10 − 3 = 7

Exercise 1

1. Find the median, the mean, the mode and the range of the following sets of numbers:
(a) 3, 12, 4, 6, 8, 5, 4
(b) 7, 21, 2, 17, 3, 13, 7, 4, 9, 7, 9
(c) 12, 1, 10, 1, 9, 3, 4, 9, 7, 9
(d) 8, 0, 3, 3, 1, 7, 4, 1, 4, 4.

2. The range for the eight numbers shown is 40.
Find the *two* possible values of the missing number.

3. The mean weight of ten people in a lift is 70 kg. The weight limit for the lift is 1000 kg. Roughly how many more people can get into the lift?

4. There were ten cowboys in a saloon. The mean age of the men was 25 and the range of their ages was 6. Write each statement below and then write next to it whether it is true, possible or false.
(a) The youngest man was 18 years old.
(b) All the men were at least 20 years old.
(c) The oldest person was 4 years older than the youngest.
(d) Every man was between 20 and 26 years old.

5. The following are the salaries of 5 employees in a small business:
    Mr A : £22,500   Mr B : £17,900   Mr C : £21,400
    Mr D : £22,500   Mr E : £85,300.
(a) Find the mean, median and mode of their salaries.
(b) Which does *not* give a fair 'average'? Explain why in one sentence.

6. A farmer has 32 cattle to sell. Their weights in kg are

81	81	82	82	83	84	84	85
85	86	86	87	87	88	89	91
91	92	93	94	96	150	152	153
154	320	370	375	376	380	381	390

[Total weight = 5028 kg]

On the telephone to a potential buyer, the farmer describes the
cattle and says the 'average' weight is 'over 157 kg'.
(a) Find the mean weight and the median weight.
(b) Which 'average' has the farmer used to describe his animals?
    Does this average describe the cattle fairly?

7. A gardening magazine sells seedlings of a plant through the post
and claims that the average height of the plants after one year's
growth will be 85 cm. A sample of 24 of the plants were measured
after one year with the following results (in cm)

6	7	7	9	34	56	85	89
89	90	90	91	91	92	93	93
93	94	95	95	96	97	97	99

[The sum of the heights is 1788 cm.]

(a) Find the mean and the median height of the sample.
(b) Is the magazine's claim about average height justified?

8. The mean weight of five men is 76 kg. The weights of four of the
men are 72 kg, 74 kg, 75 kg and 81 kg. What is the weight of the
fifth man?

9. The mean length of 6 rods is 44·2 cm. The mean length of 5 of them
is 46 cm. How long is the sixth rod?

10. (a) The mean of 3, 7, 8, 10 and x is 6. Find x.
(b) The mean of 3, 3, 7, 8, 10, x and x is 7. Find x.

11. The mean height of 12 men is 1·70 m, and the mean height of 8
women is 1·60 m. Find:
(a) the total height of the 12 men,
(b) the total height of the 8 women,
(c) the mean height of the 20 men and women.

12. The total weight of 6 rugby players is 540 kg and the mean weight
of 14 ballet dancers is 40 kg. Find the mean weight of the group of
20 rugby players and ballet dancers.

13. Write down five numbers so that:
the mean is 6
the median is 5
the mode is 4.

⌐?¬ ⌐?¬ ⌐?¬ ⌐?¬ ⌐?¬

14. Find five numbers so that the mean, median, mode and range are
all 4.

15. The numbers 3, 5, 7, 8 and N are arranged in ascending order. If the mean of the numbers is equal to the median, find N.

16. The mean of 5 numbers is 11. The numbers are in the ratio $1:2:3:4:5$. Find the smallest number.

†**17.** The median of five consecutive integers is N.
   (a) Find the mean of the five numbers.
   (b) Find the mean and the median of the squares of the integers.
   (c) Find the difference between these values.

Moving average

Here is a list of the number of children absent from a school over a 15-day period.

Day	1	2	3	4	5	6	7	8	9	10	11	12	13	14	15
Number absent	10	7	4	12	6	3	5	8	9	11	13	8	10	7	5

At the end of each day the mean number absent over *the last five days* is calculated.

So at the end of day 5, the mean over the last five days is $\dfrac{10 + 7 + 4 + 12 + 6}{5} = 7 \cdot 8$

At the end of day 6, the mean over the last five days is $\dfrac{7 + 4 + 12 + 6 + 3}{5} = 6 \cdot 4$

At the end of day 7, the mean over the last five days is $\dfrac{4 + 12 + 6 + 3 + 5}{5} = 6$

This is an example of a moving average.

Exercise 2

1. Traders on the stock market use moving averages as a guide to the performance of a company's share price. Here are the share prices, in pence, of a company over 30 days.

21	24	27	22	25	26	27	23	24	24
28	27	28	26	25	23	25	26	23	25
22	21	19	19	20	19	21	23	24	23

   (a) What was the mean price over the first 10 days?
   (b) What was the mean price over the 10 day period from day 2 to day 11?
   (c) What was the mean price over the 10-day period from day 11 to day 20?

2. Here are the prices, in pence, of shares in 'Tiger Telecom' over a
period of 20 days.

8	7	11	10	9	7	9	11	14	15
15	14	16	15	13	16	14	10	12	13

At the end of each day the mean price over the last 5 days is
calculated. So the mean price at the end of day 5 is

$$\frac{8 + 7 + 11 + 10 + 9}{5} = 9p$$

The mean price on day 6 $= \dfrac{7 + 11 + 10 + 9 + 7}{5} = 8.8p$

Work out the moving average price of the shares in this way
up to day 20 and plot the results on a graph.

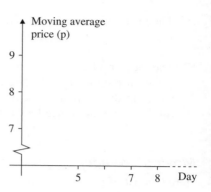

Frequency tables

A frequency table shows a number x such as a mark or a score, against
the frequency f or number of times that x occurs.
The next examples show how these symbols are used in calculating the
mean, the median and the mode.
The symbol Σ (or sigma) means 'the sum of'.

Discrete data
The marks obtained by 100 students in a test were as follows:

Mark (x)	0	1	2	3	4
Frequency (f)	4	19	25	29	23

Find:
(a) the mean mark   (b) the median mark   (c) the modal mark.

(a) Mean $= \dfrac{\Sigma xf}{\Sigma f}$

where Σxf means 'the sum of the products xf'
i.e. Σ (number $\times$ frequency)
and Σf means 'the sum of the frequencies'

Mean $= \dfrac{(0 \times 4) + (1 \times 19) + (2 \times 25) + (3 \times 29) + (4 \times 23)}{100}$

$= \dfrac{248}{100} = 2.48$

(b) The median mark is the number between the 50th and 51st
numbers. By inspection, both the 50th and 51st numbers are 3.

∴   Median $= 3$ marks

(c) The modal mark $= 3$

For grouped data, each group can be represented approximately by its mid-point. Suppose the marks of 51 students in a test were:

Mark	30–39	40–49	50–59	60–69
Frequency	7	14	21	9

We say that, for the 30–39 interval, there are 7 marks of 34·5.

Note that the mean calculated by this method is only an estimate because the raw data is not available and an assumption has been made in regard to the mid-point of each interval whereas in the example on page 296 a true mean could be found.

Exercise 3

1. A group of 50 people were asked how many books they had read in the previous year; the results are shown in the frequency table below. Calculate the mean number of books read per person.

Number of books	0	1	2	3	4	5	6	7	8
Frequency		5	5	6	9	11	7	4	2	1

2. A teacher conducted a mental arithmetic test for 26 pupils and the marks out of 10 were as follows.

Mark	3	4	5	6	7	8	9	10
Frequency	6	3	1	2	0	5	5	4

(a) Find the mean, median and mode.
(b) The teacher congratulated the class saying that "over three quarters were above average". Which 'average' justifies this statement?

3. The following tables give the distribution of marks obtained by different classes in various tests. For each table, find the mean, median and mode.

(a)

Mark	0	1	2	3	4	5	6
Frequency	3	5	8	9	5	7	3

(b)

Mark	15	16	17	18	19	20
Frequency	1	3	7	1	5	3

4. The table gives the number of words in each
   sentence of a page of writing.
   (a) Copy and complete the table.
   (b) Work out an estimate for the mean
       number of words in a sentence.

number of words	frequency f	mid-point x	fx
1–5	6	3	18
6–10	5	8	40
11–15	4		
16–20	2		
21–25	3		
Totals	20	–	

5. The results of 24 students in a test are given.

   (a) Find the mid-point of each group of marks and calculate an
       estimate of the mean mark.
   (b) Find the modal group.

Mark	Frequency
85–99	4
70–84	7
55–69	8
40–54	5

6. The number of letters delivered to the 26 houses
   in a street was as follows:

   Calculate an estimate of the mean number of
   letters delivered per house.

Number of letters delivered	Number of houses (i.e. frequency)
0 – 2	10
3 – 4	8
5 – 7	5
8 –12	3

7. The histogram shows the heights of the
   60 athletes in the 1999 British athletics team.
   (a) Calculate an estimate for the mean height
       of the 60 athletes.
   (b) Explain why your answer is an *estimate*
       for the mean height.

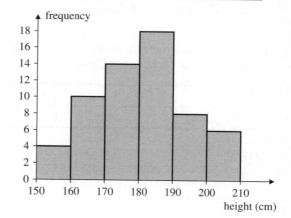

8. The number of goals scored in a series of
   football matches was as follows:

Number of goals	1	2	3
Number of matches	8	8	x

   (a) If the mean number of goals is 2·04, find x.
   (b) If the modal number of goals is 3,
       find the smallest possible value of x.
   (c) If the median number of goals is 2,
       find the largest possible value of x.

9. In a survey of the number of occupants in a number of cars, the
following data resulted.

Number of occupants	1	2	3	4
Number of cars	7	11	7	x

(a) If the mean number of occupants is $2\frac{1}{3}$, find x.

(b) If the mode is 2, find the largest possible value of x.

(c) If the median is 2, find the largest possible value of x.

10. The marks obtained by the members of a class are summarised in
the table.

Mark	x	y	z
Frequency	a	b	c

Calculate the mean mark in terms of a, b, c, x, y and z.

8.2 Data presentation

The results of a statistical investigation are known as data. Data can be
presented in a variety of different ways.

(a) Raw data

This is data in the form that it was collected, for example, the number
of peas in 40 pods.
Data in this form is difficult to interpret.

```
5  3  6  5  4  6  6  7  4  6
4  7  7  3  7  4  7  5  7  5
6  7  6  7  5  6  6  7  6  6
5  3  6  4  6  5  7  3  6  4
```

(b) Frequency tables

This presentation is in the form of a tally; the tally provides a numerical
value to the frequency of occurrence.
From the example above, there are 4 pea pods that contain 3 peas, etc.

Number of peas	Tally	Frequency (f)
3	IIII	4
4	JHT I	6
5	JHT II	7
6	JHT JHT III	13
7	JHT JHT	10

(c) Display charts

Three types of display chart are shown below; each refers to the initial data about the number of peas in a pod.

Number of peas	Angle on pie chart
3	$\frac{4}{40} \times 360 = 36°$
4	$\frac{6}{40} \times 360 = 54°$
5	$\frac{7}{40} \times 360 = 63°$
6	$\frac{13}{40} \times 360 = 117°$
7	$\frac{10}{40} \times 360 = 90°$

Pie chart

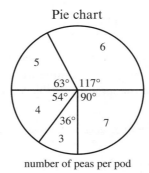

number of peas per pod

Bar chart

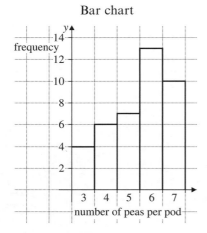

Frequency polygon

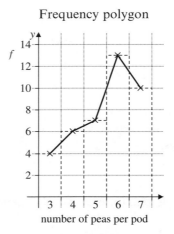

Note that in the bar chart and the frequency polygon, the vertical axis is always used to show frequency. It is the frequency that most clearly shows the mode, i.e. that more pods contained 6 peas than any other number of peas.

In the frequency polygon, the boxes of the bar chart are replaced by a line joining the tops of their mid-points.

All the above is discrete data; this is concerned solely with individually distinct number quantities. Continuous data is derived from measurements, e.g. height, weight, age, time. It can be handled as follows.

(d) Rounded data

Each measurement can be rounded and the frequency of this rounded quantity then recorded as for discrete data. For example, the lengths of 36 pea pods can be rounded to the nearest mm and then recorded as raw data — a pea pod measuring 59·2 mm is recorded as one that has a length of 59 mm.

Rounded data

52	80	65	82	77	60	72	83	63
78	84	75	53	73	70	86	55	88
85	59	76	86	73	89	91	76	92
66	93	84	62	79	90	73	68	71

(e) *Grouped data*

Each measurement can be grouped into classes with defined class
boundaries. In the table below, the same data is grouped into the classes
shown in the left-hand column.

Grouped frequency table

Length (mm)	Tally	Frequency
50–59	IIII	4
60–69	IIII I	6
70–79	IIII IIII II	12
80–89	IIII IIII	10
90–99	IIII	4

So, the pea pod measuring 59·2 mm is one of 4 recorded in the class
50–59. Because of the rounding of the data, the actual boundary of this
first class is from 49·5 to 59·5 (a length of 59·5 mm is rounded up to
60 mm). This is shown below on the horizontal axis of the bar chart for
these class boundaries.

Bar chart of grouped data

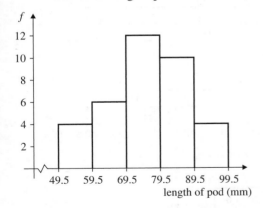

length of pod (mm)

Frequency polygon

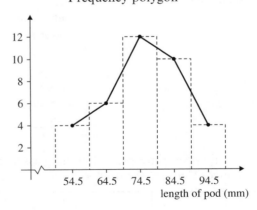

length of pod (mm)

The frequency polygon of the same data is also shown. Here the
horizontal axis records the mid-points of the classes where the
mid-point of the class 50–59 is calculated as $\dfrac{49\cdot5 + 59\cdot5}{2} = 54\cdot5$.

In some cases where there are many small classes, the mid-points of the
frequency polygon are joined with a curve. The diagram is then known
as a frequency curve.

Exercise 4

1. Karine and Jackie intend to go skiing in February. They have information about the expected snowfall in February for two possible places.

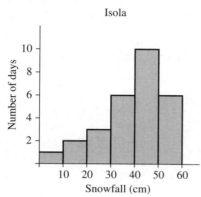

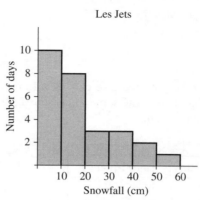

Decide where you think they should go. It doesn't matter where you decide, but you *must* say why, using the charts above to help you explain.

2. The chart shows information about people who use the internet regularly
   (a) About what percentage of boys in the 5–9 age group used the internet regularly?
   (b) In what age groups did more women than men use the internet?

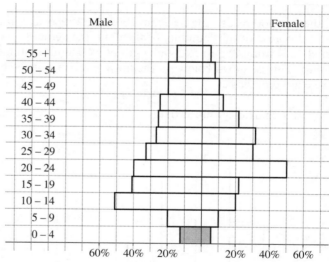

3. In a survey, the number of people in 100 cars passing a set of traffic lights was counted. Here are the results:

Number of people in car	0	1	2	3	4	5	6
Frequency	0	10	35	25	20	10	0

(a) Draw a bar chart to illustrate this data.
(b) On the same graph draw the frequency polygon.

Here the bar chart has been started. For frequency, use a scale of 1 cm for 5 units.

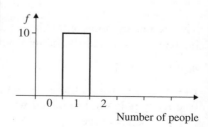

4. Two frequency polygons are shown giving the distribution of the weights of players in two different sports A and B.
   (a) How many people played sport A?
   (b) Comment on two differences between the two frequency polygons.
   (c) Either for A or for B suggest a sport where you would expect the frequency polygon of weights to have this shape. Explain in one sentence why you have chosen that sport.

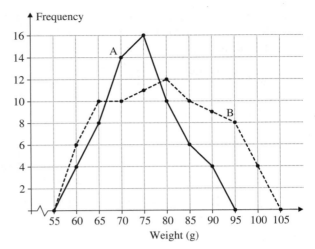

5. A scientist at an agricultural college is studying the effect of a new fertiliser for raspberries. She measures the heights of the plants and also the total weight of fruit collected. She does this for two sets of plants: one with the new fertiliser and one without it. Here are the frequency polygons:

$$\left[\begin{array}{ll} --- & \text{with fertiliser} \\ \underline{\hspace{1cm}} & \text{without fertiliser} \end{array}\right]$$

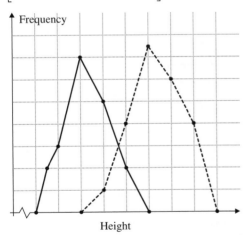

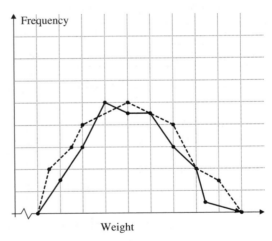

   (a) What effect did the fertiliser have on the heights of the plants?
   (b) What effect was there on the weights of fruit collected?

6. The diagram illustrates the production of apples in two countries.
   In what way could the pictorial display be regarded as misleading?

   UK
   470
   thousand
   tonnes

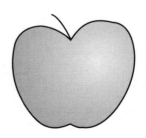

   FRANCE
   950
   thousand
   tonnes

7. The graph shows the performance of a company in the year in which a new manager was appointed. In what way is the graph misleading?

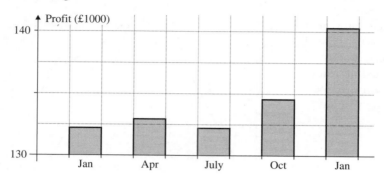

8. The pie chart illustrates the values of various goods sold by a certain shop. If the total value of the sales was £24 000, find the sales value of
   (a) toys
   (b) grass seed
   (c) records
   (d) food.

9. A quantity of scrambled eggs is made using the following recipe:

Ingredient	eggs	milk	butter	cheese	salt/pepper
Mass	450 g	20 g	39 g	90 g	1 g

Calculate the angles on a pie chart corresponding to each ingredient.

10. A firm making artificial sand sold its products in four countries.
    5% were sold in Spain
    15% were sold in France
    15% were sold in Germany
    65% were sold in U.K.
What would be the angles on a pie chart drawn to represent this information?

11. The pie chart illustrates the sales of various makes of petrol.
    (a) What percentage of sales does 'Esso' have?
    (b) If 'Jet' accounts for $12\frac{1}{2}\%$ of total sales, calculate the angles x and y.

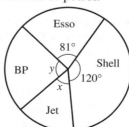

12. The cooking times for meals L, M and N are in the ratio $3 : 7 : x$. On a pie chart, the angle corresponding to L is 60°. Find x.

Stem and leaf diagrams

Data can be displayed in groups in a stem and leaf diagram.

Here are the marks of 20 girls in a science test.

54	42	61	47	24	43	55	62	30	27
28	43	54	46	25	32	49	73	50	45

We will put the marks into groups 20–29, 30–39, ... 70–79.
We will choose the tens digit as the 'stem' and the units as the 'leaf'.

The first four marks are shown [54, 42, 61, 47]

Stem (tens)	Leaf (units)
2	
3	
4	2 7
5	4
6	1
7	

The complete diagram is below ... and then with the leaves in numerical order:

stem	leaf
2	4 7 8 5
3	0 2
4	2 7 3 3 6 9 5
5	4 5 4 0
6	1 2
7	3

stem	leaf
2	4 5 7 8
3	0 2
4	2 3 3 5 6 7 9
5	0 4 4 5
6	1 2
7	3

The diagram shows the shape of the distribution. It is also easy to find the mode, the median and the range.

Back-to-back stem plots

Two sets of data can be compared using a *back-to-back stem plot*. Here are the marks of 20 boys who took the same science test as the girls above.

33	55	63	74	20	35	40	67	21	38
51	64	57	48	46	67	44	59	75	56

These marks are entered onto the back-to-back stem plot shown.

Boys	stem	Girls
1 0	2	4 5 7 8
8 5 3	3	0 2
8 6 4 0	4	2 3 3 5 6 7 9
9 7 6 5 1	5	0 4 4 5
7 7 4 3	6	1 2
5 4	7	3

It is helpful to have a key.

We can see that the boys achieved higher marks than the girls in this test.

key (boys)
1 | 5 means 51

key (girls)
2 | 4 means 24

Exercise 5

1. The marks of 24 children in a test are shown.

41	23	35	15	40	39	47	29
52	54	45	27	28	36	48	51
59	65	42	32	46	53	66	38

stem	leaf
1	
2	3
3	5
4	1
5	
6	

Draw a stem and leaf diagram. The first three entries are shown.

2. Draw a stem and leaf diagram for each set of data below.

(a)
24	52	31	55	40	37	58	61	25	46
44	67	68	75	73	28	20	59	65	39

(b)
30	41	53	22	72	54	35	47
44	67	46	38	59	29	47	28

stem	leaf
2	
3	
4	
5	
6	
7	

3. Here is the stem and leaf diagram showing the masses, in kg, of some people in a lift.
(a) Write down the range of the masses
(b) How many people were in the lift?
(c) What is the median mass?

stem (tens)	leaf (units)
3	2 5
4	1 1 3 7 8
5	0 2 5 8
6	4 8
7	1
8	2

4. In this question the stem shows the units digit and the leaf shows the first digit after the decimal point.
Draw the stem and leaf diagram using the following data:

2·4	3·1	5·2	4·7	1·4	6·2	4·5	3·3
4·0	6·3	3·7	6·7	4·6	4·9	5·1	5·5
1·8	3·8	4·5	2·4	5·8	3·3	4·6	2·8

key
3 | 7 means 3·7

stem	leaf
1	
2	
3	
4	
5	
6	

5. Here is a back-to-back stem plot sharing the pulse rates of several people.
(a) How many men were tested?
(b) What was the median pulse rate for the women?
(c) Write a sentence to describe the main features of the data.

Men		Women
5 1	4	
7 4 2	5	3
8 2 0	6	2 1
5 2	7	4 4 5 8 9
2 6	8	2 5 7
4	9	2 8

key (men)
1 | 4 means 41

key (women)
5 | 3 means 53

8.3 Scatter diagrams

Sometimes it is interesting to discover if there is a relationship (or *correlation*) between two sets of data.
Examples

- Do tall people weigh more than short people?
- If you spend longer revising for a test, will you get a higher mark?
- Do tall parents have tall children?
- If there is more rain, will there be less sunshine?
- Does the number of Olympic gold medals won by British athletes affect the rate of inflation?

If there is a relationship, it will be easy to spot if your data is plotted on a scatter diagram – that is a graph in which one set of data is plotted on the horizontal axis and the other on the vertical axis.

Each month the average outdoors temperature was recorded together with the number of therms of gas used to heat the house. The results are plotted on the scatter diagram as shown.

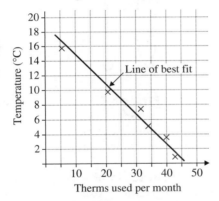

Clearly there is a high degree of *correlation* between these two figures. British Gas do in fact use weather forecasts as their main short-term predictor of future gas consumption over the whole country. A *line of best fit* has been drawn 'by eye'.
We can estimate that if the outdoor temperature was 12°C then about 17 therms of gas would be used.

Note: You can only predict within the range of values given.
If we extended the line for temperatures below zero the line of best fit predicts that about 60 therms would be used when the temperature is −4°C. But −4°C is well outside the range of the values plotted so the prediction is not valid. [Perhaps at −4°C a lot of people might stay in bed and the gas consumption would not increase by much. The point is you don't know!]
(a) The line in our example has a negative gradient and we say there is *negative correlation*.

(b) If the line of best fit has a positive gradient we say there is *positive correlation*.

(c) Some data when plotted on a scatter diagram does not appear to fit any line at all. In this case there is no correlation.

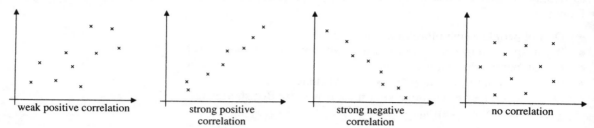

weak positive correlation   strong positive correlation   strong negative correlation   no correlation

Exercise 6

1. For this question you need to make some measurements of people in your class.
   (a) Measure everyone's height and 'armspan' to the nearest cm.

   Plot the measurements on a scatter graph. Is there any correlation?

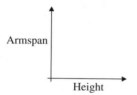

Height

Armspan

   (b) Now measure everyone's 'head circumference' just above the eyes. Plot head circumference and height on a scatter graph. Is there any correlation?
   (c) Decide as a class which other measurements [e.g. pulse rate] you can (fairly easily) take and plot these to see if any correlation exists.
   (d) Which pair of measurements gave the best correlation?

2. Plot the points given on a scatter graph, with t across the page and z up the page. Draw axes with values from 0 to 20. Describe the correlation, if any, between the values of t and z. [i.e. 'strong positive', 'weak negative' etc.]

(a)

t	8	17	5	13	19	7	20	5	11	14
z	9	16	7	13	18	10	19	8	11	15

(b)

t	4	9	13	16	17	6	7	18	10
z	5	3	11	18	6	11	18	12	16

(c)

t	12	2	17	8	3	20	9	5	14	19
z	6	13	8	15	18	2	12	9	12	6

3. Describe the correlation, if any, in these scatter graphs.

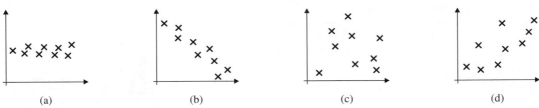

(a)   (b)   (c)   (d)

4. Plot the points given on a scatter graph, with s across the page and h up the page. Draw axes with values from 0 to 20.

s	3	13	20	1	9	15	10	17
h	6	13	20	6	12	16	12	17

(a) Draw a line of best fit..
(b) What value would you expect for h when s is 6?

5. The marks of 7 students in the two papers of a physics examination were as follows.

Paper 1	20	32	40	60	71	80	91
Paper 2	15	25	40	50	64	75	84

(a) Plot the marks on a scatter diagram, using a scale of 1 cm to 10 marks, and draw a line of best fit.
(b) A student scored a mark of 50 on Paper 1.
What would you expect her to get on Paper 2?

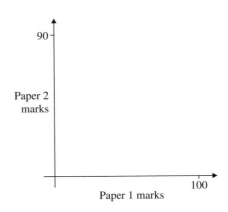

6. The table shows (i) the engine size in litres of various cars and (ii) the distance travelled in km on one litre of petrol.

Engine	0·8	1·6	2·6	1·0	2·1	1·3	1·8
Distance	13	10·2	5·4	12	7·8	11·2	8·5

(a) Plot the figures on a scatter graph using a scale of 5 cm to 1 litre across the page and 1 cm to 1 km up the page. Draw a line of best fit.
(b) A car has a 2·3 litre engine. How far would you expect it to go on one litre of petrol?

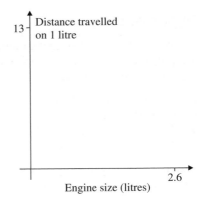

7. The data shows the latitude of 10 cities in the northern hemisphere and the average high temperatures.

City	Latitude (degrees)	Mean high temperature (°F)
Bogota	5	66
Bombay	19	87
Casablanca	34	72
Dublin	53	56
Hong Kong	22	77
Istanbul	41	64
St Petersburg	60	46
Manila	15	89
Oslo	60	50
Paris	49	59

(a) Draw a scatter diagram and draw a line of best fit. Plot latitude across the page with a scale of 2 cm to 10°. Plot temperature up the page from 40° F to 90° F with a scale of 2 cm to 10° F.

(b) Which city lies well off the line?
Do you know what factor might cause this apparent discrepancy?

(c) The latitude of Shanghai is 31°N. What do you think its mean high temperature is?

8. What sort of pattern would you expect if you took readings of the following and drew a scatter diagram?
(a) cars on roads; accident rate
(b) sales of perfume; advertising costs
(c) birth rate; rate of inflation
(d) outside temperature; sales of ice cream
(e) height of adults; age of same adults

8.4 Questionnaires, testing hypotheses

Hypotheses

Many problems of a statistical nature are made more clear when they are put in the form of a *hypothesis*.
Here are some examples of hypotheses.

- Cats prefer Kit-e-Kat.
- Boys are better at Mathematics than girls.
- Smoking damages your health.
- A university degree guarantees a higher standard of living.

A hypothesis is generally considered to be a statement which may be true but for which no proof has yet been found.

Statisticians are employed to collect and analyse data in order to obtain information from it which will prove or disprove a hypothesis. Here again are the hypotheses put in the form of questions.

- Is more Kit-e-Kat sold than other brands? — after all, you can't ask the cats!
- Do boys score higher than girls in maths tests? — you might want to write your own test for this.
- Do smokers have a shorter lifespan than non-smokers?
- Do people with a degree earn more than those without a degree?

Several factors need considering when choosing a hypothesis.

Can you test it?

In the smoker's problem, a wide variety of influences affect the lifespan of people: diet, fitness, stress, heredity, etc. How can these be eliminated so that *only* the smoking counts?

Can enough data be collected to give a reasonable result?

Think of where the data will come from. For example in the cat food problem, can you ask your local shop? Will one shop be enough? In the case of the university graduates, how can a lot of graduates (and of course non-graduates) be asked for comparison? Will they each be willing to tell you their income?

How can you know if the hypothesis has been proved or disproved?

Before collecting data, consider the criteria for proof (or disproof). In the mathematics question, what is meant by better? Do the marks of the boys have to be 5% higher (or more) on average than those of the girls?

Can the type of data to be collected be analysed?

Consider the techniques available to you: mean, median, mode, range, scatter diagrams, pie charts, frequency polygons. There are other techniques like histograms, percentiles and standard deviation which will be discussed later.

Is it interesting?

If not, the whole piece of work will be dull and tedious both for you and for others to study.

An excellent way to collect data is by means of a questionnaire. Having decided what you want to know, it is just a matter of asking the right questions.

Questionnaire design

Most surveys are conducted using questionnaires. It is very important to design the questionnaire well so that:

(a) people will cooperate and will answer the questions honestly
(b) the questions are not biased
(c) the answers to the questions can be analysed and presented for ease of understanding.

Checklist

A Provide an introduction to the sheet so that your subject knows the purpose of the questionnaire.

> 'Proposed new traffic lights'

B Make the questions easy to understand and specific to answer.
Do *not* ask vague questions like this
The answers could be:

> 'Yes, a lot'
> 'Not much'
> 'Only the best bits'
> 'Once or twice a day'

⟶ Did you see much of the Olympics on TV?

You will find it hard to analyse this sort of data.

A *better* question is:

> 'How much of the Olympic coverage did you watch?' Tick one box
>
> Not at all ☐
>
> Up to 1 hour per day ☐
>
> 1 to 2 hours per day ☐
>
> More than 2 hours per day ☐

C Make sure that the questions are not *leading* questions. It is human nature not to contradict the questioner. Remember that the survey is to find out opinions of other people, not to support your own.
Do *not* ask
> 'Do you agree that BBC has the best sports coverage?'
A better question is

> 'Which of the following has the best sports coverage?'
>
> BBC   ITV   Channel 4   Satellite TV
> ☐   ☐   ☐   ☐

You might ask for one tick or possibly numbers 1, 2, 3, 4 to show an order of preference.

D If you are going to ask sensitive questions (about age or income, for example), design the question with care so as not to offend or embarrass.
Do *not* ask:
    'How old are you?'
or   'Give your date of birth'
A better question is:

'Tick one box for your age group.'

15–17	18–20	21–30	31–50
☐	☐	☐	☐

E Do not ask more questions than necessary and put the easy questions first.

Exercise 7

Criticise the following questions and suggest a better question which overcomes the problem involved.
Write some questions with 'yes/no' answers and some questions which involve multiple responses.
Remember to word your questions simply.

1. Do you think it is ridiculous to spend money on food 'mountains' in Europe while people in Africa are starving?

2. What do you think of the new head teacher?

3. How dangerous do you think it is to fly in a single-engined aeroplane?

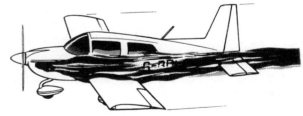

4. How much would you pay to use the new car park?
    ☐ less than £1   ☐ more than £2·50

5. Do you agree that English and Maths are the most important subjects at school?

6. Do you or your parents often hire videos from a shop?

7. Do you think that we get too much homework?

8. Do you think you would still eat meat if you had been to see the animals killed?

9. Design a questionnaire to test the truth of the statements given.
   (a) 'Most people choose to shop in a supermarket where it is easy to park a car.'
   (b) 'Children of school age watch more television than their parents.'
   (c) 'Boys prefer action films and girls prefer some sort of story.'
   (d) 'Most people who smoke have made at least one serious attempt to give it up.'

Pilot survey

Stop and consider two points.
(a) Will the questionnaire help to prove or disprove the hypothesis?
(b) How many people are needed for the survey?

Both of these questions can be answered by first doing a pilot survey, that is, a quick mini-survey of 5–10 people. Some of the questions may need changing straightaway.

Analysis

Having stated clearly what was the hypothesis being tested, display the results clearly. Diagrams like pie charts or bar charts, possibly showing percentages, are a good idea, particularly if drawn using colours. A data base or spreadsheet program on a computer may help. Draw conclusions from the results, but make sure they are justified by the evidence.

Your own work

Almost certainly the best hypotheses are your own because you are personally interested.
Here is a list of hypotheses which some students have enjoyed investigating.
(a) Young people are more superstitious than old people.
(b) Given a free choice, most girls would hardly ever choose to wear a dress in preference to something else.
(c) More babies are born in the Winter than in the Summer.
(d) First years watch more TV than fifth years.
(e) Most cars these days use unleaded petrol.

8.5 Sampling and bias

Statisticians frequently carry out surveys to investigate a characteristic of a population. A population is the set of all possible items to be observed, not necessarily people.

Surveys are done for many reasons.

- Newspapers seek to know voting intentions before an election.
- Advertisers try to find out which features of a product appeal most to the public.
- Research workers testing new drugs need to know their effectiveness.
- Government agencies conduct tests for the public's benefit such as the percentage of first class letters arriving the next day or the percentage of '999' calls answered within 5 minutes.

Census

A census is a survey in which data is collected from every member of the population of a country. A census is done in Great Britain every ten years; 1971, 1981, 1991, 2001, ... Data about age, educational qualifications, race, distance travelled to work, etc., is collected from every person in the country on the given 'Census Day'. Because no person must be omitted, it is a huge and very expensive task. Census forms take several years to prepare and many years for thorough analysis of the data collected.

Samples

If a full census is not possible or practical, then it is necessary to take a sample from the population. Two reasons for taking samples are:

- it will be much cheaper and quicker than a full census,
- the object being tested may be destroyed in the test, like testing the average lifetime of a new light bulb or the new design of a car tyre.

The choice of a sample prompts two questions.

1. Does the sample truly represent the whole population?
2. How large should the sample be?

After taking a sample, it is assumed that the result for the sample reflects the whole population. For example, if 20% of a sample of 1000 people say they will vote in a particular way, then it is assumed that 20% of the whole electorate would do so likewise, subject of course to some error.

Simple random sampling

This is a method of sampling in which every item of the population has an equal chance of selection. There are two methods of selection:

(a) Selecting *with* replacement. Here, each item selected is then replaced giving a possibility that it may be reselected.
(b) Selecting *without* replacement. Each item selected is not replaced and is therefore not available for reselection.

For large populations, there is no significant difference between these two methods, but the method chosen should always be stated clearly

Selection techniques

Out of a school of 758 students, ten are to be selected to take part in an educational experiment in which homework is banned.
How would you select these students?
First give each student a three-digit number from 001 to 758.

Method 1

Each person's number is written on a piece of card, the cards placed in a hat and mixed up. Ten cards are then drawn.

Method 2

Use a random number table. The table is shown below.

Random number table

11 74	26 93	81 44	33 93	08 72	32 79	73 31	18 22	64 70	68 50
43 36	12 88	59 11	01 64	56 23	93 00	90 04	99 43	64 07	40 36
93 80	62 04	78 38	26 80	44 91	55 75	11 89	32 58	47 55	25 71
49 54	01 31	81 08	42 98	41 87	69 53	82 96	61 77	73 80	95 27
36 76	87 26	33 37	94 82	15 69	41 95	96 86	70 45	27 48	38 80
07 09	25 23	92 24	62 71	26 07	06 55	84 53	44 67	33 84	53 20
43 31	00 10	81 44	86 38	03 07	52 55	51 61	48 89	74 29	46 47
61 57	00 63	60 06	17 36	37 75	63 14	89 51	23 35	01 74	69 93
31 35	28 37	99 10	77 91	89 41	31 57	97 64	48 62	58 48	69 19
57 04	88 65	26 27	79 59	36 82	90 52	95 65	46 35	06 53	22 54
09 24	34 42	00 68	72 10	71 37	30 72	97 57	56 09	29 82	76 50
97 95	53 50	18 40	89 48	83 29	52 23	08 25	21 22	53 26	15 87
93 73	25 95	70 43	78 19	88 85	56 67	16 68	26 95	99 64	45 69
72 62	11 12	25 00	92 26	82 64	35 66	65 94	34 71	68 75	18 67
61 02	07 44	18 45	37 12	07 94	95 91	73 78	66 99	53 61	93 78
97 83	98 54	74 33	05 59	17 18	45 47	35 41	44 22	03 42	30 00
89 16	09 71	92 22	23 29	06 37	35 05	54 54	89 88	43 81	63 61
25 96	68 82	20 62	87 17	92 65	02 82	35 28	62 84	91 95	48 83
81 44	33 17	19 05	04 95	48 06	74 69	00 75	67 65	01 71	65 45
11 32	25 49	31 42	36 23	43 86	08 62	49 76	67 42	24 52	32 45

This table is published by kind permission of the Department of Statistics
University College, London

Start anywhere in the table and go either up, down, left, or right to read numbers in groups of three digits. Starting on the bottom right-hand corner going up and then down, would give the numbers

553   108   ~~797~~   049   370   071   605   372   ~~824~~   ~~915~~   586   670   684

Why were 797, 824, 915 rejected?

So students with the above 10 numbers will take part in the experiment.

Method 3   Random number generator

Use a calculator or a computer.
On the CASIO calculator, random numbers are generated by pressing

| RAN # | after using | SHIFT | to get this function. The numbers produced are three-digit numbers between .001 and .999. Ignore the decimal point.
Although these are not true random numbers due to their method of generation, they are adequate for school use.

Random two-digit numbers are obtained simply by omitting the last digit.

Find 6 two-digit numbers between 01 and 60.

Ignoring the decimal points, a calculator gave the following.

~~819~~   453   480   ~~718~~   ~~891~~   326   050   217   ~~990~~   437

So use: 45   48   32   05   21   43

Exercise 8

1. One hundred students took a test and their results are given in the table. Each student had a two-digit identification number between 00 and 99.

Second number

		0	1	2	3	4	5	6	7	8	9
	0	53	62	48	71	41	78	64	49	82	32
	1	28	66	41	83	49	62	54	67	68	33
	2	75	81	47	35	26	70	85	93	36	35
	3	66	42	55	58	87	38	29	15	41	68
First	4	84	67	39	62	47	58	62	39	72	86
number	5	43	32	49	61	53	80	95	26	44	33
	6	77	62	41	19	26	37	56	57	40	30
	7	38	75	62	41	28	33	62	39	43	26
	8	64	52	53	68	36	42	39	38	65	70
	9	52	45	72	67	90	47	34	55	61	47

For example, Student 31 scored 42% (Row 3, Column 1) and Student 65 scored 37%.

(a) Use your judgement to select a representative sample of ten marks from the 100 given. Find the mean mark of this sample and state its range.

(b) Select several more similar samples of ten marks. Write down the range for each sample mean and find the mean of all the sample means.

2. (a) Use the random number table (page 316) to select a random sample of ten marks from the students' marks of Question 1. Find the sample mean and the range. For example, starting at 07 at the left-hand end of the sixth row of the random number table gives these random numbers and their corresponding marks:

07   09   25   23   92   24   ...
49%   32%   70%   35%   72%   26% ...

(b) Collect other sample means and their range from all the students in your class and find the mean of all the means. Compare the results from (a) and (b).

3. This example combines sampling and probability. Carefully draw
   the diagram below on graph paper with a scale of 1 cm to 1 unit.
   Points can be selected at random as follows.
   From the random number table read off two pairs, like 75 48.
   For the scale used here, this gives the point (7·5, 4·8).
   Similarly, 83 72 gives the point (8·3, 7·2).
   (a) Select 50 points in this way. Record what fraction of all the
       points lies inside the circle.

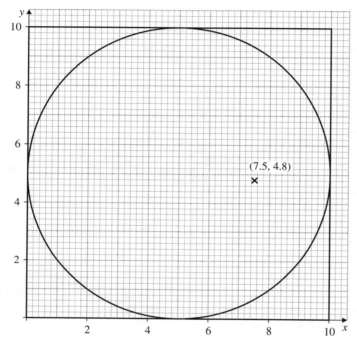

   (b) The expected probability of selecting a point inside the circle
       can be calculated using areas.

   area of circle $= \pi . 5^2 = 25\,\pi$
   area of square $= 10 \times 10 = 100\,\text{cm}^2$

   $$\text{Probability of selecting a point inside the circle} = \frac{25\pi}{100} = \frac{\pi}{4}$$

   (c) Compare the value of the fraction obtained in (a) with the
       value of the fraction $\frac{\pi}{4}$ by writing the two results in decimals.

Stratified random sampling

Some populations separate naturally into a number of sub-groups or
strata. Providing that the strata are quite distinct and that every member of
the population belongs to one and only one stratum, then the population
can be sampled using the method of stratified random sampling.

Opinion pollsters like Gallup or N.O.P. use this method when conducting polls on voting intentions. Their claim is that results are accurate to about 3 per cent.

Method

- Separate the population into suitable strata. Find what proportion of the population is in each stratum.
- Select a sample from each stratum proportional to the stratum size, by simple random sampling.

In Year 10 of a school, the 180 students are split into three groups for games; 90 play football, 55 play hockey and 35 play badminton. Use a stratified random sample of 30 students to estimate the mean weight of the 180 students.

The sample size from each of the three groups (strata) must be proportional to the stratum size. So the 30 students are made up by selecting:

from the football group, $\dfrac{90}{180} \times 30 = 15$,

from the hockey group, $\dfrac{55}{180} \times 30 = 9 \cdot 16$, (say 9)

from the badminton group, $\dfrac{35}{180} \times 30 = 5 \cdot 83$, (say 6)

The three sample means were found to be:

football 53·2 kg   hockey 49·4 kg   badminton 46·4 kg

So the mean for the
population of 180 students $= \dfrac{53 \cdot 2 + 49 \cdot 4 + 46 \cdot 4}{3}$

$= 49 \cdot 7 \, \text{kg} \ (3 \text{ s.f.})$

Exercise 9

1. The table shows the number of pupils in each year group at a new school.
   Naseem takes a stratified sample of 60 pupils from this school. Work out how many pupils from each year group should be in Naseem's sample.

Year 7	162
Year 8	161
Year 9	·157
Year 10	63
Year 11	58

2. The table gives the number of employees in five firms. A stratified sample of 100 people is taken.
   How many people should be selected from each firm?

Firm	Employees
A	398
B	1011
C	409
D	207
E	1985

Bias

Question 1(a) on page 317 required personal judgement to select a representative sample of 10 test marks taken from 100. This provided a non-random judgmental sample. The highest mark is seldom selected in this way, so this method of sampling usually gives a mean mark which is less than the true mean.

It is *not* possible to choose a random sample using personal judgement. Random samples may give results containing errors but these can be predicted and allowed for. The type of error introduced with judgmental sampling is unpredictable and thus corrections for it cannot be made. This type of unpredictable error is called *bias*.

Bias can come from a variety of sources including non-random sampling. For example, for data collected by questionnaire, non-response from subjects can introduce bias into the results. Qualities which these subjects have in common will not be represented in the collected data.

For data collected by a street survey, the time of day and location may well mean that there is a sector or sectors of the population who are not questioned.

Perhaps the most famous mistake in opinion sampling occurred in America in 1936. A magazine carried out an enormous poll in which over two million people were asked to state their preference for President, either the Democrat (F.D. Roosevelt) or the Republican (A.E. Landon). The sample was taken by selecting the two million people by simple random sampling from the telephone directories. The poll showed a large majority would vote for the Republican, but in the election the Democrat Roosevelt won.

A common cause of bias occurs when the questions asked in the survey are not clear or are leading questions. This was considered earlier in the section on questionnaires.

Exercise 10

In these questions, decide whether the method of sampling is satisfactory or not. If it is not satisfactory, suggest a better way of obtaining a sample.

1. A teacher, with responsibility for school meals, wants to hear pupils' opinions on the meals currently provided. She waits next to the dinner queue and questions the first 50 pupils as they pass.

2. To find out how satisfied customers are with the service they receive from the telephone company a person telephones 200 people chosen at random from the telephone directory.

3. John, aged 10, wants to find out the average pocket money received by children of his own age. He asks 8 of his friends how much they get.

4. An opinion pollster wants to canvas opinion about our European neighbours. He questions drivers as they are waiting to board their ferry at Dover.

5. Jim has a theory that just as many men as women do the shopping in supermarkets. To test his theory he goes to Tesco one Saturday between 09:00 am and 10:00 am and counts the numbers of men and women.

6. A journalist wants to know the views of local people about a new one-way system in the town centre. She takes the electoral roll for the town and selects a random sample of 200 people.

7. A pollster working for the BBC wants to know how many people are watching a new series which is being shown. She questions 200 people as they are leaving a supermarket between 10:00 and 12:00 one Thursday.

8. An electronics company wants to find out what percentage of people have a video at home. In a survey 1000 people are questioned throughout one week as they leave a video hire shop.

8.6 Cumulative frequency

Quartiles, interquartile range

- The range is a simple measure of spread but one extreme (very high or very low) value can have a big effect.
  The *interquartile range* is a better measure of spread.

Find the quartiles and the interquartile range for these numbers:

12   6   4   9   8   4   9   8   5   9   8   10

In order: 4   4   5 | 6   8   8 | 8   9   9 | 9   10   12

lower quartile   median   upper quartile
is 5·5   is 8   is 9

Interquartile range = upper quartile − lower quartile
= 9 − 5·5
= 3·5

Box plot

A *box plot* or *box and whisker diagram* shows the spread of a set of data.
It shows the quartiles (Q_1 and Q_3) and the median.
The 'whiskers' extend from the lowest to the highest value and show the range.

Here is a box plot.

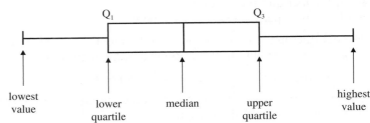

Exercise 11

For each set of data, work out: (a) the lower and upper quartile
(b) the inter-quartile range.

1. 1 1 4 4 5 8 8 8 9 10 11 11

2. 5 2 6 4 1 9 3 2 8 4 1 0

3. 7 9 11 15 18 19 23 27

4. 0 0 1 1 2 3 3 4 4 4 4 6 6 6 7 7

5. Here are three box plots. Estimate: (a) the median
(b) the interquartile range
(c) the range.

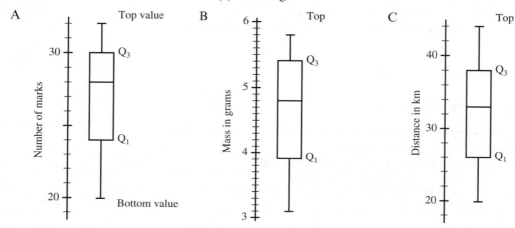

Cumulative frequency

● The total of the frequencies up to a particular value is called the
cumulative frequency

Data given in a frequency table can be used to calculate cumulative
frequencies. These new values, when plotted and joined, form a
cumulative frequency curve, sometimes called an S-shaped curve.

It is a simple matter to find the median from the halfway point of a
cumulative frequency curve.
Other points of location can also be found from this curve. The
cumulative frequency axis can be divided into 100 parts.

● The upper quartile is at the 75% point
● The lower quartile is at the 25% point

The quartiles are particularly useful in finding the central 50% of the
range of the distribution; this is known as the interquartile range.

●   interquartile range = (upper quartile) − (lower quartile)

The interquartile range is an important measure of spread in that it
shows how widely the data is spread.

Half the distribution is in the interquartile range. If the interquartile range is small, then the middle half of the distribution is bunched together.

In a survey, 200 people were asked to state their weekly earnings. The results were plotted on the cumulative frequency curve.

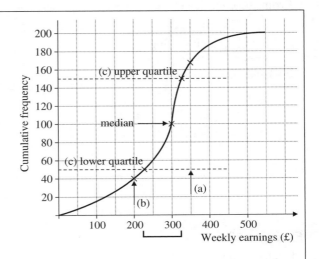

(a) How many people earned up to £350 a week?

From the curve, about 170 people earned up to £350 per week.

(b) How many people earned more than £200 a week?

About 40 people earned up to £200 per week. There are 200 people in the survey, so 160 people earned more than £200 per week.

(c) Find the interquartile range.

Lower quartile (50 People) = £225
Upper quartile (150 People) = £325
∴   Interquartile range = 325 − 225
= £100

A pet shop owner likes to weigh all his mice every week as a check on their state of health. The weights of the 80 mice are shown below.

weight (g)	frequency	cumulative frequency	weight represented by cumulative frequency
0–10	3	3	⩽ 10 g
10–20	5	8	⩽ 20 g
20–30	5	13	⩽ 30 g
30–40	9	22	⩽ 40 g
40–50	11	33	⩽ 50 g
50–60	15	48	⩽ 60 g
60–70	14	62	⩽ 70 g
70–80	8	70	⩽ 80 g
80–90	6	76	⩽ 90 g
90–100	4	80	⩽ 100 g

The table also shows the cumulative frequency.

Plot a cumulative frequency curve and hence estimate

(a) the median
(b) the interquartile range.

- Note: that the points on the graph are plotted at the upper limit of each group of weights.

From the cumulative frequency curve,

$$\begin{aligned}
\text{median} &= 55\,\text{g} \\
\text{lower quartile} &= 36\,\text{g} \\
\text{upper quartile} &= 68\,\text{g} \\
\text{interquartile range} &= (68 - 36)\,\text{g} \\
&= 32\,\text{g}
\end{aligned}$$

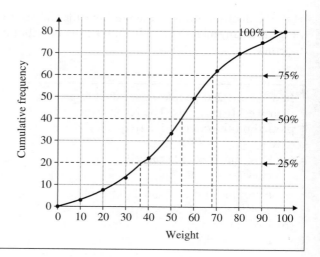

Exercise 12

1. The graph shows the cumulative frequency curve for the marks of 40 students in an examination.
   From the graph estimate:
   (a) the median mark
   (b) the mark at the lower quartile and at the upper quartile
   (c) the interquartile range
   (d) the pass mark if three-quarters of the students passed.

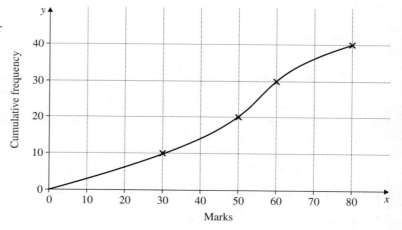

2. The graph shows the cumulative frequency curve for the marks of 60 students in an examination.
   From the graph estimate:
   (a) the median mark
   (b) the mark at the lower quartile and at the upper quartile
   (c) the interquartile range
   (d) the pass mark if two-thirds of the students passed.

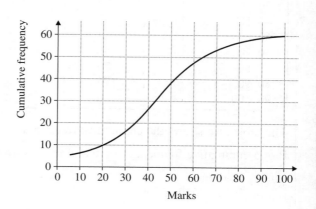

3. The lifetimes of 500 electric light
   bulbs were measured in a
   laboratory. The results are
   shown in the cumulative
   frequency diagram.
   (a) How many bulbs had a
       lifetime of 1500 hours or less?
   (b) How many bulbs had a
       lifetime of between 2000
       and 3000 hours?
   (c) After how many hours were
       70% of the bulbs dead?
   (d) What was the shortest
       lifetime of a bulb?

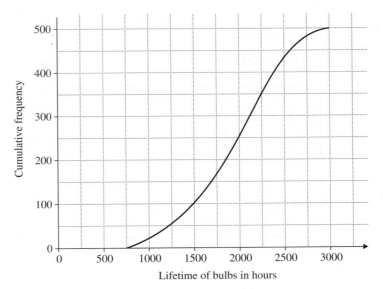

4. A photographer measures all the snakes required for a scene in a
   film involving a snake pit.

   (a) Draw a cumulative frequency curve for the results below.

length (cm)	frequency	cumulative frequency	upper limit
0–10	0	0	⩽ 10
10–20	2	2	⩽ 20
20–30	4	6	⩽ 30
30–40	10	16	⩽ 40
40–50	17		⋮
50–60	11		⋮
60–70	3		⋮
70–80	3		⋮

   Use a scale of 2 cm for 10 units across the page for the lengths and
   2 cm for 10 units up the page for the cumulative frequency.
   Remember to plot points at the *upper* end of the classes
   (10, 20, 30 etc).
   (b) Find: (i) the median   (ii) the interquartile range.

5. As part of a medical inspection, a nurse measures the heights of 48 pupils in a school.

(a) Draw a cumulative frequency curve for the results below.

height (cm)	frequency	cumulative frequency
$140 \leqslant h < 145$	2	2 [$\leqslant 145$ cm]
$145 \leqslant h < 150$	4	6 [$\leqslant 150$ cm]
$150 \leqslant h < 155$	8	14 [$\leqslant 155$ cm]
$155 \leqslant h < 160$	9	
$160 \leqslant h < 165$	12	
$165 \leqslant h < 170$	7	
$170 \leqslant h < 175$	4	
$175 \leqslant h < 180$	2	

Use a scale of 2 cm for 5 units across the page and 2 cm for 10 units up the page.

(b) Find: (i) the median
          (ii) the interquartile range.

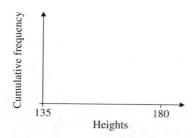

6. Hugo and Boris are brilliant darts players. They recorded their scores over 60 throws.

Here are Hugo's scores:

Score (x)	$30 < x \leqslant 60$	$60 < x \leqslant 90$	$90 < x \leqslant 120$	$120 < x \leqslant 150$	$150 < x \leqslant 180$
Frequency	10	4	13	23	10

Draw a cumulative frequency curve.

Use a scale of 2 cm for 20 points across the page
           and 2 cm for 10 throws up the page.

For Hugo, find: (a) his median score
                (b) the interquartile range of his scores.

For his 60 throws, Boris had a median score of 105 and an interquartile range of 20.

Which of the players is more consistent? Give a reason for your answer.

7. The 'life' of a Mickey Mouse photocopying machine is tested by
timing how long it works before breaking down. The results for 50
machines are.

Time, t (hours)	$2 < t \leqslant 4$	$4 < t \leqslant 6$	$6 < t \leqslant 8$	$8 < t \leqslant 10$	$10 < t \leqslant 12$
Frequency	9	6	15	15	5

(a) Draw a cumulative frequency graph.
   [t across the page, 1 cm = 1 hour; C.F. up the page]
(b) Find: (i) the median life of a machine
           (ii) the interquartile range.
(c) The makers claim that their machines will work for at
   least 10 hours. What percentage of the machines do
   not match this description?

8. In an international competition 60 children from Britain and France
did the same science test.

Marks	Britain frequency	France frequency	Britain cum. freq.	France cum. freq.
1–5	1	2	1 [$\leqslant$ 5·5]	2 [$\leqslant$ 5·5]
6–10	2	5	3 [$\leqslant$ 10·5]	7 [$\leqslant$ 10·5]
11–15	4	11	7 [$\leqslant$ 15·5]	
16–20	8	16		
21–25	16	10		
26–30	19	8		
31–35	10	8		

Note: The upper class boundaries for the marks are 5·5, 10·5, 15·5
etc.
The cumulative frequency graph should be plotted for values $\leqslant$ 5·5,
$\leqslant$ 10·5, $\leqslant$ 15·5 and so on.
(a) Using the same axes, draw the cumulative frequency curves for
   the British and French results.
   Use a scale of 2 cm for 5 marks across the page
               and 2 cm for 10 people up the page.
(b) Find the median mark for each country.
(c) Find the interquartile range for the British results.
(d) Describe in one sentence the main difference between the two
   sets of results.

9. A new variety of plant is tested by planting a sample of seeds and measuring the heights of the plants after six weeks. The results were:

height (cm)	0–	2–	4–	6–	8–	10–	12–	14–	16 and over
frequency	12	0	2	4	10	12	10	8	0

Note: The height interval '0–' means $0 \leqslant$ height < 2. The class boundaries are 0 and 2.

(a) Draw a cumulative frequency curve for the data.
    Use a scale of 1 cm to 1 cm across the page and 2 cm to 10 units up the page.
(b) Find the median height of the plants.
(c) Why do you think there are so many plants in the '0–' interval?
(d) The 'average' height of the plants after six weeks is to be printed on the packet.
    The actual mean height of the plants is 9·0 cm.
    Which 'average' height gives the better indication of the likely performance of the plants: the mean or the median?
    Explain your answer.

†10. The age distribution of the populations of two countries, A and B, is shown below

Age	Number of people in A (millions)	Number of people in B (millions)
Under 10	15	2
10–19	11	3
20–39	18	5
40–59	7	13
60–79	3	14
80–99	1	7

(a) Copy and complete the cumulative frequency tables below

Age	Country A	Country B
Under 10	15	
Under 20	26	
Under 40		
Under 60		
Under 80		
Under 100		

(b) Using the same axes draw cumulative frequency curves for countries A and B.
    Use a scale of 2 cm to 20 years across the page and 2 cm to 10 million up the page.
(c) State the population of country B.
(d) State the median age for the two countries.
(e) Describe the main difference in the age distribution of the two countries.

8.7 Histograms

In a histogram, the frequency of the data is shown by the *area* of each bar. Histograms resemble bar charts but are not to be confused with them: in bar charts the frequency is shown by the height of each bar. Histograms often have bars of varying widths. Because the area of the bar represents frequency, the height must be adjusted to correspond with the width of the bar. The vertical axis is not labelled frequency but frequency density.

$$\text{frequency density} = \frac{\text{frequency}}{\text{class width}}$$

Histograms can be used to represent both discrete data and continuous data, but their main purpose is for use with continuous data.

Draw a histogram from the table shown for the distribution of ages of passengers travelling on a flight to New York.

Note that the data has been collected into class intervals of different widths.

To draw the histogram, the heights of the bars must be adjusted by calculating frequency density.

Ages	Frequency
$0 \leqslant x < 20$	28
$20 \leqslant x < 40$	36
$40 \leqslant x < 50$	20
$50 \leqslant x < 70$	30
$70 \leqslant x < 100$	18

Ages	Frequency	Frequency density (f.d.)
$0 \leqslant x < 20$	28	$28 \div 20 = 1 \cdot 4$
$20 \leqslant x < 40$	36	$36 \div 20 = 1 \cdot 8$
$40 \leqslant x < 50$	20	$20 \div 10 = 2$
$50 \leqslant x < 70$	30	$30 \div 20 = 1 \cdot 5$
$70 \leqslant x < 100$	18	$18 \div 30 = 0 \cdot 6$

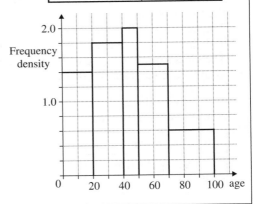

Exercise 13

1. The lengths of 20 copper nails were measured. The results are shown in the frequency table.
   Calculate the frequency densities and draw the histogram as started below.

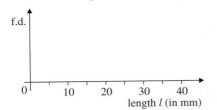

Length l (in mm)	Frequency	Frequency density (f.d.)
$0 \leqslant L < 20$	5	$5 \div 20 = 0 \cdot 25$
$20 \leqslant L < 25$	5	
$25 \leqslant L < 30$	7	
$30 \leqslant L < 40$	3	

2. The volumes of 55 containers were measured
and the results presented in a frequency
table as shown.
Calculate the frequency densities and draw
the histogram.

Volume (mm³)	Frequency
$0 \leqslant V < 5$	5
$5 \leqslant V < 10$	3
$10 \leqslant V < 20$	12
$20 \leqslant V < 30$	17
$30 \leqslant V < 40$	13
$40 \leqslant V < 60$	5

3. Thirty students in a class are weighed on the first day of term. Draw
a histogram to represent this data.

Note that the weights do not start at zero. This can be shown on
the graph as follows.

Weight (kg)	Frequency
30–40	5
40–45	7
45–50	10
50–55	5
55–70	3

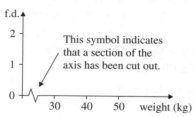

This symbol indicates
that a section of the
axis has been cut out.

4. The ages of 120 people passing through a
turnstyle were recorded and are shown in
the frequency table.
The notation –10 means '0 < age ≤ 10'
and similarly –15 means '10 < age ≤ 15'.
The class boundaries are 0, 10, 15, 20, 30, 40.
Draw the histogram for the data.

Age (yrs)	Frequency
–10	18
–15	46
–20	35
–30	13
–40	8

5. The daily profit made at a beach resort sailing
club is shown in the table.

Profit ($)	Frequency
0–40	16
40–80	48
80–100	30
100–120	40
120–160	36
160–240	16

Draw a histogram for the data.

6. Another common notation is used here for the masses of plums
picked in an orchard, shown in the table below.

Mass (g)	20–	30–	40–	60–	80–
Frequency	11	18	7	5	0

The initial 20– means 20 g ≤ mass < 30 g.
Draw a histogram with class boundaries at 20, 30, 40, 60, 80.

7. The heights of 50 Olympic athletes were measured as shown in the table below.

Height (cm)	170–174	175–179	180–184	185–194
Frequency	8	17	14	11

These values were rounded off to the nearest cm. For example, an athlete whose height h is 181 cm could be entered anywhere in the class $180{\cdot}5\,\text{cm} \leqslant h < 181{\cdot}5\,\text{cm}$. So the table is as follows.

Height	169·5–174·5	174·5–179·5	179·5–184·5	184·5–194·5
Frequency	8	17	14	11

Draw a histogram with class boundaries at 169·5, 174·5, 179·5, . . .

Finding frequencies from histograms

If the data is already presented in a histogram, it can be useful to draw up the relevant frequency table. The actual size of the sample tested can then be found as well as the frequencies in each class. However, if a sample size is too small, the results of the analysis may not truly represent the distribution which is being examined.

Given the histogram of children's heights shown, draw up the relevant frequency table and find the size of the sample which was tested.

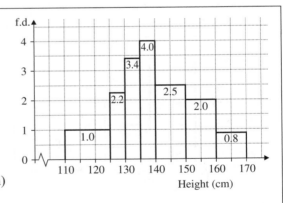

Use   frequency density $= \dfrac{\text{frequency}}{\text{class width}}$

So   frequency $=$ (frequency density) $\times$ (class width)

Remember:
the area of each bar represents frequency.

Height (cm)	f.d.	Frequency
$110 \leqslant x < 125$	1·0	$1{\cdot}0 \times 15 = 15$
$125 \leqslant x < 130$	2·2	$2{\cdot}2 \times 5 = 11$
$130 \leqslant x < 135$	3·4	$3{\cdot}4 \times 5 = 17$
$135 \leqslant x < 140$	4·0	$4{\cdot}0 \times 5 = 20$
$140 \leqslant x < 150$	2·5	$2{\cdot}5 \times 10 = 25$
$150 \leqslant x < 160$	2·0	$2{\cdot}0 \times 10 = 20$
$160 \leqslant x < 170$	0·8	$0{\cdot}8 \times 10 = 8$
		Total frequency $= 116$

So there were 116 children in the sample measured.

Exercise 14

1. For the histogram shown:
   (a) How many of the lengths are in
   these intervals?
   (i) 40–60 cm   (ii) 30–40 cm
   (b) What is the total frequency?

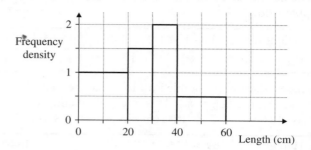

2. For the histogram shown, find the
total frequency.

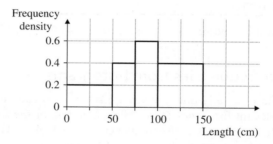

3. The histogram shows the ages of the trees in a small wood. How
many trees were in the wood altogether?

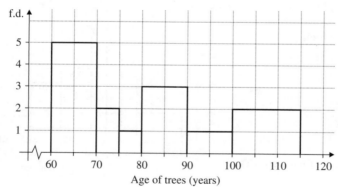

4. One day a farmer weighs all the hens'
eggs which he collects. The results are
shown in this histogram.
There were 8 eggs in the class 50–60 g.
Work out the numbers on the frequency
density axis and hence find the total
number of eggs collected.

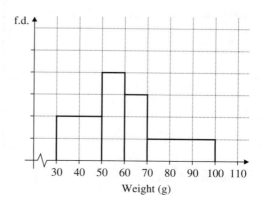

5. This histogram shows the number of letters delivered at some houses in a street. Notice that the data is discrete and that, for example, the class boundaries for 1–2 letters are 0·5–2·5.
Given that 3 letters were delivered to 8 houses, how many letters were delivered altogether?

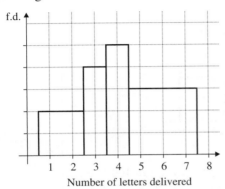

Number of letters delivered

6. Here are the percentage marks obtained by 44 pupils in an exam.

41	61	11	46	51	21	63	48	25	87	47
53	52	40	55	15	70	54	74	43	85	38
47	23	52	18	75	47	73	45	80	59	88
52	46	76	49	31	57	27	58	5	70	42

Make a tally chart and a frequency table using groups of marks 0–19, 20–39, 40–49, 50–59, 60–89.
Work out the frequency densities and hence draw a histogram.

7. The manager of a small company found that the number of sick days taken by his 50 employees in 1999 was as follows.

10	31	17	22	7	1	9	36	5	13	26	2	8
6	0	24	30	11	20	28	8	18	38	3	16	12
23	15	7	21	12	8	37	22	11	19	40	18	9
6	13	30	27	3	28	4	29	41	40	14		

(a) Draw up a frequency table and construct a histogram for this set of data after grouping into classes of equal width.
(b) Repeat using more suitable class widths.
(c) In 2000, the manager introduced piped music, carpeting and better illumination in the workplace. The number of sick days taken by his employees in 2000 was as follows.

4	6	11	4	12	5	18	3	10	4	9	3	13
8	15	0	6	7	1	7	28	5	23	1	39	14
33	2	7	2	16	30	13	1	17	7	41	3	22
1	18	24	38	8	3	9	26	41	0	43		

Draw up a frequency table using the same classes as in part (b) and a histogram. Comment, in one sentence, on the two comparable histograms.

8. The histogram shows the distribution of marks obtained by students in a test.

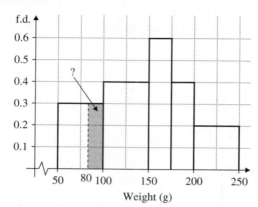

To find the number of people with a mark from, say, 50 to 100, use 0.3×50 to get 15. This is really the area of the rectangle with base from 50 to 100.

The number of people with a mark from 80 to 100 is given by the area of the shaded rectangle which is $0.3 \times 20 = 6$. So 6 people have marks from 80 to 100.

Similarly, use the areas of rectangles to find the number of people with marks in the following ranges.

(a) 100–130   (b) 90–150   (c) 90–160   (d) 160–230

8.8 Standard deviation

So far, we have met two measures of the way in which the data is spread. These are the range and, from a cumulative frequency curve, the interquartile range in which the middle 50% of the data lies. Standard deviation is the third measure of spread (or dispersion) about the mean. Standard deviation gives a more detailed picture of the way in which the data is dispersed about the mean as the centre of the distribution. Its main use is to compare two sets of data.

$$\text{Standard deviation (s.d.)} = \sqrt{\frac{\Sigma \ (x - \bar{x})^2}{n}}$$

where Σ means 'the sum of'
        x represents an item of data
and n is the number of items of data.

The symbol $\bar{x}$ is the mean where $\bar{x} = \dfrac{\Sigma x}{n}$

Sometimes the Greek symbol σ (sigma) is used for standard deviation instead of s.d.

Find the mean and standard deviation of the set of numbers
3, 4, 7, 9, 12.

Mean $\bar{x} = \dfrac{\Sigma x}{n} = \dfrac{3 + 4 + 7 + 9 + 12}{5} = 7$

Make a table

x	$(x - \bar{x})$	$(x - \bar{x})^2$
3	−4	16
4	−3	9
7	0	0
9	2	4
12	5	25
		$\Sigma(x - \bar{x})^2 = 54$

s.d. $= \sqrt{\dfrac{\Sigma(x - \bar{x})^2}{n}} = \sqrt{\dfrac{54}{5}} = 3\cdot286\ 335\ 345$

So $\bar{x} = 7$,   s.d. $= 3\cdot29$ (to 3 s.f.)

Find the mean and standard deviation of the frequency distribution.

x	0	2	3	6	8	10
f	2	2	1	1	3	1

$$\bar{x} = \frac{\Sigma fx}{\Sigma f}$$

$$\bar{x} = \frac{2(0) + 2(2) + 1(3) + 1(6) + 3(8) + 1(10)}{10} = 4\cdot7$$

x	f	$(x - \bar{x})$	$(x - \bar{x})^2$	$f(x - \bar{x})^2$
0	2	−4·7	22·09	44·18
2	2	−2·7	7·29	14·58
3	1	−1·7	2·89	2·89
6	1	1·3	1·69	1·69
8	3	3·3	10·89	32·67
10	1	5·3	28·09	28·09
	$n = \Sigma f = 10$			$\Sigma f(x - \bar{x})^2 = 124\cdot1$

For a frequency distribution, use the formula

s.d. $= \sqrt{\dfrac{\Sigma f(x - \bar{x})^2}{n}} = \sqrt{\dfrac{124\cdot1}{10}} = 3\cdot522\ 782\ 991$

So, $\bar{x} = 4\cdot7$
    s.d. $= 3\cdot52$ (to 3 s.f.)

Compare the results of ten students in two subjects as follows.

Mathematics
52   18   31   68   78   16   94   40   75   64
History
42   39   60   54   61   58   46   49   60   67

The mean mark for each subject is the same (53·6).
Yet the marks in mathematics are more widely spread with a range
94–16 compared to 63–39 for history.
It can be shown that the standard deviation for the mathematics
marks was 25·2 and that for history was 8·7.

Note:
When finding the standard deviation of grouped data, the mid-point of
each interval is used to represent that interval. The working then
follows that of the second example.
For example, the grouped data on the left would be replaced by the
data on the right.

Interval	f
1–5	2
6–10	1
11–15	2

Mid-point	f
3	2
8	1
13	2

Exercise 15

1. Find the mean and standard deviation of the following sets of
   numbers.
   (a) 1, 4, 8, 9, 10   (b) 3·2, 4·7, 5·1, 5·2, 6·3
   (c) 103, 109, 110, 112, 125, 131   (d) −5, −2, 0, 1, 2, 3

2. A production engineer can use either of two methods to make car
   batteries. He takes a random sample of 8 batteries made by each
   method and measures the lifetime of each battery. The lifetimes in
   months are given below.

Made by Method A	21	24	22	23	25	23	26	20
Made by Method B	23	27	20	22	26	20	27	19

   Find the mean and standard deviation for each sample. Which
   method would you recommend the engineer to use in future?
   Give your reason.

3. A professional golfer has two sets of clubs and is not sure which set
   gives better results. The last six scores playing with the 'Mizuno' set
   were 68, 71, 72, 68, 67, 74. The last six scores playing with the
   'Lynx' set were 70, 65, 74, 72, 75, 64. Find the mean and standard

deviation of the scores using each set of clubs.
Which set of clubs gives the more consistent results?

4. The children in two schools took the same test in mathematics and
their results are shown.

St Mary's School	Birchwood School
mean mark = 52%	mean mark = 51·8%
s.d = 7·2	s.d = 11·2

What can you say about these two sets of results?

5. As part of a health improvement programme, people from one town
and from one village in Gambia were measured. Here are the results.

People in town	People in village
mean height = 171 cm	mean height = 163 cm
s.d = 8·4	s.d = 3·7

What can you say about these two sets of results?

6. Find the mean and standard deviation of the frequency distributions.

(a)
x	4	8	12	16	20
f	4	2	3	1	3

(b)
x	3	4	5	7	9
f	1	5	4	2	3

7. Find the estimated mean and standard deviation of the frequency
distributions of the following grouped data. Note that it does not
matter if the intervals are of varying widths.

(a)
Interval	10–20	20–30	30–40	40–50
f	2	3	3	5

(b)
Interval	0–6	7–13	14–20	21–27
f	2	4	3	5

(c)
Interval	0–5	6–11	12–15	16–20
f	2	4	1	3

8. The weekly earnings of 30 people are shown in
the histogram.
(a) Find the number of people in each interval
and make a frequency distribution.
(b) Calculate the mean and standard deviation
of the weekly earnings for the whole group.

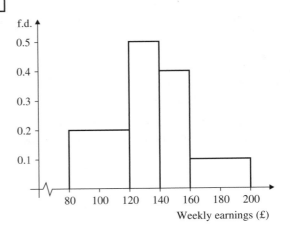

Mean and standard deviation on a calculator

These instructions are for a CASIO scientific calculator with the statistical functions. Most CASIO calculators work in the same way. For another make of calculator, read the instruction handbook. The process will be similar.

[A] Set the calculator into SD mode by pressing | MODE | | 3 |
   This gives the functions marked in blue.

[B] Clear the memories by pressing | SHIFT | | KAC |

   This is essential as most calculators retain data in this mode even after switch off.

[C] Enter the data (use data from the first example on page 335).

3		DATA
4		DATA
7		DATA
9		DATA
12		DATA

[D] The answers have now been worked out!

   To get: $\bar{x}$ press | SHIFT | | 1 | $\bar{x} = 7$

   s.d. $(x\sigma_n)$ press | SHIFT | | 2 | s.d. $= 3.29$

To enter data from the frequency distribution in the second example on page 335 (remember to clear memory first):

0	×	2	DATA		6	×	1	DATA
2	×	2	DATA		8	×	3	DATA
3	×	1	DATA		10	×	1	DATA

Note two things:

1. Enter | x | | × | | f |. Do not reverse the order.
2. Do not press | = |.

Try some of the previous questions using the calculator in this way. Let the machine do the 'number crunching'!

Exercise 16

1. A farmer records the weights of a batch of sugar beet.

   Calculate the mean and standard deviation for the weight of the sugar beet.

Weight (grams)	Frequency
204–216	7
217–229	18
230–242	30
243–255	25
256–268	10

2. The farmer in Question **1** discovered that his scales were incorrect and that each sugar beet had been weighed 10 grams less than its actual weight.
   Draw up a corrected frequency distribution and calculate the mean and standard deviation of this new distribution.
   (a) How does the mean compare to the old mean?
   (b) How does the standard deviation compare to the old standard deviation?

3. The following are the diameters in cm of 12 flowers produced by an experimental botanist.
   3·8   3·3   2·9   1·3   1·0   0·3   0·4   0·9   1·0   1·8   2·2   3·0
   Find the mean and standard deviation of the measurements.

4. The botanist found a new fertiliser which doubled the diameter of the flowers in Question **3**. Write down the new set of measurements and find the new mean and standard deviation.
   (a) How does the new mean compare to the original mean?
   (b) How does the new standard deviation compare?

Summary of mean and standard deviation rules

The four questions of the last exercise illustrate two rules which apply to the mean and to standard deviation.

Rule 1 If a constant is added to each member of a set of data, the mean is also increased by that constant, but the standard deviation (i.e. the spread) is unchanged.

Rule 2 If each member of a distribution is multiplied by a constant, both the mean and the standard deviation are multiplied by that constant.

9 PROBABILITY

9.1 Relative frequency

To work out the probability of a drawing pin landing point up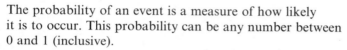
we can conduct an experiment in which a drawing pin is dropped many
times. If the pin lands 'point up' on x occasions out of a total number
of N trials, the *relative frequency* of landing 'point up' is $\dfrac{x}{N}$.

When an experiment is repeated many times we can use the relative
frequency as an estimate of the probability of the event occurring.
The graph shows how the ratio $\dfrac{x}{N}$ settles down to a consistent

value as N becomes large (say 1000 or 10 000 trials). So the
relative frequency is a more reliable guide to the
probability of the event occurring when the number
of trials is large.

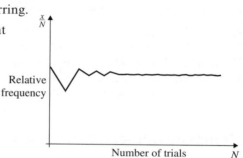

The probability of an event is a measure of how likely
it is to occur. This probability can be any number between
0 and 1 (inclusive).

There are four different ways of estimating probabilities.

Method A   Use symmetry (the theoretical probability).
- The probability of rolling a 3 on a fair dice is $\frac{1}{6}$.
  This is because all the scores, 1, 2, 3, 4, 5, 6 are equally likely.
- Similarly the probability of getting a head when tossing a fair
  coin is $\frac{1}{2}$.

Method B   Conduct an experiment or survey to collect data
- The dice shown has five faces, not all the same shape. To estimate
  the probability of the dice landing with a '1' showing, I could
  conduct an experiment to see what happened in, say, 500 trials.

- I might want to know the probability that the next car going past
  the school gates is driven by a woman.
  I could conduct a survey in which the drivers of cars are recorded
  over a period of time.

Method C   Look at past data
- If I wanted to estimate the probability of my plane crashing as it
  lands at Heathrow airport I could look at accident records at
  Heathrow over the last five years or so.

Method D   Make a subjective estimate

We have to use this method when the event is not repeatable. It is not really a 'method' in the same sense as are methods A, B, C.

- We might want to estimate the probability of England beating France in a soccer match next week. We could look at past results but these could be of little value for all sorts of reasons. We might consult 'experts' but even they are notoriously inaccurate in their predictions.

Exercise 1

In Questions **1** to **10** state which method A, B, C or D you would use to estimate the probability of event given.

1. The probability that a person chosen at random from a class will be left-handed.

2. The probability that there will be snow in the ski resort to which a school party is going in February next year.

3. The probability of drawing an 'ace' from a pack of playing cards.

4. The probability that you hole a six foot putt when playing golf.

5. The probability that the world record for running 1500 m will be under 3 min 20 seconds by the year 2020.

6. The probability of winning the National Lottery.

7. The probability of rolling a 3 using a dice which is suspected of being biased.

8. The probability that a person selected at random would vote 'Labour' in a general election tomorrow.

9. The probability that a train will arrive within ten minutes of its scheduled arrival time.

10. The probability that the current Wimbledon Ladies Champion will successfully defend her title next year.

11. Saddam rolls a dice and records the number of ones that he gets.

Number of rolls	10	25	50	100	300	1000	2000
Number of ones	1	3	5	12	54	160	338

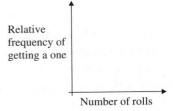

Plot a graph of relative frequency of getting one against the number of rolls.

12. Conduct an experiment where you cannotpredict the result.
    You could roll a dice with a piece of 'blu-tack' stuck to it.
    Or make a spinner where the axis is not quite in the centre.
    Or drop a drawing pin.
    Conduct the experiment many times and
    work out the relative frequency of a 'success' after every
    10 or 20 trials.
    Plot a relative frequency graph to see if the results
    'settle down' to a consistent value.

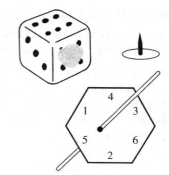

13. The spinner has an equal chance of giving
    any digit from 0 to 9.
    Four friends did an experiment when they
    spun the pointer a different number of
    times and recorded the number of zeros
    they got.
    Here are their results.

	Number of spins	Number of zeros	Relative frequency
Steve	10	2	0·2
Nick	150	14	0·093
Mike	200	41	0·205
Jason	1000	104	0·104

One of the four recorded his results incorrectly. Say who you think
this was and explain why.

9.2 Working out probabilities

The probability of an event occurring can be calculated using
symmetry. We argue, for example, that when we toss a coin we have an
equal chance of getting a 'head' or a 'tail'. So the probability of
spinning a 'head' is a half.
We write 'p (spinning a head) $=\frac{1}{2}$'.

A single card is drawn from a pack of 52 playing cards.
Find the probability of the following results:
(a) the card is a Queen,
(b) the card is a Club,
(c) the card is the Jack of Hearts.

There are 52 equally likely outcomes of the 'trial' (drawing a card).
(a) p (Queen) $=\frac{4}{52}=\frac{1}{13}$
(b) p (Club) $=\frac{13}{52}=\frac{1}{4}$
(c) p (Jack of Hearts) $=\frac{1}{52}$.

13 Spades
13 Hearts
13 Diamonds
13 Clubs

Total   52

Exercise 2

1. If one card is picked at random from a pack of 52 playing cards, what is the probability that it is:
   (a) a King,
   (b) the Ace of Clubs,
   (c) a Heart?

2. Nine counters numbered 1, 2, 3, 4, 5, 6, 7, 8, 9 are placed in a bag. One is taken out at random. What is the probability that it is:
   (a) a '5',   (b) divisible by 3,
   (c) less than 5,   (d) divisible by 4?

3. A bag contains 5 green balls, 2 red balls and 4 yellow balls. One ball is taken out at random. What is the probability that it is:
   (a) green,
   (b) red,
   (c) yellow?

4. A cash bag contains two 20p coins, four 10p coins, five 5p coins, three 2p coins and three 1p coins. Find the probability that one coin selected at random is:
   (a) a 10p coin,
   (b) a 2p coin,
   (c) a silver coin.

5. One card is selected at random from those shown. Find the probability of selecting:
   (a) a Heart,
   (b) an Ace,
   (c) the 10 of Clubs,
   (d) a Spade,
   (e) a Heart or a Diamond.

6. A pack of playing cards is well shuffled and a card is drawn. Find the probability that the card is:
   (a) a Jack.
   (b) a Queen or a Jack,
   (c) the ten of Hearts,
   (d) a Club higher than the 9 (count the Ace as high).

7. One ball is selected at random from those opposite. Find the probability of selecting:
   (a) a white ball,
   (b) a yellow or a black ball,
   (c) a ball which is not red.

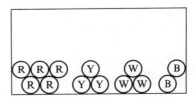

R = red
Y = yellow
W = white
B = black

8. (a) A bag contains 5 red balls, 6 green balls and 2 black balls. Find the probability of selecting:
   (i) a red ball   (ii) a green ball.
   (b) One black ball is removed from the bag. Find the new probability of selecting:
   (i) a red ball   (ii) a black ball.

9. A pack of playing cards is split so that all the picture cards (Kings, Queens, Jacks) are in Pile A and all the other cards are in Pile B. Find the probability of selecting:
   (a) the Queen of Clubs from pile A.
   (b) the seven of Spades from pile B.
   (c) any Heart from pile B.

pile A

pile B

10. One ball is selected at random from a bag containing 12 balls of which x are white.
    (a) What is the probability of selecting a white ball?
    When a further 6 white balls are added the probability of selecting a white ball is doubled.
    (b) Find x.

11. A large firm employs 3750 people.
    One person is chosen at random.
    What is the probability that that person's birthday is on a Monday in the year 2000?

12. The numbering on a set of 28 dominoes is as follows:

| 6/6 | 6/5 | 6/4 | 6/3 | 6/2 | 6/1 | 6/0 | 5/5 | 5/4 | 5/3 |

| 5/2 | 5/1 | 5/0 | 4/4 | 4/3 | 4/2 | 4/1 | 4/0 | 3/3 | 3/2 |

| 3/1 | 3/0 | 2/2 | 2/1 | 2/0 | 1/1 | 1/0 | 0/0 |

(a) What is the probability of drawing a domino from a full set with
    (i) at least one six on it?
    (ii) at least one four on it?
    (iii) at least one two on it?
(b) What is the probability of drawing a 'double' from a full set?
(c) If I draw a double five which I do not return to the set, what is the probability of drawing another domino with a five on it?

Expectation

A fair dice is rolled 240 times. How many times would you expect to roll a number greater than 4?

We can roll a 5 or a 6 out of the six equally likely outcomes.

$\therefore$ p (number greater than 4) $= \frac{2}{6} = \frac{1}{3}$

Expected number of successes = (probability of a success) × (number of trials)

Expected number of scores greater than $4 = \frac{1}{3} \times 240$
$$= 80$$

Exercise 3

1. A fair dice is rolled 300 times. How many times would you expect to roll:
   (a) an even number
   (b) a 'six'?

2. The spinner shown has four equal sectors. How many 3's would you expect in 100 spins?

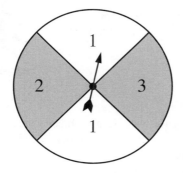

3. About one in eight of the population is left-handed. How many left-handed people would you expect to find in a firm employing 400 people?

4. A bag contains a large number of marbles of which one in five is red. If I randomly select one marble on 200 occasions how many times would I expect to select a red marble?

5. The spinner shown is used for a simple game. A player pays 10p and then spins the pointer, winning the amount indicated.
   (a) What is the probability of winning nothing?
   (b) If the game is played by 200 people how many times would you expect the 50p to be won?

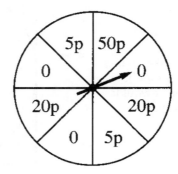

ne numbered cards are shuffled and put into a pile.

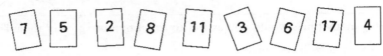

One card is selected at random and not replaced. A second card is then selected.
(a) If the first card was the '11' find the probability of selecting an even number with the second draw.
(b) If the first card was an odd number, find the probability of selecting another odd number.

7. When the ball is passed to Vinnie, the probability that he kicks it is 0·2 and the probability that he heads it is 0·1. Otherwise he will miss the ball completely, fall over and claim a foul.
In one game his ever-optimistic team mates pass the ball to Vinnie 150 times.
(a) How often would you expect him to head the ball?
(b) How often would you expect him to miss the ball?

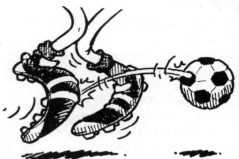

8. A small pack of 20 cards consists of the Ace, King, Queen, Jack and ten of all four suits. One card is selected and then replaced. This procedure is repeated 100 times.
How many times would you expect to select:
(a) an Ace,
(b) the Queen of Spades,
(c) a red card,
(d) any King or Queen?

† 9. A point is chosen at random from inside the square.
(a) What is the probability of choosing a point which lies outside the circle? Give your answer as a decimal to 4 s.f.
(b) In a computer simulation 5000 points are chosen using a random number generator. How many points would you expect to be chosen which lie outside the circle?

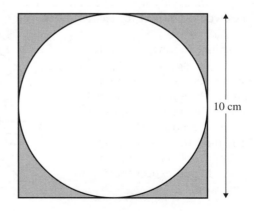

10 cm

Combined events

When a 10p coin and a 50p coin
are tossed together we have two
events occurring:

- tossing the 10p coin
- tossing the 50p coin.

We can list all the possible outcomes as shown.

Note that a *sample space* is a list of all the possible
outcomes and a *sample space diagram* is a table or
diagram which shows all the possible outcomes.

(10p first) head, head
head, tail
tail, head
tail, tail.

A red dice and a black dice are thrown together. Show all the
possible outcomes.

We could list them in pairs with the red dice first:

(1,1), (1,2), (1,2), ...(1,6)
(2,1), (2,2), ...
and so on.

In this example it is easier to see all
the possible outcomes when
they are shown on a grid.

The X shows 6 on the red dice
and 2 on the black.

The ◯ shows 3 on the red dice
and 5 on the black.

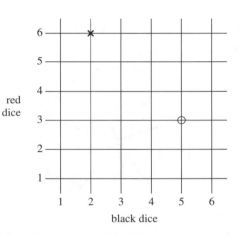

Exercise 4

1. Three coins (10p, 20p, 50p)
   are tossed together.
   (a) List all the possible ways in which they could land.
   (b) What is the probability of getting three heads?

2. List all the possible outcomes when four coins are tossed together.
   How many are there altogether?

3. A black dice and a white dice are thrown together
   (a) Draw a grid to show all the possible outcomes.
       [See the box above.]
   (b) How many ways can you get a total of nine on the two dice?
   (c) What is the probability that you get a total of nine?

..e two spinners shown are spun together.

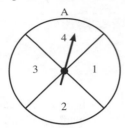 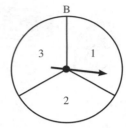

List all the possible outcomes. Find the probability of obtaining:
(a) a total of 4,   (b) the same number on each spinner.

5. Four friends, Wayne, Xavier, Yves and Zara, each write their name
on a card and the four cards are placed in a hat. Two cards are
chosen to decide who does the maths homework that night.
List all the possible combinations.
What is the probability that the names drawn are Xavier and Yves?

6. The spinner is spun and the dice is thrown at the same time.

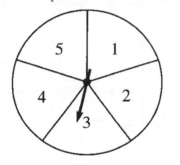 

(a) Draw a grid to show all the possible outcomes.
(b) A 'win' occurs when the number on the spinner is greater than or
equal to the number on the dice. Find the probability of a 'win'.

7. A red dice and a blue dice are thrown at the same time. Show all
the possible outcomes in a systematic way. Find the probability of
obtaining:
(a) a total of 10,   (b) a total of 12,
(c) a total less than 6,   (d) the same number on both dice.

8. A dice is thrown; when the result has been recorded, the dice is
thrown a second time. Display all the possible outcomes of the two
throws. Find the probability of obtaining:
(a) a total of 4 from the two throws,
(b) a total of 8 from the two throws,
(c) a total between 5 and 9 inclusive from the two throws,
(d) a number on the second throw which is double the number on
the first throw,
(e) a number on the second throw which is four times the number
on the first throw.

9. Two cards are selected at random from the 3, 4, 5 and 6 of Hearts.
Find the probability that the total of the two cards is more than 9.

10. (a) How many possible outcomes are there when six coins are tossed together?
(b) What is the probability of tossing six heads?

11. Shirin and Dipika are playing a game in which three coins are tossed. Shirin wins if there are no heads or one head. Dipika wins if there are either two or three heads. Is the game fair to both players?

12. Pupils X, Y and Z play a game in which four coins are tossed.

X wins if there is 0 or 1 head.
Y wins if there are 2 heads.
Z wins if there are 3 or 4 heads.

Is the game fair to all three players?

13. Four cards numbered 2, 4, 5 and 7 are mixed up and placed face down.

In a game you pay 10p to select two cards. You win 25p if the total of the two cards is nine.
How much would you expect to win or lose if you played the game 12 times?

†**14.** Two dice and two coins are thrown at the same time. Find the probability of obtaining:
(a) two heads and a total of 12 on the dice,
(b) a head, a tail and a total of 9 on the dice,

...se 5 A practical exercise

1. The ⟨RAN #⟩ button on a calculator generates random numbers between ·000 and ·999. It can be used to simulate tossing three coins. We could say any *odd* digit is a *tail* and any *even* digit is a *head*.

So the number ·568 represents THH
and ·605 represents HHT

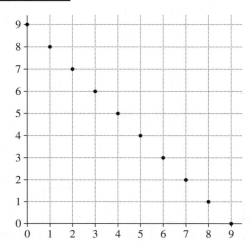

Use the ⟨RAN #⟩ button to simulate the tossing of three coins.

'Toss' the three coins 32 times and work out the relative frequencies of (a) three heads and (b) two heads and a tail.

Compare your results with the values that you would expect to get theoretically.

2. A calculator can be used to simulate throwing imaginary dice with 10 faces numbered 0, 1, 2, 3, 4, 5, 6, 7, 8, 9.

The ⟨RAN #⟩ button gives a random 3 digit number between 0·000 and 0·999. We can use the first 2 digits after the point to represent the numbers on two ten-sided dice.

So ⟨0·763⟩ means 7 on one dice and 6 on the second.
[Ignore the '3']

and ⟨0·031⟩ means 0 on one dice and 3 on the second.
[Ignore the '1']

We add the numbers on the two 'dice' to give the total.

So for ⟨0·763⟩ the total is 13 and for ⟨0·031⟩ the total is 3.

Press the ⟨RAN #⟩ button lots of times and record the totals in a tally chart.

Total	0	1	2	. . .	18
Tally		||	||||	. . .	

This grid can be used to predict the result. There are 100 equally likely outcomes like 3,7 or 4,1. The ten results shown each give a total of 9 which is the most likely score. So in 100 throws you could expect to get a total of 9 on ten occasions.

Plot your results on a frequency graph. Compare your results with the values you would expect.

3. In Gibson Academy there are six forms (A, B, C, D, E, F) in Year 10 and the Mathematics Department has been asked to work out a schedule so that, over 5 weeks, each team can play each of the others in a basketball competition.

The games are all played at lunch time on Wednesdays and three games have to be played at the same time.

For example in Week 1 we could have

> Form A *v* Form B
> Form C *v* Form D
> Form E *v* Form F

Work out a schedule for the remaining four weeks so that each team plays each of the others. Check carefully that each team plays every other team just once.

9.3 Exclusive events

Events are *mutually exclusive* if they cannot occur at the same time.

Examples
- Selecting an ace ⎫ from a
  Selecting a ten ⎬ pack of cards

- Tossing a 'head'
  Tossing a 'tail'

- The total sum of the probabilities of mutually exclusive events is 1.
- The probability of something happening is 1 minus the probability of it not happening.

The 'OR' rule : the addition rule

- For exclusive events A and B

    $p(\text{A or B}) = p(\text{A}) + p(\text{B})$

Many questions involving exclusive events can be done *without* using the addition rule.

...rd is selected at random from a pack of 52 playing cards.
What is the probability of selecting any King or Queen?

(a) Count the number of ways in which a King *or* Queen can occur.
That is 8 ways.
p (selecting a King or a Queen) $= \frac{8}{52} = \frac{2}{13}$

(b) Since 'selecting a King' and 'selecting a Queen' are exclusive events, we could use the addition law.

p (selecting a King) $= \frac{4}{52}$

p (selecting a Queen) $= \frac{4}{52}$

So, p (selecting a King or a Queen) $= \frac{4}{52} + \frac{4}{52} = \frac{8}{52}$ as before.

You can decide for yourself which method is easier in any given question.

Non-exclusive events

If the events are not exclusive, the addition rule *cannot* be used.

Here is a spinner with numbers and colours.
The events 'spinning a red' and 'spinning a 1' are *not* exclusive because a red and a 1 can occur at the same time.

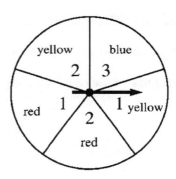

Exercise 6

1. State whether or not these pairs of events are mutually exclusive.
   (a) Drawing a Club from a pack of cards – Drawing a red card.
   (b) Drawing an ace from a pack of cards – Drawing a Diamond.
   (c) [Two coins are tossed] Getting at least one head – Getting two tails
   (d) [Two dice are thrown] Getting the same on both dice – Getting two numbers less than 4.
   (e) [A single dice is thrown] Getting a prime number – Getting an even number.

2. A dice is thrown. Pair the following events so that each pair is mutually exclusive.
   Event A getting an odd number
   Event B getting a 6
   Event C getting a prime number
   Event D getting a number greater than 4
   Event E getting a factor of 4
   Event F getting a 2

3. A bag contains 10 red balls, 5 blue balls and 7 green balls. Find the probability of selecting at random:
(a) a red ball,
(b) a green ball,
(c) a blue *or* a red ball,
(d) a red *or* a green ball.

4. A roulette wheel has the numbers 1 to 36 once only. What is the probability of spinning either a 10 or a 20?

5. A bag contains a large number of balls coloured red, white, black or green. The probabilities of selecting each colour are as follows:

colour	red	white	black	green
probability	0·3	0·1		0·3

Find the probability of selecting a ball
(a) which is black
(b) which is not white.

6. A fair dice is rolled. What is the probability of rolling either a 1 or a 6?

7. From a pack of cards I have already selected the King, Queen, Jack and ten of Diamonds. What is the probability that on my next draw I will select either the Ace or the nine of Diamonds?

8. In an opinion poll people were asked to state the party for which they are going to vote.

If one person is chosen at random from the sample, what is the probability that he or she would vote either Conservative or Labour?

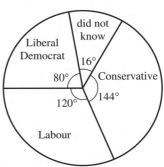

9. In a survey the number of people in cars is recorded.
When a car passes the school gates the probability of having 1, 2, 3, ... occupants is as follows.

number of people	1	2	3	4	more than 4
probability	0·42	0·23		0·09	0·02

(a) Find the probability that the next car past the school gates contains   (i) three people   (ii) less than 4 people.
(b) One day 2500 cars passed the gates. How many of the cars would you expect to have 2 people inside?

...ty pupils were asked to state the activities they enjoyed from ...wimming (S), tennis (T) and hockey (H). The numbers in each set are shown.
One pupil is randomly selected.
(a) Which of the following pairs of events are exclusive?
    (i) 'selecting a pupil from S', 'selecting a pupil from H'.
    (ii) 'selecting a pupil from S', 'selecting a pupil from T'.
(b) What is the probability of selecting a pupil who enjoyed either hockey or tennis?

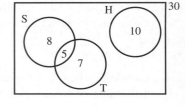

11. The spinner shown has four equal sectors.
    (a) Which of the following pairs of events are exclusive?
        (i) 'spinning 1', 'spinning green',
        (ii) 'spinning 3', 'spinning 2'.
        (iii) 'spinning blue', 'spinning 1'.
    (b) What is the probability of spinning either a 1 or a green?

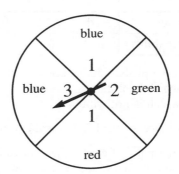

12. Here is a spinner with unequal sectors. When the pointer is spun the probability of getting each colour and number is as follows.

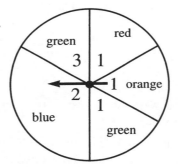

red	0·1
blue	0·3
orange	0·2
green	0·4

1	0·5
2	0·3
3	0·2

(a) What is the probability of spinning either 1 or 2?
(b) What is the probability of spinning either blue or green?
(c) Why is the probability of spinning either a 1 or a green *not* $0.5 + 0.4$?

13. According to a weather forecaster in Norway, 'The probability of snow on Saturday is 0·6 and the probability of snow on Sunday is 0·5 so it is certain to snow over the weekend.' Explain why this statement is wrong.

9.4 Independent events, tree diagrams

Two events are *independent* if the occurrence of one event is unaffected by the occurrence of the other.
So, obtaining a head on one coin and a tail on another coin when the coins are tossed at the same time are independent.

The 'AND' rule: the multiplication rule

- For independent events A and B
  $p(\text{A and B}) = p(\text{A}) \times p(\text{B})$

The multiplication rule only works for *independent* events.

Two coins are tossed and the results listed.

  HH HT TH TT

The probability of tossing two heads is $\frac{1}{4}$.

By the multiplication rule for independent events:

$p(\text{two heads}) = p(\text{head on first coin}) \times p(\text{head on second coin})$

$\therefore \quad p(\text{two heads}) = \frac{1}{2} \times \frac{1}{2} = \frac{1}{4}$ as before.

A fair coin is tossed and a fair dice is rolled. Find the probability of obtaining a 'head' and a 'six'.

The two events are independent.

$p(\text{head } and \text{ six}) = p(\text{head}) \times p(\text{six}) = \frac{1}{2} \times \frac{1}{6} = \frac{1}{12}$

When the two spinners are spun, what is the probability of getting a B on the first and a 3 on the second?

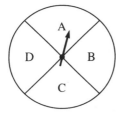 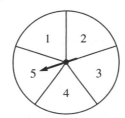

The events 'B on the first spinner' and '3 on the second spinner' are independent.

$\therefore \quad p(\text{spinning B and 3}) = p(\text{B}) \times p(3) = \frac{1}{4} \times \frac{1}{5} = \frac{1}{20}$

Exercise 7

1. A card is drawn from a pack of playing cards and a dice is thrown.
   Events A and B are as follows:
   A: 'a Jack is drawn from the pack'
   B: 'a three is thrown on the dice'.
   (a) Write down the values of $p(\text{A})$, $p(\text{B})$.
   (b) Write down the value of $p(\text{A and B})$.

2. A coin is tossed and a dice is thrown. Write down the probability of obtaining:
   (a) a 'head' on the coin,
   (b) an odd number on the dice,
   (c) a 'head' on the coin and an odd number on the dice.

A contains 3 red balls and 3 white balls. Box B contains 1 red and 4 white balls.

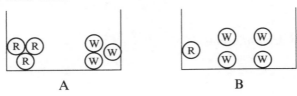

A   B

One ball is randomly selected from Box A and one from Box B. What is the probability that both balls selected are red?

4. In an experiment, a card is drawn from a pack of playing cards and a dice is thrown.
   Find the probability of obtaining:
   (a) a card which is an ace and a six on the dice,
   (b) the king of clubs and an even number on the dice,
   (c) a heart and a 'one' on the dice.

5. A card is taken at random from a pack of playing cards and replaced. After shuffling, a second card is selected. Find the probability of obtaining:
   (a) two cards which are clubs,
   (b) two Kings,
   (c) two picture cards.

6. A ball is selected at random from a bag containing 3 red balls, 4 black balls and 5 green balls. The first ball is replaced and a second is selected. Find the probability of obtaining:
   (a) two red balls,   (b) two green balls.

7. The letters of the word 'INDEPENDENT' are written on individual cards and the cards are put into a box. A card is selected and then replaced and then a second card is selected. Find the probability of obtaining:
   (a) the letter 'P' twice,   (b) the letter 'E' twice.

8. A fruit machine has three independent reels and pays out a Jackpot of £100 when three apples are obtained.

   Each reel has 15 pictures. The first reel has 3 apples, the second has 4 apples and the third has 2 apples.
   Find the probability of winning the Jackpot.

9. Three coins are tossed and two dice are thrown at the same time.
   Find the probability of obtaining:
   (a) three heads and a total of 12 on the dice,
   (b) three tails and a total of 9 on the dice.

10. A coin is biased so that it shows 'Heads' with a probability of $\frac{2}{3}$.
The same coin is tossed three times. Find the probability of
obtaining:
(a) two tails on the first two tosses,
(b) a head, a tail and a head (in that order).

11. A fair dice and a biased dice are thrown together. The probabilities
of throwing the numbers 1 to 6 are shown for the biased dice.

Fair   Biased

$p(6) = \frac{1}{4}; \ \ p(1) = \frac{1}{12}$
$p(2) = p(3) = p(4) = p(5) = \frac{1}{6}$

Find the probability of obtaining a total of 12 on the two dice.

12. Philip and his sister toss a coin to decide who does the washing up.
If it's heads Philip does it. If it's tails his sister does it.
What is the probability that Philip does the washing up every day
for a week (7 days)?

13. Here is the answer sheet for five questions in a
multiple choice test.
What is the probability of getting all five correct
by guessing?

Answer sheet			
1. (A)	(B)	(C)	
2. (A)	(B)		
3. (A)	(B)	(C)	
4. (A)	(B)	(C)	(D)
5. (A)	(B)		

14. This spinner has six equal sectors.

Find the probability of getting:
(a) 20 Qs in 20 trials.
(b) No Qs in n trials
(c) At least one Q in n trials.

.agrams

A bag contains 5 red balls and 3 green balls.
A ball is drawn at random and then replaced.
Another ball is drawn.

What is the probability that both balls are green?

The branch marked * involves the selection
of a green ball twice.
The probability of this event is obtained by
simply multiplying the fractions on the
two branches.

$\therefore \quad p(\text{two green balls}) = \frac{3}{8} \times \frac{3}{8} = \frac{9}{64}$

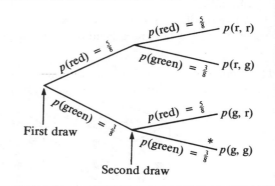

$p(\text{red}) = \frac{5}{8}$ — $p(\text{r, r})$
$p(\text{red}) = \frac{5}{8}$
$p(\text{green}) = \frac{3}{8}$ — $p(\text{r, g})$

First draw
$p(\text{green}) = \frac{3}{8}$
$p(\text{red}) = \frac{5}{8}$ — $p(\text{g, r})$
$p(\text{green}) = \frac{3}{8}$ * — $p(\text{g, g})$

Second draw

A bag contains 5 red balls and 3 green balls.
A ball is selected at random and *not* replaced.
A second ball is then selected.
Find the probability of selecting
(a) two green balls
(b) one red ball and one green ball.

Notice that since the first ball is not
replaced there are only 7 balls in the bag
for the second selection.

(a) $p(\text{two green balls}) = \frac{3}{8} \times \frac{2}{7}$

$\quad = \frac{3}{28}$

(b) $p(\text{one red, one green}) = (\frac{5}{8} \times \frac{3}{7}) + (\frac{3}{8} \times \frac{5}{7})$

$\quad = \frac{15}{28}$

$p(\text{red}) = \frac{4}{7}$ — $p(\text{r, r}) = \frac{5}{8} \times \frac{4}{7}$
$p(\text{red}) = \frac{5}{8}$
$p(\text{green}) = \frac{3}{7}$ — $p(\text{r, g}) = \frac{5}{8} \times \frac{3}{7}$

First draw
$p(\text{green}) = \frac{3}{8}$
$p(\text{red}) = \frac{5}{7}$ — $p(\text{g, r}) = \frac{3}{8} \times \frac{5}{7}$
$p(\text{green}) = \frac{2}{7}$ — $p(\text{g, g}) = \frac{3}{8} \times \frac{2}{7}$

Second draw

Notice that we can add here because the events 'red then green' and 'green then red' are exclusive.

As a check all the fractions at the ends of the branches should add up to one.

So $(\frac{5}{8} \times \frac{4}{7}) + (\frac{5}{8} \times \frac{3}{7}) + (\frac{3}{8} \times \frac{5}{7}) + (\frac{3}{8} \times \frac{2}{7}) = \frac{20}{56} + \frac{15}{56} + \frac{15}{56} + \frac{6}{56}$

$= 1$

Exercise 8

1. A bag contains 10 discs; 7 are black and 3 white. A disc is selected,
   and then replaced. A second disc is selected. Copy and complete the
   tree diagram showing all the probabilities and outcomes.

   Find the probability of the following:
   (a) both discs are black
   (b) both discs are white.

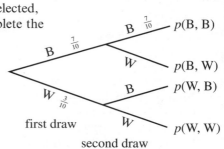

B $\frac{7}{10}$ — $p(\text{B, B})$
B $\frac{7}{10}$
W — $p(\text{B, W})$
W $\frac{3}{10}$
B — $p(\text{W, B})$

first draw
W — $p(\text{W, W})$

second draw

2. A bag contains 5 red balls and 3 green balls.
A ball is drawn and then replaced before a ball is drawn again.
Draw a tree diagram to show all the possible outcomes. Find the
probability that:
(a) two green balls are drawn
(b) the first ball is red and the second is green.

3. A bag contains 7 green discs and 3 blue discs. A disc is drawn and
not replaced.
A second disc is drawn. Copy and complete the tree diagram.

Find the probability that:
(a) both discs are green
(b) both discs are blue.

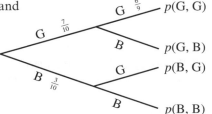

4. A bag contains 5 red balls, 3 blue balls
and 2 yellow balls. A ball is drawn and
not replaced. A second ball is drawn.

Find the probability of drawing:
(a) two red balls
(b) one blue ball and one yellow ball (in any order)
(c) two yellow balls.

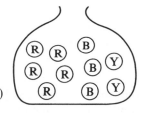

5. A bag contains 4 red balls, 2 green balls
and 3 blue balls. A ball is drawn and not
replaced. A second ball is drawn.

Find the probability of drawing:
(a) two blue balls
(b) two red balls
(c) one red ball and one blue ball (in any order)
(d) one green ball and one red ball (in any order).

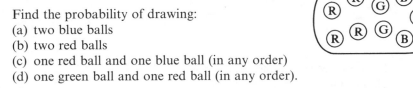

6. A six-sided dice is thrown three times. Complete the tree diagram,
showing at each branch the two events:
'three' and 'not three' (written $\bar{3}$).

What is the probability of throwing a total of
(a) three threes,
(b) no threes,
(c) one three,
(d) at least one three (use part (b))?

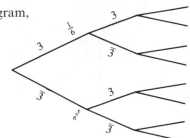

7. A card is drawn at random from a pack of 52 playing cards. The
card is replaced and a second card is drawn. This card is replaced
and a third card is drawn. What is the probability of drawing:
(a) three hearts?
(b) at least two hearts?
(c) exactly one heart?

...contains 6 red marbles and 4 yellow marbles. A marble is
...wn at random and *not* replaced. Two further draws are made,
again without replacement. Find the probability of drawing:
(a) three red marbles,
(b) three yellow marbles,
(c) no red marbles,
(d) at least one red marble. [Hint: Use part (c).]

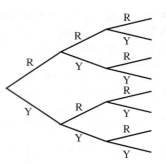

9. When a cutting is taken from a geranium the probability that it
grows is $\frac{3}{4}$. Two cuttings are taken. What is the probability that:
(a) both cuttings grow
(b) neither of them grows?

10. A dice has its six faces marked 0, 1, 1, 1, 6, 6.
Two of these dice are thrown together and
the total score is recorded.
Draw a tree diagram.

(a) How many different totals are possible?
(b) What is the probability of obtaining
a total of 7?

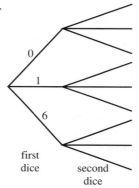

11.

Louise loves the phone! When the
phone rings $\frac{2}{3}$ of the calls are for
Louise, $\frac{1}{4}$ are for her brother Herbert
and the rest are for me.
On Christmas Day I answered the
phone twice. Find the probability that:
(a) both calls were for Louise
(b) one call was for me and one was
for Herbert.

12. Bag A contains 3 red balls and 1 white ball.
Bag B contains 2 red balls and 3 white balls.

A ball is chosen at random from each bag in turn.
Find the probability of taking:
(a) a white ball from each bag
(b) two balls of the same colour.

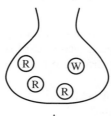

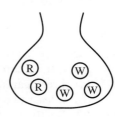

A

B

13. A teacher decides to award exam grades A, B or C by a new fairer method. Out of 20 children, three are to receive A's, five B's and the rest C's. She writes the letters A, B and C on 20 pieces of paper and invites the pupils to draw their exam result, going through the class in alphabetical order. Find the probability that:
(a) the first three pupils all get grade 'A'
(b) the first three pupils all get grade 'B'
(c) the first four pupils all get grade 'B'.
(Do not cancel down the fractions.)

There are 10 boys and 12 girls in a class. Two pupils are chosen at random. What is the probability that one boy and one girl are chosen in either order?

Here two pupils are chosen. It is like choosing one pupil without replacement and then choosing a second. Here is the tree diagram.

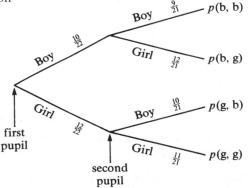

$$p(b, g) + p(g, b) = (\tfrac{10}{22} \times \tfrac{12}{21}) + (\tfrac{12}{22} \times \tfrac{10}{21})$$
$$= \tfrac{40}{77}$$

● Remember:
When a question says '2 balls are drawn' or '3 people are chosen', remember to draw a tree diagram *without* replacement.

Exercise 9

1. There are 1000 components in a box of which 10 are known to be defective. Two components are selected at random. What is the probability that:
(a) both are defective   (b) neither are defective
(c) just one is defective?
(Do *not* simplify your answers.)

2. There are 10 boys and 15 girls in a class. Two children are chosen at random. What is the probability that:
(a) both are boys   (b) both are girls
(c) one is a boy and one is a girl?

3. Two similar spinners are spun together.
(a) What is the most likely total score on the two spinners?
(b) What is the probability of obtaining this score on three successive spins of the two spinners?

...ontains 3 red, 4 white and 5 green balls. Three balls are ...cted without replacement.

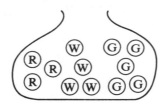

Find the probability that the three balls chosen are:

(a) all red   (b) all green   (c) one of each colour.

(d) If the selection of the three balls was carried out 1100 times, how often would you expect to choose three red balls?

5. The diagram represents 15 students in a class and shows the following.

G   represents those who are girls

S   represents those who are swimmers

F   represents those who believe in Father Christmas.

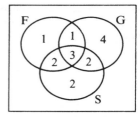

A student is chosen at random. Find the probability that the student:

(a) can swim   (b) is a girl swimmer

(c) is a boy swimmer who believes in Father Christmas.

Two students are chosen at random. Find the probability that:

(d) both are boys,   (e) neither can swim.

6. There are 500 ball bearings in a box of which 100 are known to be undersize. Three ball bearings are selected at random. What is the probability that:

(a) all three are undersize   (b) none are undersize?

Give your answers as decimals correct to three significant figures.

7. There are 9 boys and 15 girls in a class. Three children are chosen at random. What is the probability that:

(a) all three are boys   (b) all three are girls

(c) one is a boy and two are girls?

Give your answers as fractions.

8. A box contains x milk chocolates and y plain chocolates. Two chocolates are selected at random. Find, in terms of x and y, the probability of choosing:

(a) a milk chocolate on the first choice

(b) two milk chocolates

(c) one of each sort

(d) two plain chocolates.

9. A pack of z cards contains x 'winning cards'. Two cards are selected at random. Find, in terms of x and z, the probability of choosing:

(a) a 'winning' card on the first choice

(b) two 'winning cards' in the two selections

(c) exactly one 'winning' card in the pair.

10. When a golfer plays any hole, he will take 3, 4, 5, 6, or 7 strokes with probabilities of $\frac{1}{10}$, $\frac{1}{5}$, $\frac{2}{5}$, $\frac{1}{5}$, and $\frac{1}{10}$ respectively. He never takes more than 7 strokes. Find the probability of the following events:
   (a) scoring 4 on each of the first three holes
   (b) scoring 3, 4 and 5 (in that order) on the first three holes
   (c) scoring a total of 28 for the first four holes
   (d) scoring a total of 10 for the first three holes.

11. Assume that births are equally likely on each of the seven days of the week. Two people are selected at random. Find the probability that:
   (a) both were born on a Sunday
   (b) both were born on the same day of the week.

12. (a) A playing card is drawn from a standard pack of 52 and then replaced. A second card is drawn. What is the probability that the second card is the same as the first?
   (b) A card is drawn and *not* replaced. A second card is drawn. What is the probability that the second card is *not* the same as the first?

9.5 Conditional probability

The performance of a climber is affected by the weather. When it rains, he falls on a mountain with a probability of $\frac{1}{10}$, whereas in dry weather he falls with a probability of only $\frac{1}{50}$. In the climbing seasons, the probability of rain on any day is $\frac{1}{4}$.

What is the probability that he falls on the mountain on a day chosen at random?

Draw a tree diagram.

$p\,(\text{he falls}) = p(\text{r, f}) + p(\bar{\text{r}}, \text{f})$

$\qquad = (\frac{1}{4} \times \frac{1}{10}) + (\frac{3}{4} \times \frac{1}{50}) = \frac{1}{25}$

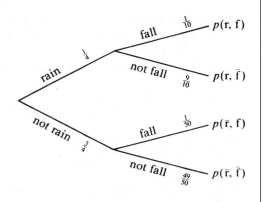

0

...g A contains 3 red balls and 3 blue balls. Bag B contains
1 red ball and 3 blue balls.

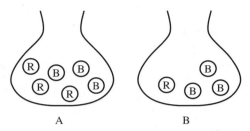

A ball is taken at random from Bag A and placed in Bag B. A ball
is then chosen from Bag B. What is the probability that the ball
taken from B is red?

2. On a Monday or a Thursday. Mr Gibson paints a 'masterpiece'
   with a probability of $\frac{1}{5}$. On any other day, the probability of
   producing a 'masterpiece' is $\frac{1}{100}$. In common with other great
   painters, Gibson never knows what day it is. Find the probability
   that on one day chosen at random, he will in fact paint a
   masterpiece.

3. If a hedgehog crosses a certain road before 7.00 a.m., the
   probability of being run over is $\frac{1}{10}$. After 7.00 a.m., the
   corresponding probability is $\frac{3}{4}$. The probability of the hedgehog
   waking up early enough to cross before 7.00 a.m. is $\frac{4}{5}$.

   What is the probability of the following events:
   (a) the hedgehog waking up too late to reach the road before
       7.00 a.m.,
   (b) the hedgehog waking up early and crossing the road in safety,
   (c) the hedgehog waking up late and crossing the road in safety,
   (d) the hedgehog waking up early and being run over,
   (e) the hedgehog crossing the road in safety?

4. In Stockholm the probability of snow falling on a day in January is
   0·2. If it snows on one day, the probability of snow on the following
   day increases to 0·35.
   What is the probability that for two consecutive days:
   (a) it will snow on both days
   (b) it will not snow on either day
   (c) it will snow on just one of the two days?

5. Two boxes are shown containing red and white balls.
   One ball is selected from Box A and placed in Box B.
   A ball is then selected from Box B and placed in Box A.
   What is the probability that Box A now contains 4 red
   balls?

Box A

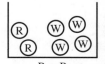

Box B

6. Michelle has been revising for a multiple choice exam in history but she had only had time to learn 70% of the facts being tested. If there is a question on any of the facts she has revised she will get that question right. Otherwise she will simply guess one of the five possible answers.
  (a) A question is chosen randomly from the paper. What is the probability that she will get it right?
  (b) If the paper has 50 questions what mark do you expect her to get?

7. Surgeons can operate to cure Pythagoratosis but the success rate at the first attempt is only 65%. If the first operation fails the operation can be repeated but this time the success rate is only 20%. After a second failure there is so little chance of success that surgeons will not operate again.

There is an outbreak of the disease at Gibson College and 38 students contract the disease.
How many of these students can we expect to be saved after both operations?

10 USING AND APPLYING MATHEMATICS

10.1 Conjectures

A conjecture (or hypothesis) is simply a statement which may or may not be true. In mathematics people are constantly looking for rules or formulas to make calculations easier. Suppose you carry out an investigation and after looking at the results you spot what you think might be a rule.

You could make the conjecture that:

$\quad$ 'x^2 is always greater than x'

or that 'the angle in a semi circle is always $90°$'

or that 'the expression $x^2 + x + 41$ gives a prime number for all
$\qquad$ integer values of x'

There are three possibilities for any conjecture. It may be
(a) true, $\qquad$ (b) false, $\qquad$ (c) not proven.

(a) True

To show that a conjecture is true it is not enough to simply find lots of examples where it is true. We have to *prove* it for *all* values. We return to proof later in this section.

(b) False

To show that a conjecture is false it is only necessary to find one *counter-example*.

Consider the conjecture 'x^2 is always greater than x'

It is true that $\quad 2^2 > 2; \quad 3 \cdot 1^2 > 3 \cdot 1; \quad (-5)^2 > 5$

But $(\frac{1}{2})^2$ is *not* greater than $\frac{1}{2}$.

This is a counter-example so the conjecture is false.

We have *disproved* the conjecture.

Consider the conjecture:

'the expression $x^2 + x + 41$ gives a prime number for all integer values of x'

If we try the first few values of x, we get these results.

x	1	2	3	4	5	6
$x^2 + x + 41$	43	47	53	61	71	83

The expression does give prime numbers for these values of x and indeed, it does so for all values of x up to 39.

But when $x = 40$, $x^2 + x + 41 = 1681$
$$= 41 \times 41$$
So we do not always obtain a prime number.

The conjecture is, therefore, false.

Notice that it takes only *one* counter-example to disprove a conjecture.

(c) Not proven

Suppose you have a conjecture for which you cannot find a counter-example but which you also cannot prove. In this case the conjecture is *not proven*.

Until recently, the most famous example of a conjecture not proven was 'Fermat's Last Theorem' which stated that there are no whole numbers a, b, c for which $a^3 + b^3 = c^3$ [or further, that $a^n + b^n = c^n$ $(n > 2)$].

For 358 years no mathematician could prove the conjecture but no one could find a counter-example. In 1994 Andrew Wiles, a British mathematician, finally proved the theorem after devoting himself to it for seven years. The proof is extremely long (100 pages) and it is said that only a handful of people in the world understand it.

Exercise 1

In Questions **1** to **12**, consider the conjecture given. Some are true and some are false. Write down a counter-example where the conjecture is false. If you cannot find a counter-example, state that the conjecture is 'not proven' as you are *not* asked to prove it.

1. $(n + 1)^2 = n^2 + 1$ for all values of n.

2. $\dfrac{1}{x}$ is always less than x.

3. $\sqrt{n + 1}$ is always smaller than $\sqrt{n} + 1$.

4. For all values of n, $2n$ is greater than $n - 2$.

5. For any set of numbers, the median is always smaller than the mean.

6. $\sqrt{x}$ is always less than x.

7. The diagonals of a parallelogram never cut at right angles.

8. The product of two irrational numbers is always another irrational number.

9. If n is even $5n^2 + n + 1$ is odd.

10. All numbers in the sequence 31, 331, 3331, 33 331, 333 331, 3 333 331, 33 333 331, 333 333 331, ... are prime.
[Hint: Divide by 17.]

11. $n^2 + n > 0$ for all values of n.

12. The expression $a^n - a$ is always a multiple of n.

13. Here is a circle with 2 points on its circumference and 2 regions.   This circle has 3 points ($n = 3$) and 4 regions ($r = 4$).

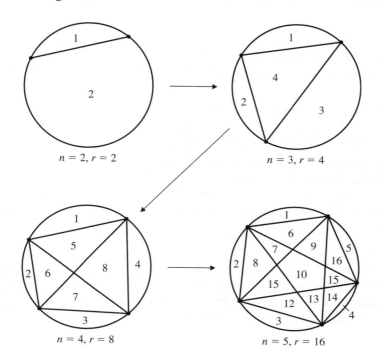

$n = 2, r = 2$

$n = 3, r = 4$

$n = 4, r = 8$

$n = 5, r = 16$

Each time, every point on the circumference is joined to all the others.

(a) Do you think a conjecture is justified at this stage?
   Consider:
   'The circle with 6 points on its circumference will have 32 regions'.
   Or again:
   'With n points on the circle the number of regions, r, is given by $r = 2^{n-1}$
   Draw the next two circles in the sequence with $n = 6$ and $n = 7$.
   Be careful to avoid this sort of thing:

where three lines pass through a single point.

Under a magnifying glass we might see this:
and a region might be lost.
What can you say about your conjecture at
this stage?

(b) When you have the results for $n = 6$ and $n = 7$ you might get
somewhere (although it is quite difficult) by looking at
differences in the results.

10.2 Logical argument

- Here is a statement

    Triangle PQR is equilateral $\Rightarrow$ angle QRP $= 60°$.

    The statement can be read:

    If triangle PQR is equilateral, then angle QRP equals $60°$.

    or

    Triangle PQR is equilateral implies that angle QRP equals $60°$.
- Here is another statement.

    $a > b \Rightarrow a - b > 0$

    This can be read:

    If a is greater than b, then a less b is greater than zero

    or

    a is greater than b implies that a less b is greater than zero.

Note $x = 4 \Rightarrow x^2 = 16$ is true
but $x^2 = 16 \Rightarrow x = 4$ is not true because x could be -4.

Exercise 2

1. Write the following as sentences.
    (a) Today is Monday $\Rightarrow$ tomorrow is Tuesday.
    (b) It is raining $\Rightarrow$ there are clouds in the sky.
    (c) Abraham Lincoln was born in 1809 $\Rightarrow$ Abraham Lincoln is dead.

2. State whether the following statements are true or false.
    (a) $x - 4 = 3 \Rightarrow x = 7$
    (b) n is even $\Rightarrow n^2$ is even
    (c) $a = b \Rightarrow a^2 = b^2$
    (d) $x > 2 \Rightarrow x = 3$
    (e) $a + b$ is odd $\Rightarrow ab$ is even (a and b are whole numbers).
    (f) $x^2 = 4 \Rightarrow x = 2$
    (g) $x = 45° \Rightarrow \tan x = 1$
    (h) $5^x = 1 \Rightarrow x = 0$
    (i) $pq = 0 \Rightarrow p = 0$
    (j) a is odd $\Rightarrow a$ can be written in the form $2k + 1$, where k is an
    integer.

3. Think of pairs of statements like those in Question **1**. Link the
statements with the $\Rightarrow$ sign.

10.3 Proof

- In the field of science a theory, like Newton's theory of gravitation, can never be proved. It can only be considered highly likely using all the evidence available at the time. The history of science contains many examples of theories which were accepted at the time but were later shown to be untrue when more accurate observation was possible.
- A mathematical proof is far more powerful. Once a theorem is proved mathematically it will *always* be true. Pythagoras proved his famous theorem over 2500 years ago and when he died he knew it would never be disproved.

  A proof starts with simple facts which are accepted. The proof then argues logically to the result which is required.
- It is most important to realise that a result *cannot* be proved simply by finding thousands or even millions of results which support it.

  In 1738 Euler made the conjecture that there were no solutions to the equation $a^4 + b^4 + c^4 = d^4$.

  For 250 years this result remained unproved but it was then shown that $2\,682\,440^4 + 15\,365\,639^4 + 18\,796\,760^4 = 20\,615\,673^4$, thereby showing that Euler's conjecture was false.

Algebraic proof

Prove that, if a and b are odd numbers, then $a + b$ is even.

If a is odd, there is remainder 1 when a is divided by 2.
$\therefore$ a may be written in the form $(2m + 1)$ where m is a whole number.
Similarly b may be written in the form $(2n + 1)$.

$\therefore$ $a + b = 2m + 1 + 2n + 1$
$\qquad\quad = 2(m + n + 1)$

$\therefore$ $a + b$ is even, as required.

Prove that the answer to every line of the pattern below is 8.

$3 \times 5 - 1 \times 7$
$4 \times 6 - 2 \times 8$
$5 \times 7 - 3 \times 9$
$\quad\vdots \qquad \vdots$

Proof
The nth line of the pattern is $(n + 2)(n + 4) - n(n + 6)$
$$= n^2 + 6n + 8 - (n^2 + 6n)$$
$$= 8$$

The result is proved.

[Much harder but very interesting]

Prove that the product of 4 consecutive numbers is always one less than a square number.

e.g. $1 \times 2 \times 3 \times 4 = 24$ or $3 \times 4 \times 5 \times 6 = 360$
$= 25 - 1$ $= 19^2 - 1$

Proof

Let the first number be n.

Then the next three consecutive numbers are $(n + 1)$, $(n + 2)$, $(n + 3)$

Find the product

$n(n + 1)(n + 2)(n + 3)$
$= (n^2 + n)(n^2 + 5n + 6)$
$= n^4 + 6n^3 + 11n^2 + 6n$

Write the products as $(n^4 + 6n^3 + 11n^2 + 6n + 1) - 1$.
Note that this does not alter the value of the product.

The expression in the brackets factorises as

$(n^2 + 3n + 1)(n^2 + 3n + 1)$

$\therefore$ The product is $(n^2 + 3n + 1)^2 - 1$

Since $(n^2 + 3n + 1)^2$ is a square number, the result is proved.

Exercise 3

1. If a is odd and b is even, prove that ab is even.

2. If a and b are both odd, prove that ab is odd.

3. Prove that the sum of two odd numbers is even.

4. Prove that the square of an even number is divisible by 4.

5. Prove that the answer to every line of the pattern below is 3.

$2 \times 4 - 1 \times 5 =$
$3 \times 5 - 2 \times 6 =$
$4 \times 6 - 3 \times 7 =$
$\vdots \qquad \vdots$

[Hint: see the second example on page 370.]

6. In any three consecutive numbers prove that the product of the first and the third number is one less than the square of the middle number.

7. Prove that the product of four consecutive numbers is divisible by 4.

8. Prove that the sum of the squares of any two consecutive integers is an odd number.

9. Prove that the sum of the squares of five consecutive numbers is divisible by 5.

e.g. $2^2 + 3^2 + 4^2 + 5^2 + 6^2 = 90 = 5 \times 18$

Begin by writing the middle number as n, so the other numbers are $n - 2, n - 1, n + 1, n + 2$.

10. Find the sum of the squares of three consecutive numbers and then subtract 2. Prove that the result is always 3 times a square number.

11. A proof of Pythagoras' Theorem.

Four right-angled triangles with sides a, b and c are drawn around the tilted square as shown.

Area of large square $= (a + b)^2$ ①

Also area of large square = area of 4 triangles + area of tilted square.   ②

Equate ① and ② and hence complete the proof of Pythagoras' Theorem.

12. Here is the 'proof' that $1 = 2$.
Let $a = b$

$\Rightarrow ab = b^2$ [multiply by b]

$\Rightarrow ab - a^2 = b^2 - a^2$ [subtract a^2]

$\Rightarrow a(b - a) = (b + a)(b - a)$ [factorise]

$\Rightarrow a = b + a$ [divide by $(b - a)$]

$\Rightarrow a = a + a$ [from top line]

$\Rightarrow 1 = 2$

Which step in the argument is not allowed?

Geometric proof

Over 2000 years ago Greek mathematicians were very keen on proofs. Euclid wrote a best seller called 'The Elements' in which he first of all stated some very basic assumptions called *axioms*. He then proceeded to prove one result after another, basing each new result on a previous result which had already been proved. Examples of his axioms are 'parallel lines do not meet in either direction' and 'only one straight line can be drawn between two points'.

When you are writing a geometric proof of your own you do not need to go right back to Euclid's axioms. What you should do is state clearly after each line what facts you have used.
For example 'angles in a triangle add up to $180°$'
or 'opposite angles are equal.'

Prove that the angle at the centre of a circle is twice the angle at the circumference.

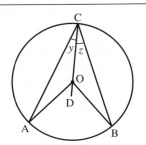

Draw the straight line COD.
Let $A\hat{C}O = y$ and $B\hat{C}O = z$.

In triangle AOC,

$\quad\quad AO = OC \quad\quad$ (radii)

$\therefore\quad O\hat{C}A = O\hat{A}C \quad$ (isosceles triangle)

$\therefore\quad C\hat{O}A = 180 - 2y \quad$ (angle sum of triangle)

$\therefore\quad A\hat{O}D = 2y \quad\quad$ (angles on a straight line)

Similarly from triangle COB, we find

$\quad\quad D\hat{O}B = 2z$

Now $\quad A\hat{C}B = y + z$

and $\quad A\hat{O}B = 2y + 2z$

$\therefore\quad\quad A\hat{O}B = 2 \times A\hat{C}B$ as required.

Prove that opposite angles in a cyclic quadrilateral add up to 180°.

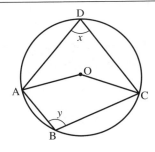

Draw radii OA and OC.
Let $A\hat{D}C = x$ and $A\hat{B}C = y$.

$\quad\quad A\hat{O}C$ obtuse $= 2x$ (angle at the centre)

$\quad\quad A\hat{O}C$ reflex $= 2y$ (angle at the centre)

$\therefore\quad\quad\quad 2x + 2y = 360°$ (angles at a point)

$\therefore\quad\quad\quad x + y = 180°$ as required.

Exercise 4

1. Triangle ABC is isosceles.

Prove that $C\hat{B}D = 2 \times C\hat{A}B$.

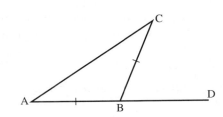

2. Triangle PQR is isosceles and

$P\hat{S}Q = 90°$.

Prove that PS bisects $Q\hat{P}R$.

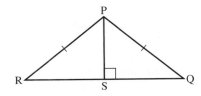

3. Two chords of a circle AB and CD intersect at X.

   Use similar triangles to prove that AX.BX = CX.DX. This result is called the Intersecting Chords Theorem.

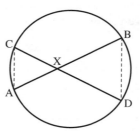

4. Line ATB touches a circle at T and TC is a diameter. AC and BC cut the circle at D and E respectively. Prove that the quadrilateral ADEB is cyclic.

5. Prove that the angle in a semicircle is a right angle.

6. TC is a tangent to a circle at C and BA produced meets this tangent at T.

   Show that triangles TCA and TBC are similar and hence prove that $TC^2 = TA \times TB$.

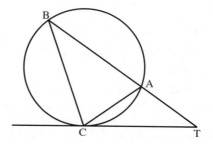

7. Given that BOC is a diameter and that $A\widehat{D}C = 90°$, prove that AC bisects $B\widehat{C}D$.

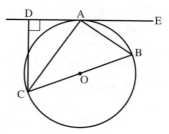

10.4 Coursework tasks

There are a large number of possible starting points for investigations here so it may be possible to allow students to choose investigations which appeal to them. On other occasions the same investigation may be set to a whole class.

Here are a few guidelines for pupils:
(a) If the set problem is too complicated try an easier case;
(b) Draw your own diagrams;
(c) Make tables of your results and be systematic;
(d) Look for patterns;
(e) Is there a rule or formula to describe the results?
(f) Can you *predict* further results?
(g) Can you *prove* any rules which you may find?
(h) Where possible extend the task further by asking questions like 'what happens if . . .'

1 Opposite corners

Here the numbers are arranged in 9 columns.

In the 2 × 2 square . . .

$6 \times 16 = 96$

$7 \times 15 = 105$

. . . the difference between them is 9.

In the 3 × 3 square . . .

$22 \times 42 = 924$

$24 \times 40 = 960$

. . . the difference between them is 36.

1	2	3	4	5	6	7	8	9
10	11	12	13	14	15	16	17	18
19	20	21	22	23	24	25	26	27
28	29	30	31	32	33	34	35	36
37	38	39	40	41	42	43	44	45
46	47	48	49	50	51	52	53	54
55	56	57	58	59	60	61	62	63
64	65	66	67	68	69	70	71	72
73	74	75	76	77	78	79	80	81
82	83	84	85	86	87	88	89	90

2 × 2 square:

6	7
15	16

3 × 3 square:

22	23	24
31	32	33
40	41	42

Investigate to see if you can find any rules or patterns connecting the size of square chosen and the difference.

If you find a rule, use it to *predict* the difference for larger squares.

Test your rule by looking at squares like 8 × 8 or 9 × 9.

Can you *generalise* the rule?

[What is the difference for a square of size $n \times n$?]

Can you *prove* the rule?

Hint:

In a 3 × 3 square . . .

x	?
?	?

What happens if the numbers are arranged in six columns or seven columns?

1	2	3	4	5	6
7	8	9	10	11	12
13	14	15	16	17	18
19					

	1	2	3	4	5
6	7	8	9	10	11
12	13	14	15	16	17
18	19	20	21		
22					

2 Hiring a car

You are going to hire a car for one week (7 days).
Which of the firms below should you choose?

Gibson car hire	Snowdon rent-a-car	Hav-a-car
£170 per week unlimited mileage	£10 per day 6·5 p per mile	£60 per week 500 miles without charge 22p per mile over 500 miles.

Work out as detailed an answer as possible.

3 Half-time score

The final score in a football match was 3–2. How many different scores were possible at half-time?

Investigate for other final scores where the difference between the teams is always one goal. [1–0, 5–4, etc.]. Is there a pattern or rule which would tell you the number of possible half-time scores in a game which finished 58–57?
Suppose the game ends in a draw. Find a rule which would tell you the number of possible half-time scores if the final score was 63–63.
Investigate for other final scores [3–0, 5–1, 4–2, etc.].
Find a rule which gives the number of different half-time scores for *any* final score (say $a - b$).

4 An expanding diagram

Look at the series of diagrams.

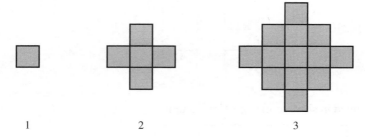

Each time new squares are added all around the outside of the previous diagram.
Draw the next few diagrams in the series and count the number of squares in each one.
How many squares are there in diagram number 15 or in diagram number 50?
What happens if we work in three dimensions? Instead of adding squares we add cubes all around the outside. How many cubes are there in the fifth member of the series or the fifteenth?

5 Maximum box

(a) You have a square sheet of card 24 cm by 24 cm.
   You can make a box (without a lid) by cutting squares from the corners and folding up the sides.
   What size corners should you cut out so that the volume of the box is as large as possible?
   Try different sizes for the corners and record the results in a table.

Length of the side of the corner square (cm)	Dimensions of the open box (cm)	Volume of the box (cm³)
1	22 × 22 × 1	484
2	—	—

24 cm

24 cm

Now consider boxes made from different sized cards:
15 cm × 15 cm and 20 cm by 20 cm.
What size corners should you cut out this time so that the volume of the box is as large as possible?
Is there a connection between the size of the corners cut out and the size of the square card?

(b) Investigate the situation when the card is not square. Take rectangular cards where the length is twice the width (20 × 10, 12 × 6, 18 × 9 etc.)
Again, for the maximum volume is there a connection between the size of the corners cut out and the size of the original card?

6 Timetabling

(a) Every year a new timetable has to be written for the school. We will look at the problem of writing the timetable for one department (mathematics). The department allocates the teaching periods as follows:

U6	2 sets (at the same times); 8 periods in 4 doubles.
L6	2 sets (at the same times); 8 periods in 4 doubles.
Year 5	6 sets (at the same times); 5 single periods.
Year 4	6 sets (at the same times); 5 single periods.
Year 3	6 sets (at the same times); 5 single periods.
Year 2	6 sets (at the same times); 5 single periods.
Year 1	5 mixed ability forms; 5 single periods not necessarily at the same times.

Here are the teachers and the maximum number of maths periods which they can teach.

A	33	F	15	(Must be Years 5, 4, 3)
B	33	G	10	(Must be Years 2, 1)
C	33	H	10	(Must be Years 2, 1)
D	20	I	5	(Must be Year 3)
E	20			

Furthermore, to ensure some continuity of teaching, teachers B and C must teach the U6 and teachers A, B, C, D, E, F must teach year 5.

A timetable form which has been started is shown overleaf.

M	5					U6 B, C	U6 B, C		
Tu		5	U6 B, C	U6 B, C					
W					5				
Th						5	U6 B, C	U6 B, C	
F	U6 B, C	U6 B, C		5					

Your task is to write a complete timetable for the mathematics department subject to the restrictions already stated.

(b) If that was too easy, here are some changes.

U6 and L6 have 4 sets each (still 8 periods)
Two new teachers:
   J 20 periods maximum
   K 15 periods maximum but cannot teach on Mondays.

Because of games lessons: A cannot teach Wednesday afternoon
                             B cannot teach Tuesday afternoon
                             C cannot teach Friday afternoon
Also: A, B, C and E must teach U6
      A, B, C, D, E, F must teach year 5
For the pupils, games afternoons are as follows:
Monday year 2; Tuesday year 3; Wednesday year 5 L6, U6;
Thursday year 4; Friday year 1.

7 Alphabetical order

A teacher has four names on a piece of paper which are in no particular order (say Smith, Jones, Biggs, Eaton). He wants the names in alphabetical order.
One way of doing this is to interchange each pair of names which are clearly out of order.

So he could start like this; S J B E
the order becomes: J S B E
He would then interchange S and B.

Using this method, what is the *largest* number of interchanges he could possibly have to make?
What if he had thirty names, or fifty?

8 What shape tin?

We need a cylindrical tin which will contain a volume of $600 \, \text{cm}^3$ of drink.

What shape should we make the tin so that we use the minimum amount of metal?
In other words, for a volume of $600 \, \text{cm}^3$, what is the smallest possible surface area?

Hint: Make a table.

r	h	A
2	?	?
3	?	?
⋮		

What shape tin should we design to contain a volume of $1000 \, \text{cm}^3$?

9 Painting cubes

The large cube on the right consists of 27 unit cubes.

All six faces of the large cube are painted green.

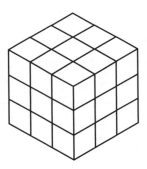

- How many unit cubes have 3 green faces?

- How many unit cubes have 2 green faces?

- How many unit cubes have 1 green face?

- How many unit cubes have 0 green faces?

Answer the four questions for the cube which is $n \times n \times n$.

10 Discs

(a) You have five black discs and five white discs which are arranged in a line as shown.

We want to get all the black discs to the right-hand end and all the white discs to the left-hand end.

The only move allowed is to interchange two neighbouring discs.

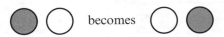   becomes

How many moves does it take?
How many moves would it take if we had fifty black discs and fifty white discs arranged alternately?

(b) Suppose the discs are arranged in pairs

 ... etc.

How many moves would it take if we had fifty black discs and fifty white discs arranged like this?
[Hint: In both cases work with a smaller number of discs until you can see a pattern].

(c) Now suppose you have three colours black, white and green arranged alternately.

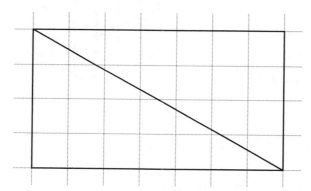 ... etc.

You want to get all the black discs to the right, the green discs to the left and the white discs in the middle.
How many moves would it take if you have 30 discs of each colour?

11 Diagonals

In a 4 × 7 rectangle, the diagonal passes through 10 squares.

Draw rectangles of your own choice and count the number of squares through which the diagonal passes.
A rectangle is 640 × 250. How many squares will the diagonal pass through?

12 Chess board

Start with a small board, just 4 × 4.
How many squares are there? [It is not just 16!]
How many squares are there on an 8 × 8 chess board?
How many squares are there on an $n \times n$ chess board?

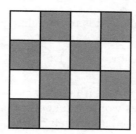

13 Find the connection

Work through the flow diagram several times, using a calculator.

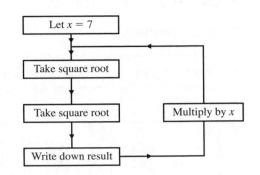

What do you notice?
Try different numbers for x (suggestions: 11, 5, 8, 27)
What do you notice?

What happens if you take the square root three times?

Suppose in the flow diagram you change 'Multiply by x' to 'Divide by x'. What happens now?

Suppose in the flow diagram you change 'Multiply by x' to 'Multiply by x^2'. What happens now?

You can vary the number of times you take the square root and whether you 'Multiply by x', 'Multiply by x^2', 'Divide by x' etc.

14 Spotted shapes

For this investigation you need dotted paper. If you have not got any, you can make your own using a felt tip pen and squared paper.

The rectangle in Diagram 1 has
10 dots on the perimeter ($p = 10$)
and 2 dots inside the shape ($i = 2$).
The area of the shape is 6 square units ($A = 6$)

Diagram 1

The triangle in Diagram 2 has
9 dots on the perimeter ($p = 9$)
and 4 dots inside the shape ($i = 4$).
The area of the triangle is $7\frac{1}{2}$ square
units ($A = 7\frac{1}{2}$)

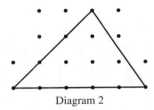

Diagram 2

Draw more shapes of your own design and record the values for p, i and A in a table. Make some of your shapes more difficult like the one in Diagram 3.

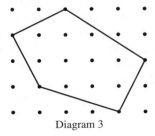

Diagram 3

This is quite difficult.
Be systematic, recording the values for i, p, A in a table. Start by drawing several shapes with $i = 0$. Find a connection between p and A.
Now increase i by one at a time.
Try to find a formula connecting p, i and A.

10.5 Puzzles and games

1 Crossnumbers

(a) Copy out the crossnumber pattern.
(b) Fit all the given numbers into the correct spaces. Tick off the numbers from the lists as you write them in the square.

1.

2 digits	3 digits	4 digits	5 digits	6 digits
11	121	2104	14700	216841
17	147	2356	24567	588369
18	170	2456	25921	846789
19	174	3714	26759	861277
23	204	4711	30388	876452
31	247	5548	50968	
37	287	5678	51789	
58	324	6231	78967	
61	431	6789	98438	
62	450	7630		*7 digits*
62	612	9012		6645678
70	678	9921		
74	772			
81	774			
85	789			
94	870			
99				

2.

2 digits		3 digits	4 digits	5 digits	6 digits
12	47	129	2096	12641	324029
14	48	143	3966	23449	559641
16	54	298	5019	33111	956782
18	56	325	5665	33210	
20	63	331	6462	34509	
21	67	341	7809	40551	
23	81	443	8019	41503	
26	90	831	8652	44333	*7 digits*
27	91	923		69786	1788932
32	93			88058	5749306
38	98			88961	
39	99			90963	
46				94461	
				99654	

2 Estimating game

This is a game for two players. On squared paper draw an answer grid
with the numbers shown.

Answer grid

891	7047	546	2262	8526	429
2548	231	1479	357	850	7938
663	1078	2058	1014	1666	3822
1300	1950	819	187	1050	3393
4350	286	3159	442	2106	550
1701	4050	1377	4900	1827	957

The players now take turns to choose two numbers from the question
grid below and multiply them on a calculator.

Question grid

11	26	81
17	39	87
21	50	98

The game continues until all the numbers in the answer grid have been
crossed out. The object is to get four answers in a line (horizontally,
vertically or diagonally). The winner is the player with most lines of
four. A line of *five* counts as *two* lines of four. A line of *six* counts as
three lines of four.

3 The chess board problem

(a) On the 4 × 4 square, four objects have been placed, subject to the
restriction that nowhere are there two objects on the same row,
column or diagonal.

Subject to the same restrictions:
  (i) find a solution for a 5 × 5 square, using five objects,
 (ii) find a solution for a 6 × 6 square, using six objects,
(iii) find a solution for a 7 × 7 square, using seven objects,
(iv) find a solution for a 8 × 8 square, using eight objects.

It is called the chess board problem because the objects could be 'Queens' which can move any number of squares in any direction.

(b) Suppose we remove the restriction that no two Queens can be on the same row, column or diagonal. Is it possible to attack every square on an 8×8 chess board with less than eight Queens?
Try the same problem with other pieces like knights or bishops.

4 Creating numbers

Using only the numbers 1, 2, 3 and 4 once each and the operations $+, -, \times, \div, !$ create every number from 1 to 100.
You can use the numbers as powers and you must use all of the numbers 1, 2, 3 and 4.
[4! is pronounced 'four factorial' and means $4 \times 3 \times 2 \times 1$ (i.e. 24)
similarly $3! = 3 \times 2 \times 1 = 6$
    and $5! = 5 \times 4 \times 3 \times 2 \times 1 = 120$]

Examples: $1 = (4 - 3) \div (2 - 1)$
$20 = 4^2 + 3 + 1$
$68 = 34 \times 2 \times 1$
$100 = (4! + 1)(3! - 2!)$

11 REVISION

11.1 Revision exercises (non-calculator)

Revision exercise 1

1. In the diagram, the equations of the lines are $y = 3x$, $y = 6$, $y = 10 - x$ and $y = \frac{1}{2}x - 3$.

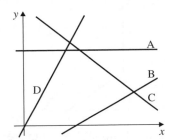

Find the equation corresponding to each line.

2. Find x.

(a) $2^x = 32$ 　　　　(b) $\dfrac{4}{x} = 7$

3. Find the median of: 6, 3, 8, 2, 9.

4. (a) A lies on a bearing of 040° from B. Calculate the bearing of B from A.
   (b) The bearing of X from Y is 115°. Calculate the bearing of Y from X.

5. Given $a = 3$, $b = 4$ and $c = -2$, evaluate:
   (a) $2a^2 - b$ 　　(b) $a(b - c)$ 　　(c) $2b^2 - c^2$

6. Work out $6578 \div 26$.

7. Increase £5000 by 5%.

8. Convert $\frac{3}{8}$ to a decimal.

9. Work out
   (a) $\frac{3}{5} \times \frac{3}{4}$ 　　　　　(b) $\frac{7}{12} - \frac{1}{8}$

10. $a = \frac{1}{2}$, $b = \frac{1}{4}$. Which one of the following has the greatest value?

    (i) ab 　　(ii) $a + b$ 　　(iii) $\dfrac{a}{b}$

    (iv) $\dfrac{b}{a}$ 　　(v) $(ab)^2$

11. (a) On a map, the distance between two points is 16 cm. Calculate the scale of the map if the actual distance between the points is 8 km.
    (b) On another map, two points appear 1·5 cm apart and are in fact 60 km apart. Calculate the scale of the map.

12. Given that $s - 3t = rt$, express:
    (a) s in terms of r and t
    (b) r in terms of s and t.

13. Work out an estimate of the value of $\dfrac{61\cdot51 \times 4\cdot93}{31\cdot33}$ correct to one significant figure.

14. The shaded region A is formed by the lines $y = 2$, $y = 3x$ and $x + y = 6$. Write down the three inequalities which define A.

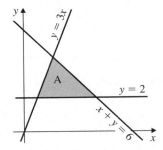

Revision exercise 2

1. A camera is bought for £300 and then sold for £315. Work out the percentage profit.

2. The sides of a square are increased by 10%. By what percentage is the area increased?

3. Work out x, if $4^x = 16^3$.

4. Twenty-seven small wooden cubes fit exactly inside a cubical box without a lid. How many of the cubes are touching the sides or the bottom of the box?

5. Given that $x = 4$, $y = 3$, $z = -2$, evaluate:
(a) $2x(y + z)$ (b) $(xy)^2 - z^2$
(c) $x^2 + y^2 + z^2$ (d) $(x + y)(x - z)$
(e) $\sqrt{[x(1 - 4z)]}$ (f) $\dfrac{xy}{z}$

6. When two dice are thrown simultaneously, what is the probability of obtaining the same number on both dice?

7. A target consists of concentric circles of radii 3 cm and 9 cm.

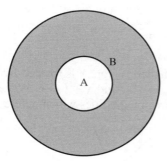

(a) Find the area of A, in terms of π.

(b) Find the ratio $\dfrac{\text{area of B}}{\text{area of A}}$.

8. A motorist travelled 200 miles in five hours. Her average speed for the first 100 miles was 50 m.p.h. What was her average speed for the second 100 miles?

9. The perimeter of this rectangle is 40 cm. Find x.

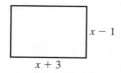
$x - 1$
$x + 3$

10.

Marks	3	4	5	6	7	8
Number of pupils	2	3	6	4	3	2

The table shows the number of pupils in a class who scored marks 3 to 8 in a test. Find:
(a) the mean mark
(b) the modal mark
(c) the median mark.

11. In the diagram, triangles ABC and EBD are similar but DE is *not* parallel to AC. Given that AD = 5 cm, DB = 3 cm and BE = 4 cm, calculate the length of BC.

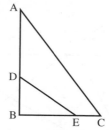

12. Nadia said: 'I thought of a number, multiplied it by 6, then added 15. My answer was less than 200'.
(a) Write down Nadia's statement in symbols, using x as the starting number.
(b) Nadia actually thought of a prime number. What was the largest prime number she could have thought of?

Revision exercise 3

1. (a) A piece of meat, initially weighing 2·4 kg, is cooked and subsequently weighs 1·9 kg. What is the percentage loss in weight?
(b) An article is sold at a 6% loss for £225·60. What was the cost price?

2. In the diagram, the equations of the lines are $2y = x - 8$, $2y + x = 8$, $4y = 3x - 16$ and $4y + 3x = 16$.

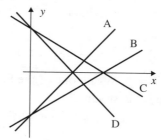

Find the equation corresponding to each line.

3. (a) Calculate the speed (in metres per second) of a slug which moves a distance of 30 cm in 1 minute.
   (b) Calculate the time taken for a bullet to travel 8 km at a speed of 5000 m/s.
   (c) Calculate the distance flown, in a time of four hours, by a pigeon which flies at a speed of 12 m/s.

4. Solve the simultaneous equations
$$7c + 3d = 29$$
$$5c - 4d = 33$$

5. Describe the single transformation which maps
   (a) $\triangle ABC$ onto $\triangle DEF$
   (b) $\triangle ABC$ onto $\triangle PQR$
   (c) $\triangle ABC$ onto $\triangle XYZ$

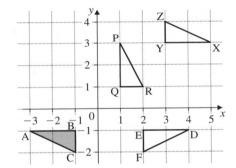

6. The letters r, h, a represent lengths. For each of the following formulas, state whether z is a length, an area, a volume or an impossible expression.
   (a) $z = r^2 + h^2$
   (b) $z = \pi h^2 a + rha$
   (c) $z = 3a^2 + h$
   (d) $z = \pi(r + h)$

7. Sketch the curve $y = \sin x$ for x from $0°$ to $360°$. If $\sin 28° = 0.469$, give another angle whose sine is 0.469.

8. Sketch the curve $y = \tan x$ for x from $0°$ to $360°$. Find two solutions of the equation $\tan x = 1$.

9. Given that $y = \dfrac{k}{k + w}$
   (a) Find the value of y when $k = \frac{1}{2}$ and $w = \frac{1}{3}$.
   (b) Express w in terms of y and k.

10. It is given that $y = \dfrac{k}{x}$ and that $1 \leqslant x \leqslant 10$.
   (a) If the smallest possible value of y is 5, find the value of the constant k.
   (b) Find the largest possible value of y.

Revision exercise 4

1. What values of x satisfy the inequality $15 - 4x > 12$?

2. Work out $30\,000 \times 5$ million and write the answer in standard form.

3. Factorise: (a) $x^2 + 8x + 15$
   (b) $x^2 + x - 6$
   (c) $5x^2 - 30x$

4. In the diagram the area of the smaller square is $10\,\text{cm}^2$. Find the area of the larger square.

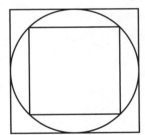

5. A bag contains 8 balls of which 2 are red and 6 are white. A ball is selected and not replaced. A second ball is selected. Find the probability of obtaining:
   (a) two red balls,   (b) two white balls,
   (c) one ball of each colour.

6. Several pupils took a science test and the results are shown in the histogram below.

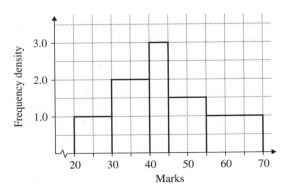

(a) If the pass mark was 45, how many pupils passed the test?

(b) How many took the test altogether?

7. Draw a histogram for the data below giving the ages of people at a disco.

Ages	Frequency
$14 \leqslant x < 16$	10
$16 \leqslant x < 17$	18
$17 \leqslant x < 18$	26
$18 \leqslant x < 21$	30
$21 \leqslant x < 26$	40

8. The shaded region B is formed by the lines $x = 0$, $y = x - 2$ and $x + y = 7$

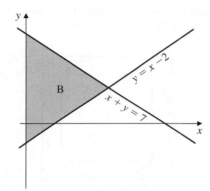

Write down the three inequalities which define B.

9. Estimate the answer correct to one significant figure. Do not use a calculator.

(a) $(612 \times 52) \div 49 \cdot 2$

(b) $(11 \cdot 7 + 997 \cdot 1) \times 9 \cdot 2$

(c) $\sqrt{\left(\dfrac{91 \cdot 3}{10 \cdot 1} \right)}$ (d) $\pi \sqrt{(5 \cdot 2^2 + 18 \cdot 2^2)}$

10. In the quadrilateral PQRS, $PQ = QS = QR$, PS is parallel to QR and $Q\widehat{R}S = 70°$. Calculate:

(a) $R\widehat{Q}S$

(b) $P\widehat{Q}S$.

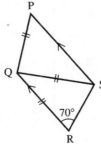

11. The triangle PQR is rolled clockwise along the line AB.

Draw the locus of P as the triangle rolls around Q and then R.

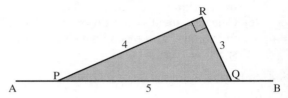

Revision exercise 5

1. Solve these equations by factorising.

(a) $x^2 = 2x + 15$

(b) $x^2 + 12 = 8x$

2. One solution of the equation $2x^2 - 7x + k = 0$ is $x = -\frac{1}{2}$. Find the value of k.

3. The diagram is the speed–time graph of a car.

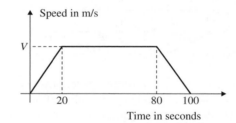

Given that the total distance travelled is $2 \cdot 4$ km, calculate:

(a) the value of the maximum speed V,

(b) the distance travelled in the first 30 seconds of the motion.

4. The radii of two spheres are in the ratio $2 : 5$. The volume of the smaller sphere is $16 \, \text{cm}^3$. Calculate the volume of the larger sphere.

5. The surface areas of two similar jugs are $50 \, \text{cm}^2$ and $450 \, \text{cm}^2$ respectively.

(a) If the height of the larger jug is $10 \, \text{cm}$, find the height of the smaller jug.

(b) If the volume of the smaller jug is $60 \, \text{cm}^3$, find the volume of the larger jug.

6. Find the angles marked with letters. (O is the centre of the circle.)

(a)   (b)

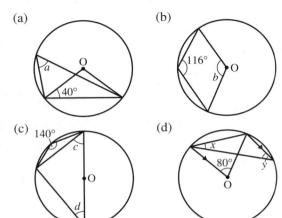

(c) 140°   (d)

7. Estimate the probability that the next President of Russia was born in April.

8. The probability that it will be wet today is $\frac{1}{6}$. If it is dry today, the probability that it will be wet tomorrow is $\frac{1}{8}$. What is the probability that both today and tomorrow will be dry?

9. Two dice are thrown. What is the probability that the *product* of the numbers on top is:
(a) 12   (b) 4   (c) 11?

10. In a mixed school there are twice as many boys as girls and ten times as many girls as teachers. Using the letters b, g, t to represent the number of boys, girls and teachers, find an expression for the total number of boys, girls and teachers. Give your answer in terms of b only.

Revision exercise 6

1. The diagram is the speed–time graph of a bus.

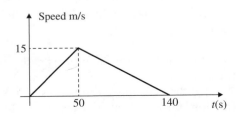

Calculate:
(a) the acceleration during the first 50 seconds
(b) the total distance travelled
(c) how long it takes before it is moving at 12 m/s for the first time.

2. A car is an enlargement of a model, the scale factor being 10.
(a) If the windscreen of the model has an area of 100 cm², find the area of the windscreen on the actual car (answer in m²).
(b) If the capacity of the boot of the car is 1 m³, find the capacity of the boot on the model (answer in cm³).

3. Solve the equations:
(a) $4(y + 1) = \dfrac{3}{1 - y}$
(b) $4(2x - 1) - 3(1 - x) = 0$

4. A coin is tossed four times. What is the probability of obtaining at least three 'heads'?

5. ABCD is a parallelogram and AE bisects angle A. Prove that DE = BC.

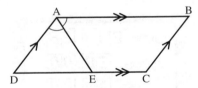

6. Without using a calculator, work out:
(a) $9^{-\frac{1}{2}} + (\frac{1}{8})^{\frac{1}{3}} + (-3)^0$
(b) $(1000)^{-\frac{1}{3}} - (0.1)^2$

7. (a) Given that $x - z = 5y$, express z in terms of x and y.
(b) Given that $mk + 3m = 11$, express m in terms of k.
(c) For the formula $T = C\sqrt{z}$, express z in terms of T and C.

8. Here is the graph of $y = f(x)$.

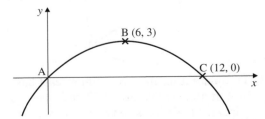

Draw three sketches to show:
(a) $y = f(x) - 3$
(b) $y = f(x - 3)$
(c) $y = f(3x)$

Give the new coordinates of A, B and C on each sketch.

9. The sides of a right-angled triangle have lengths $(x - 3)$ cm, $(x + 11)$ cm and $2x$ cm, where $2x$ is the hypotenuse. Find x.

10. In the parallelogram OABC, M is the mid-point of AB and N is the mid-point of BC.
If $\overrightarrow{OA} = \mathbf{a}$ and $\overrightarrow{OC} = \mathbf{c}$, express in terms of $\mathbf{a}$ and $\mathbf{c}$:

(a) $\overrightarrow{CA}$ (b) $\overrightarrow{ON}$ (c) $\overrightarrow{NM}$

Describe the relationship between CA and NM.

11. Solve the equations:
(a) $\dfrac{x + 3}{x} = 2$
(b) $x^2 = 5x$
(c) $(7^x)^2 = 1$

12. Draw the graph of $y = \dfrac{5}{x} + 2x - 3$, for $\frac{1}{2} \leqslant x \leqslant 7$, taking 2 cm to one unit for x and 1 cm to one unit for y.
Use the graph to find approximate solutions to the equation $\dfrac{5}{x} + 2x = 9$.

11.2 Revision exercises (calculator allowed)

Revision exercise 7

1. The pump shows the price of petrol in a garage.

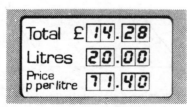

One day I buy £20 worth of petrol: How many litres do I buy?

2. Given that OA = 10 cm and AÔB = 70° (where O is the centre of the circle), calculate:
(a) the arc length AB
(b) the area of minor sector AOB.

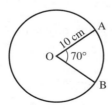

3. Throughout his life Mr Cram's heart has beat at an average rate of 72 beats per minute. Mr Cram is sixty years old. How many times has his heart beat during his life? Give the answer in standard form correct to two significant figures.

4. The edges of a cube are all increased by 10%. What is the percentage increase in the volume?

5. A cylinder of radius 8 cm has a volume of 2 litres. Calculate the cylinder height.

6. The dimensions of the rectangle are correct to the nearest cm.

Give the maximum and minimum values for the area of the rectangle consistent with this data.

7. The fraction $\frac{139}{99}$ gives an approximate value for $\sqrt{2}$.
What is the percentage error in using this fraction? Give your answer correct to 2 s.f.

8. Evaluate the following using a calculator: (answers to 4 sig. fig.)

(a) $\dfrac{0.74}{0.81 \times 1.631}$

(b) $\sqrt{\left(\dfrac{9.61}{8.34 - 7.41}\right)}$

(c) $\left(\dfrac{0.741}{0.8364}\right)^4$

(d) $\dfrac{8.4 - 7.642}{3.333 - 1.735}$

9. The mean of four numbers is 21.
(a) Calculate the sum of the four numbers.
Six other numbers have a mean of 18.
(b) Calculate the mean of the ten numbers.

10. Given BD = 1 m, calculate the length AC.

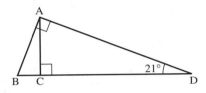

11. Use the method of trial and improvement to find a solution of the equation $x^5 = x^3 + 1$, giving your answer correct to 2 decimal places.

12. Given that x is an acute angle and that $3 \tan x - 2 = 4 \cos 35.3°$ calculate:
(a) $\tan x$
(b) the value of x in degrees correct to 1 d.p.

Revision exercise 8

1. Sainsburys sell their 'own-label' raspberry jam in two sizes.

Which jar represents the better value for money? You are given that 1 kg = 2.20 lb.

2. Sketch the curve $y = \cos x$ for x from 0° to 360°.
(a) If $\cos 70° = 0.342$, find another angle whose cosine is 0.342.
(b) Find two values of x if $\cos x = 0.5$.

3. Evaluate the following and give the answers to 3 significant figures:
(a) $\sqrt[3]{(9.61 \times 0.0041)}$

(b) $\left(\dfrac{1}{9.5} - \dfrac{1}{11.2}\right)^3$

(c) $\dfrac{15.6 \times 0.714}{0.0143 \times 12}$

(d) $\sqrt[4]{\left(\dfrac{1}{5 \times 10^3}\right)}$

4. Calculate the side or angle marked with a letter.

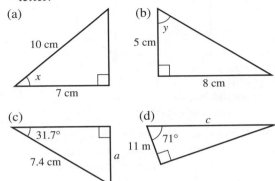

5. A cube of side 10 cm is melted down and made into five identical spheres. Calculate the radius of each sphere.

6. In figure 1 a circle of radius 4 cm is inscribed in a square. In figure 2 a square is inscribed in a circle of radius 4 cm.
Calculate the shaded area in each diagram.

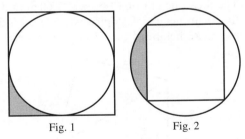

Fig. 1   Fig. 2

7. The dimensions of the cylinder are accurate to the nearest mm.

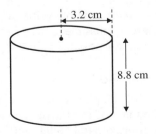

Work out the minimum possible volume of the cylinder. Give your answer to 3 s.f.

8. The figure shows a cube of side 10 cm.

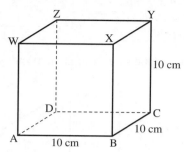

Calculate:
(a) the length of AC
(b) the angle YAC.

9. A bag contains x green discs and 5 blue discs. A disc is selected and replaced. A second disc is drawn. Find, in terms of x, the probability of selecting:
(a) a green disc on the first draw,
(b) a green disc on the first and second draws.

10. A copper pipe has external diameter 18 mm and thickness 2 mm. The density of copper is 9 g/cm^3 and the price of copper is £150 per tonne. What is the cost of the copper in a length of 5 m of this pipe?

11. The mean height of 10 boys is 1·60 m and the mean height of 15 girls is 1·52 m. Find the mean height of the 25 boys and girls.

12. Find x.

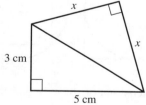

13. If the values of p are $-2 \leqslant p \leqslant 4$, find:
(a) the smallest possible value of p^2
(b) the largest possible value of p^2
(c) the smallest possible value of p^3.

Revision exercise 9

1. The mass of the planet Jupiter is about 350 times the mass of the Earth. The mass of the earth is approximately $6·03 \times 10^{21}$ tonnes. Give an estimate correct to 2 significant figures for the mass of Jupiter.

2. Work out the difference between one ton and one tonne.

| 1 tonne = 1000 kg |
| 1 ton = 2240 lb |
| 1 lb = 454 g |

Give your answer to the nearest kg.

3. A regular octagon of side length 20 cm is to be cut out of a square card.

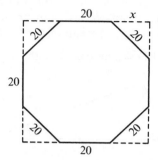

(a) Find the length x and hence find the size of the smallest square card from which this octagon can be cut.
(b) Calculate the area of the octagon, correct to 3 s.f.

4. Solve these equations, correct to 2 d.p.
(a) $3x^2 - 5x - 11 = 0$
(b) $(x - 2)^2 = 2x$

5. A formula for z is $z = ut - x^2$. The values of u, t and x are 5·2, 8·8 and 6·3 respectively, all correct to one decimal place. Work out the minimum possible value of z consistent with this data.

6. Calculate the length of AB.

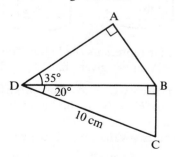

7. Find the mean and standard deviation of these numbers.
5, 17, 23, 11, 4, 8, 15, 32, 9, 12.

8. Two lighthouses A and B are 25 km apart and A is due West of B. A submarine S is on a bearing of 137° from A and on a bearing of 170° from B. Find the distance of S from A and the distance of S from B.

9. The diagram shows a rectangular block. AY = 12 cm, AB = 8 cm, BC = 6 cm.

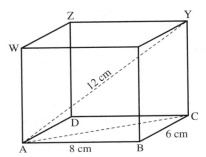

Calculate
(a) the length YC
(b) the angle YÂZ

10. The square has sides of length 3 cm and the arcs have centres at the corners. Find the shaded area.

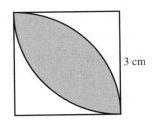

11. Given cos AĈB = 0·6, AC = 4 cm, BC = 5 cm and CD = 7 cm, find the length of AB and AD.

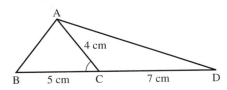

12. A sphere of radius 5 cm is melted down and made into a solid cube. Find the length of a side of the cube.

13. If AB = AC, find angle x, to the nearest degree.

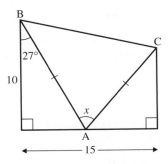

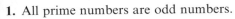

Revision exercise 10

Which of the following statements are true?

1. All prime numbers are odd numbers.

2. Every positive integer greater than 10 has an even number of factors.

3. If $x^2 = x$, then x must be the number 1.

4. The translation $\begin{pmatrix} 0 \\ 0 \end{pmatrix}$ is the only transformation that leaves *any* shape completely unchanged.

5. All non-negative numbers are positive.

6. If we add a given number to both the numerator and denominator of a fraction, then the new fraction is equivalent to the original fraction.

7. If both the numerator and denominator of a fraction are squared, then the new fraction is equivalent to the original fraction.

8. If we multiply both the numerator and denominator of a fraction by a given non-zero number, then the new fraction is equivalent to the original fraction.

9. $a \times (b \times c) = (a \times b) \times c$

10. $a \div (b \div c) = (a \div b) \div c$

11. All mathematical curves cross the x-axis or the y-axis or both.

12. If a quadrilateral has exactly 2 lines of symmetry, then it must be a rectangle.

13. Except for 1, no cube number is also a square number.

14. x^2 is never equal to $5x + 14$.

15. If n is a positive integer, then $n^2 + n + 5$ is a prime number.

16. If m and n are positive integers, then $6m + 4n + 13$ is an odd number.

17. No square number differs from a cube number by exactly 2.

18. A polygon having all its sides equal is a *regular* polygon.

19. If $a^2 = 7^2$, then a must be 7.

20. $\dfrac{a+b}{c} = \dfrac{a}{c} + \dfrac{b}{c}$

21. $\dfrac{a}{b+c} = \dfrac{a}{b} + \dfrac{a}{c}$

22. An enlargement always changes the area of a shape, unless the scale factor of the enlargement is 1.

23. $a \times (b + c) = ab + ac$

24. $a \times (b \times c) = ab \times ac$

25. $(a + b)^3 = a^3 + b^3$

26. $\sqrt{x + y} = \sqrt{x} + \sqrt{y}$

27. x^2 is never less than x.

28. $(a + b)(c + d) = ac + bd$

29. 2^x is always positive.

30.

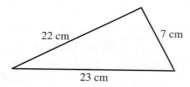

This triangle (not drawn to scale) is a *right-angled* triangle.

31. Suppose that we accurately draw the straight line $y = x$ on a set of axes. Then if we take a protractor and measure the acute angle between the line and the x-axis, it will be found to be $45°$.

32. If we have an unlimited supply of 5p, 7p and 11p stamps, then we can make up any amount above 13p from just these stamps.

33. If n is a positive integer greater than 1, then $2^n < n^3$.

34. There is exactly one point which lies on *both* of the straight lines $y = \frac{1}{2}x + 5$ and $x - 2y = 3$.

35. If n is a positive integer, then $n^4 - 10n^3 + 35n^2 - 48n + 24$ is equal to $2n$.

36. If the number x is multiplied by 78·39, and the result is then divided by 78·39, then the final answer is x.

37. If the number x is increased by 8·3%, and the result is then decreased by 8·3%, then the final answer is x.

38. Written as a fraction, π is $\frac{22}{7}$.

39. If triangle ABC is isosceles, then $\widehat{ABC} = \widehat{ACB}$.

40. If the product of two numbers is 8, then one of the numbers must be 8.

41. If the product of two numbers is 0, then one of the numbers must be 0.

42. A cuboid has 6 faces, 12 edges, and 8 vertices.

43. A pyramid has 5 faces, 8 edges, and 5 vertices.

44. The number 133! (i.e. the number obtained by working out the product $1 \times 2 \times 3 \times 4 \times 5 \times \ldots\ldots \times 132 \times 133$) ends in exactly 26 noughts.

11.3 Multiple choice tests [non-calculator]

Test 1

1. How many mm are there in 1 m 1 cm?

   A 1001
   B 1110
   C 1010
   D 1100

2. The circumference of a circle is 16π cm. The radius, in cm, of the circle is:

   A 2
   B 4
   C $\frac{4}{\pi}$
   D 8

3. In the triangle below the value of $\cos x$ is:

   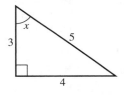

   A 0·8
   B 1·333
   C 0·75
   D 0·6

4. The line $y = 2x - 1$ cuts the x-axis at P. The coordinates of P are:

   A $(0, -1)$
   B $(\frac{1}{2}, 0)$
   C $(-\frac{1}{2}, 0)$
   D $(-1, 0)$

5. The formula $b + \dfrac{x}{a} = c$ is rearranged to make x the subject. What is x?

   A $a(c - b)$
   B $ac - b$
   C $\dfrac{c - b}{a}$
   D $ac + ab$

6. The mean weight of a group of 11 men is 70 kg. What is the mean weight of the remaining group when a man of weight 90 kg leaves?

   A 80 kg
   B 72 kg
   C 68 kg
   D 62 kg

7. Find x if $2^{x+2} = 16^{x-7}$

   A 8
   B 9
   C 10
   D 11

8. In standard form the value of $2000 \times 80\,000$ is:

   A 16×10^6
   B $1{\cdot}6 \times 10^9$
   C $1{\cdot}6 \times 10^7$
   D $1{\cdot}6 \times 10^8$

9. The solutions of the equation $(x - 3)(2x + 1) = 0$ are:

   A $-3, \frac{1}{2}$
   B $3, -2$
   C $3, -\frac{1}{2}$
   D $-3, -2$

10. In the triangle the size of angle x is:

    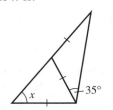

    A 35°
    B 70°
    C 110°
    D 40°

11. A man paid tax on £9000 at 30%. He paid the tax in 12 equal payments. Each payment was:

    A £2·25
    B £22·50
    C £225
    D £250

12. The approximate value of $\dfrac{3{\cdot}96 \times (0{\cdot}5)^2}{97{\cdot}1}$ is:

    A 0·01
    B 0·02
    C 0·04
    D 0·1

13. Given that $\dfrac{3}{n} = 5$, then $n =$

    A 2
    B -2
    C $1\frac{2}{3}$
    D 0·6

14. Cube A has side 2 cm. Cube B has side 4 cm. $\left(\dfrac{\text{Volume of B}}{\text{Volume of A}}\right) =$

    A 2
    B 4
    C 8
    D 16

15. How many tiles of side 50 cm will be needed to cover the floor shown?

    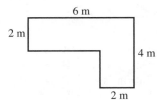

    A 16
    B 32
    C 64
    D 84

16. The equation
$ax^2 + x - 6 = 0$ has a
solution $x = -2$ What is
a?

    A 1
    B -2
    C $\sqrt{2}$
    D 2

17. Which of the following
is/are correct?
1. $\sqrt{0.16} = 0.4$
2. $0.2 \div 0.1 = 0.2$
3. $\frac{4}{7} > \frac{3}{5}$

    A **1** only
    B **2** only
    C **3** only
    D **1** and **2**

18. How many prime
numbers are there
between 30 and 40?

    A 0
    B 1
    C 2
    D 3

19. A man is paid £180 per
week after a pay rise of
20%. What was he paid
before?

    A £144
    B £150
    C £160
    D £164

20. A car travels for 20
minutes at 45 m.p.h. and
then for 40 minutes at
60 m.p.h. The average
speed for the whole
journey is:

    A $52\frac{1}{2}$ m.p.h.
    B 50 m.p.h
    C 54 m.p.h.
    D 55 m.p.h.

21. The point $(3, -1)$ is
reflected in the line
$y = 2$. The new
coordinates are:

    A $(3, 5)$
    B $(1, -1)$
    C $(3, 4)$
    D $(0, -1)$

22. Two discs are randomly
taken from a bag
containing 3 red discs
and 2 blue discs. What is
the probability of taking
2 red discs?

    A $\frac{9}{25}$
    B $\frac{1}{10}$
    C $\frac{3}{10}$
    D $\frac{2}{5}$

23. The shaded area, in cm^2,
is:

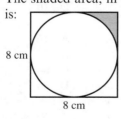

    A $16 - 2\pi$
    B $16 - 4\pi$
    C $\frac{4}{\pi}$
    D $64 - 8\pi$

24. Given the equation
$5^x = 120$, the best
approximate solution is
$x =$

    A 2
    B 3
    C 4
    D 25

25. What is the sine of 45°?

    A 1
    B $\frac{1}{2}$
    C $\frac{1}{\sqrt{2}}$
    D $\sqrt{2}$

Test 2

1. What is the value of the
expression
$(x - 2)(x + 4)$ when
$x = -1$?

    A 9
    B -9
    C 5
    D -5

2. The perimeter of a
square is 36 cm. What is
its area?

    A 36 cm^2
    B 324 cm^2
    C 81 cm^2
    D 9 cm^2

3. AB is a diameter of the
circle. Find the angle
BCO.

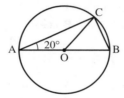

    A 70°
    B 20°
    C 60°
    D 50°

4. The gradient of the line
$2x + y = 3$ is:

    A 3
    B -2
    C $\frac{1}{2}$
    D $-\frac{1}{2}$

5. A firm employs 1200
people, of whom 240 are
men. The percentage of
employees who are men
is:

    A 40%
    B 10%
    C 15%
    D 20%

6. A car is travelling at a
constant speed of
30 m.p.h. How far will
the car travel in 10
minutes?

    A $\frac{1}{3}$ mile
    B 3 miles
    C 5 miles
    D 6 miles

7. What are the
coordinates of the point
$(1, -1)$ after reflection in
the line $y = x$?

    A $(-1, 1)$
    B $(1, 1)$
    C $(-1, -1)$
    D $(1, -1)$

8. $\frac{1}{3} + \frac{2}{5} =$

A $\frac{2}{8}$

B $\frac{3}{8}$

C $\frac{3}{15}$

D $\frac{11}{15}$

9. In the triangle the size of the largest angle is:

A 30°

B 90°

C 120°

D 80°

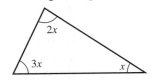

10. 800 decreased by 5% is:

A 795

B 640

C 760

D 400

11. Which of the statements is (are) true
1. $\tan 60° = 2$
2. $\sin 60° = \cos 30°$
3. $\sin 30° > \cos 30°$

A **1** only

B **2** only

C **3** only

D **2** and **3**

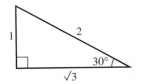

12. Given $a = \frac{3}{5}$, $b = \frac{1}{3}$, $c = \frac{1}{2}$ then

A $a < b < c$

B $a < c < b$

C $a > b > c$

D $a > c > b$

13. The *larger* angle between South-West and East is:

A 225°

B 240°

C 135°

D 315°

14. In a triangle PQR, $P\widehat{Q}R = 50°$ and point X lies on PQ such that $QX = XR$. Calculate $Q\widehat{X}R$.

A 100°

B 50°

C 80°

D 65°

15. What is the value of $1 - 0.05$ as a fraction?

A $\frac{1}{20}$

B $\frac{9}{10}$

C $\frac{19}{20}$

D $\frac{5}{100}$

16. Find the length x.

A 5

B 6

C 8

D $\sqrt{50}$

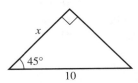

17. Given that $m = 2$ and $n = -3$, what is mn^2?

A -18

B 18

C -36

D 36

18. The graph of $y = (x - 3)(x - 2)$ cuts the y-axis at P. The coordinates of P are:

A (0, 6)

B (6, 0)

C (2, 0)

D (3, 0)

19. £240 is shared in the ratio $2 : 3 : 7$. The largest share is:

A £130

B £140

C £150

D £160

20. Adjacent angles in a parallelogram are $x°$ and $3x°$. The smallest angles in the parallelogram are each:

A 30°

B 45°

C 60°

D 120°

21. When the sides of a square are increased by 10% the area is increased by:

A 10%

B 20%

C 21%

D 15%

22. The volume, in cm³, of the cylinder is:

A 9π

B 12π

C 600π

D 900π

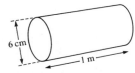

23. A car travels for 10 minutes at 30 m.p.h. and then for 20 minutes at 45 m.p.h. The average speed for the whole journey is:

A 40 m.p.h.

B $37\frac{1}{2}$ m.p.h.

C 20 m.p.h.

D 35 m.p.h.

24. Four people each toss a coin. What is the probability that the fourth person will toss a 'tail'?

A $\frac{1}{2}$

B $\frac{1}{4}$

C $\frac{1}{8}$

D $\frac{1}{16}$

25. A rectangle 8 cm by 6 cm is inscribed inside a circle. What is the area, in cm², of the circle?

- **A** 10π
- **B** 25π
- **C** 49π
- **D** 100π

Test 3

1. The price of a T.V. changed from £240 to £300. What is the percentage increase?

- **A** 15%
- **B** 20%
- **C** 60%
- **D** 25%

2. Find the length x.

- **A** 6
- **B** 5
- **C** $\sqrt{44}$
- **D** $\sqrt{18}$

3. The bearing of A from B is 120°. What is the bearing of B from A?

- **A** 060°
- **B** 120°
- **C** 240°
- **D** 300°

4. Numbers m, x and y satisfy the equation $y = mx^2$. When $m = \frac{1}{2}$ and $x = 4$ the value of y is:

- **A** 4
- **B** 8
- **C** 1
- **D** 2

5. A school has 400 pupils, of whom 250 are boys. The ratio of boys to girls is:

- **A** 5 : 3
- **B** 3 : 2
- **C** 3 : 5
- **D** 8 : 5

6. A train is travelling at a speed of 30 km per hour. How long will it take to travel 500 m?

- **A** 2 minutes
- **B** $\frac{3}{50}$ hour
- **C** 1 minute
- **D** $\frac{1}{2}$ hour

7. The approximate value of $\dfrac{9\cdot65 \times 0\cdot203}{0\cdot0198}$ is:

- **A** 99
- **B** 9·9
- **C** 0·99
- **D** 180

8. Which point does *not* lie on the curve $y = \dfrac{12}{x}$?

- **A** (6, 2)
- **B** ($\frac{1}{2}$, 24)
- **C** (−3, −4)
- **D** (3, −4)

9. $t = \dfrac{c^3}{y}$, $y =$

- **A** $\dfrac{t}{c^3}$
- **B** $c^3 t$
- **C** $c^3 - t$
- **D** $\dfrac{c^3}{t}$

10. The largest number of 1 cm cubes which will fit inside a cubical box of side 1 m is:

- **A** 10^3
- **B** 10^6
- **C** 10^8
- **D** 10^{12}

11. The nth term of a sequence is $u_n = n(n - 2)$. Find the largest value of n for which $u_n < 2000$.

- **A** 50
- **B** 2001
- **C** 44
- **D** 45

12. Which of the following has the largest value?

- **A** $\sqrt{100}$
- **B** $\sqrt{\dfrac{1}{0\cdot1}}$
- **C** $\sqrt{1000}$
- **D** $\dfrac{1}{0\cdot01}$

13. Two dice numbered 1 to 6 are thrown together and their scores are added. The probability that the sum will be 12 is:

- **A** $\frac{1}{6}$
- **B** $\frac{1}{12}$
- **C** $\frac{1}{18}$
- **D** $\frac{1}{36}$

14. The length, in cm, of the minor arc is:

- **A** 3π
- **B** π
- **C** 6π
- **D** $13\frac{1}{2}\pi$

15. Metal of weight 84 kg is made into 40 000 pins. What is the weight, in kg, of one pin?

- **A** 0·0021
- **B** 0·0036
- **C** 0·021
- **D** 0·21

16. What is the value of x which satisfies both equations?
$3x + y = 1$
$x - 2y = 5$

- **A** −1
- **B** 1
- **C** −2
- **D** 2

17. What is the new fare when the old fare of £250 is increased by 8%?

A £258
B £260
C £270
D £281·25

18. What is the area of this triangle?

A $12x^2$
B $15x^2$
C $16x^2$
D $30x^2$

19. What values of x satisfy the inequality $2 - 3x > 1$?

A $x < -\frac{1}{3}$
B $x > -\frac{1}{3}$
C $x > \frac{1}{3}$
D $x < \frac{1}{3}$

20. A right-angled triangle has sides in the ratio $5:12:13$. The tangent of the smallest angle is:

A $\frac{12}{5}$
B $\frac{12}{13}$
C $\frac{5}{13}$
D $\frac{5}{12}$

21. The area of △ABE is $4\,\text{cm}^2$. The area of △ACD is:

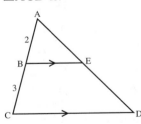

A $10\,\text{cm}^2$
B $6\,\text{cm}^2$
C $25\,\text{cm}^2$
D $16\,\text{cm}^2$

22. Given $2^x = 3$ and $2^y = 5$, the value of 2^{x+y} is:

A 15
B 8
C 4
D 125

23. The probability of an event occurring is 0·35. The probability of the event *not* occurring is:

A $\dfrac{1}{0\cdot35}$
B 0·65
C 0·35
D 0

24. How long is the perimeter of the rectangle?

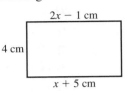

A 20 cm
B 30 cm
C 24·5 cm
D can't be found

25. On a map a distance of 36 km is represented by a line of 1·8 cm. What is the scale of the map?

A 1 : 2000
B 1 : 20 000
C 1 : 200 000
D 1 : 2 000 000

Test 4

1. What is the value of x satisfying the simultaneous equations
$3x + 2y = 13$
$x - 2y = -1$?

A 7
B 3
C $3\frac{1}{2}$
D 2

2. A straight line is 4·5 cm long. $\frac{2}{5}$ of the line is:

A 0·4 cm
B 1·8 cm
C 2 cm
D 0·18 cm

3. The mean of four numbers is 12. The mean of three of the numbers is 13. What is the fourth number?

A 9
B 12·5
C 7
D 1

4. How many whole numbers satisfy both of the inequalities below?
$5x > 4$ and $3x - 4 < 32$

A 12
B none
C 11
D 36

For Questions **5** to **7** use the diagram below.

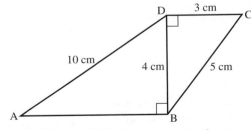

5 The length of AB, in cm, is:

A 6
B $\sqrt{116}$
C 8
D $\sqrt{84}$

6. The sine of angle DCB is:

- **A** 0·8
- **B** 1·25
- **C** 0·6
- **D** 0·75

7. The tangent of angle CBD is:

- **A** 0·6
- **B** 0·75
- **C** 1·333
- **D** 1·6

8. The value of 4865·355 correct to 2 significant figures is:

- **A** 4865·36
- **B** 4865·35
- **C** 4900
- **D** 49

9. What is the surface area of a cube of volume $1000\,\text{cm}^3$?

- **A** $800\,\text{cm}^2$
- **B** $60\,\text{cm}^2$
- **C** $480\,\text{cm}^2$
- **D** $600\,\text{cm}^2$

10. The area of a circle is $100\pi\,\text{cm}^2$. The radius, in cm, of the circle is:

- **A** 50
- **B** 10
- **C** $\sqrt{50}$
- **D** 5

11. If $f(x) = x^2 - 3$, then $f(3) - f(-1) =$

- **A** 5
- **B** 10
- **C** 8
- **D** 9

12. In the triangle BE is parallel to CD. What is x?

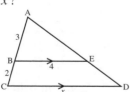

- **A** $6\frac{2}{3}$
- **B** 6
- **C** $7\frac{1}{2}$
- **D** $5\frac{3}{4}$

13. Simplify $\dfrac{n+n+n+n}{n}$

- **A** $4n$
- **B** 4
- **C** n^3
- **D** can't be done

14. In a group of 20 people, 5 cannot swim. If two people are selected at random, what is the probability that neither of them can swim?

- **A** $\frac{1}{16}$
- **B** $\frac{1}{19}$
- **C** $\frac{35}{76}$
- **D** $\frac{12}{19}$

15. Given $16^x = 4^4$, what is x?

- **A** -2
- **B** $-\frac{1}{2}$
- **C** $\frac{1}{2}$
- **D** 2

16. What is the area, in m^2, of a square with each side 0·02 m long?

- **A** 0·0004
- **B** 0·004
- **C** 0·04
- **D** 0·4

17. I start with x, then square it, multiply by 3 and finally subtract 4. The final result is:

- **A** $(3x)^2 - 4$
- **B** $(3x - 4)^2$
- **C** $3x^2 - 4$
- **D** $3(x - 4)^2$

18. n people buy x drinks at z pence each and divide the cost equally between them. How much does each person pay?

- **A** $\dfrac{z+x}{n}$
- **B** $\dfrac{xz}{n}$
- **C** $\dfrac{x+n}{z}$
- **D** $\dfrac{xn}{z}$

19. What are the coordinates of the point $(2, -2)$ after reflection in the line $y = -x$?

- **A** $(-2, 2)$
- **B** $(2, -2)$
- **C** $(-2, -2)$
- **D** $(2, 2)$

20. The area of a circle is $36\pi\,\text{cm}^2$. The circumference, in cm, is:

- **A** 6π
- **B** 18π
- **C** $12\sqrt{\pi}$
- **D** 12π

21. The gradient of the line $2x - 3y = 4$ is:

- **A** $\frac{2}{3}$
- **B** $1\frac{1}{2}$
- **C** $-\frac{4}{3}$
- **D** $-\frac{3}{4}$

22. When all three sides of a triangle are trebled in length, the area is increased by a factor of:

- **A** 3
- **B** 6
- **C** 9
- **D** 27

23. $a = \sqrt{\left(\dfrac{m}{x}\right)}$

$x =$

- **A** $a^2 m$
- **B** $a^2 - m$
- **C** $\dfrac{m}{a^2}$
- **D** $\dfrac{a^2}{m}$

12 EXAMINATION QUESTIONS

Examination exercise 1: Number (non calculator)

1. In each of the following, put in a decimal point so that the measurement is reasonable.
   (a) Diameter of a 2p coin.   2750 cm
   (b) Weight of an average woman   5714 kg
   (c) Capacity of a petrol tank on a moped   7698 litres
   (d) Length of a football pitch   1076 m   [M]

2. The scale of a map was 1 : 2000. The map is reduced in size so that it uses a quarter of the original area of paper. What is the scale of the new map?   [S]

3. (a) Find the exact value of

$$\left[\left(\frac{2}{3}\right)^2 + \frac{5}{6}\right] \div 11\frac{1}{2}$$

   (a) Given that $a = 64$, evaluate

$$a^{\frac{1}{2}} + a^{\frac{2}{3}}$$ [M]

4. A metal rivet fits into a hole as shown.
   The diameter of the rivet is $\frac{3''}{5}$.
   The diameter of the hole is $\frac{7''}{8}$.
   Calculate the distance a, which you should express as
   (a) a fraction of an inch (b) a decimal.   [N]

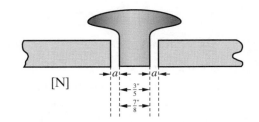

5. Write down the prime factors of
   (a) 273,
   (b) 2004.
   Hence find the highest common factor of 273 and 2002.   [S]

6. This item appeared in a Sunday newspaper.

   > Rubik's New Puzzle
   >
   > A new version of the Rubik Cube Puzzle has been invented. It has 256 times the number of arrangements of the original one. The original Rubik Cube had 43×10^{30} different arrangements.

   Calculate the number of arrangements of the new puzzle, giving your answer in standard form, correct to one place of decimals. [N]

7. The time, T seconds, that an experiment takes depends on the temperature, t degrees, and is given by the formula

$$T = t^4 - t + 1$$

Calculate the time the experiment takes when the temperature is -3 degrees.   [N]

8. A cassette tape is played at $1\frac{7}{8}$ inches per second.
 (a) Work out the length of tape played in $2\frac{1}{2}$ seconds.
 (b) Work out how long it takes for $9\frac{3}{8}$ inches of tape to be played.   [L]

9. (a) Express 9800 as a product of prime factors.
 (b) Find the smallest square number which is a multiple of 9800.

10. Jack shares £180 between his two children Ruth and Ben. The ratio of Ruth's share to Ben's share is $5:4$.
 (a) Work out how much each child is given.
 Ben then gives 10% of his share to Ruth.
 (b) Work out the percentage of the £180 that Ruth now has.   [N.I.]

11. Imagine you do not have a calculator with you.
 Show how you would prove that $13\,000^{\frac{1}{4}}$ lies between 10 and 11 without the use of a calculator.   [N.I.]

12. (a) Explain what is meant by the statement

'$\sqrt{8}$ is an irrational number.'

n is a **rational** number.

$$k \times \sqrt{8} = n$$

 (b) Write down a non-zero value of k and a non-zero value of n which satisfy this equation.
 (c) Write $\sqrt{8}$ in the form 2^c, where c is a **rational** number.

$$4^{d+1} = \sqrt{8}$$

 (d) Find the value of d.   [L]

Examination exercise 2: Number (calculators allowed)

1. (a) 35% of the population of a country are city-dwellers. Express the ratio of city-dwellers to non city-dwellers in its simplest form.
 (b) (i) The population of another country is 73 million people. Write this number in standard form.
   (ii) The ratio of under 18s to over 18s is $4:11$. Calculate the number of over 18s in the population, giving your answer to the nearest thousand.   [N.I.]

2. Nesta invests £508 in a bank account paying compound interest at a rate of 10% per annum.
Calculate the total amount in Nesta's bank account after 2 years.
[L]

3. Jane is going on holiday to France and needs to buy some films for her camera. If she buys them in England before she goes, they will cost her £2·25 each. If she waits until she gets to France and buys them there, they will cost her 20·16 francs each.
If the exchange rate is £1 to 9·60 francs, find out how much, in English currency, she can save on each film by buying them in France.

4. The approximate area of the world is 135 085 000 km^2.
(a) Express the number 135 085 000 in standard form.
The approximate area of Africa is 30 132 000 km^2.
(b) Calculate the area of Africa as an approximate percentage of the area of the world. (Give your answer correct to one decimal place.)
[L]

5. A sum of money was invested at a fixed rate of compound interest added annually.
At the end of the first year it had amounted to £990 and at the end of the second year it had amounted to £1089.
(a) Calculate the yearly rate of interest.
(b) Calculate the original sum of money which was invested.   [N]

6. Aziz wants to make a plastic mug in the shape of a cylinder. The volume, V, of the mug must be 510 cubic centimetres and the height, h, 9·6 centimetres, Aziz has to work out the radius, r, in centimetres. He knows that r is given by the formula

$$r = \sqrt{\frac{V}{\pi H}}.$$

(a) Taking $\pi = 3\cdot14$, calculate the value of r, correct to 1 decimal place.
(b) Explain clearly how Aziz could estimate the value of r, correct to 1 significant figure, if he did not have a calculator or mathematical tables.
[M]

7. Supergrowth Unit Trust claims that the value of its units is likely to grow by 21% compound interest per annum. Assuming that this claim is true, calculate the value, after 5 years, of an investment of £1000 in Supergrowth Unit Trust.
[M]

8. In a General Election a candidate loses his deposit if he does not obtain at least 5% of the total votes cast in the constituency for which he is seeking election.
(a) In a certain constituency, three candidates, A, B and C, had put up for election.

Out of the 42 560 votes cast, 21 523 people voted for candidate A, 18 862 voted for candidate B, and the rest voted for candidate C.

Decide whether, and by how many votes, candidate C lost or saved his deposit.

(b) In another constituency, there were just two candidates, R and S. The winner, who was candidate R, received 16 017 votes. By letting x be the number of people who voted for candidate S, or otherwise, calculate the least number of votes candidate S would have to obtain in order not to lose his deposit.   [N]

9.

It's a fact!

– SOMEONE IN THE UK IS BURGLED EVERY 66 SECONDS –

(a) Assuming that this statement is true, find correct to 2 significant figures, the number of burglaries in the UK in 1 year.

(b) Write your answer to part (a) in standard form.   [M]

10. (a) Write down a rational number x such that $\sqrt{42} < x < \sqrt{44}$.

(b) Write down an irrational number y such that $6 < y < 7$.

(c) Write down a rational number whose decimal form is never-ending.   [N.I.]

11.
$$1 \, \text{m}^3 = 220 \text{ gallons}$$
$$1 \, \text{m}^3 = 10^6 \, \text{cm}^3$$

(a) How many m^3 are equal to one gallon?

Write your answer in standard form, correct to 3 significant figures.

The petrol tank of a small car holds 6 gallons when it is 80% full.

(b) What is the capacity of the petrol tank in cm^3?

Give your answer correct to 3 significant figures.   [L]

12.
$$F = \frac{ab}{a - b}$$

Imran uses this formula to calculate the value of F.

Imran estimates the value of F without using a calculator.

$a = 49{\cdot}8$ and $b = 30{\cdot}6$.

(a)   (i) Write down approximate values for a and b that Imran could use to estimate the value of F.

(ii) Work out the estimate for the value of F that these approximations give.

Imran works out the value of F with two new values for a and b.

(iii) Use your calculator to work out the accurate value for F.
Use $a = 49\cdot8$ and $b = 30\cdot6$.
Write down all the figures on your calculator display.

(b) Calculate the value of F when
$a = 9\cdot6 \times 10^{12}$ and $b = 4\cdot7 \times 10^{11}$
Give your answer in standard form, correct to two significant
figures.   [L]

13. The world record for the 100 metres sprint is $9\cdot83$ seconds.
(a) Calculate the average speed of the athlete in km/h.
(b) A newspaper report said that the athlete was running at 40 km/h.
Calculate the percentage error of this estimate.   [M]

14. The area of a rectangular field is 28 500 square metres and its length
is 195 metres, both measurements being correct to three significant
figures.
(a) Find:
(i) the greatest possible breadth of the field
(ii) the smallest possible breadth of the field.
(b) What is the breadth of the field correct to two significant
figures?   [N]

15. As part of a project, some children are timing cars as they travel
along a section of road which they have measured as 100 metres.
One car is timed at 6 seconds.
(a) Assuming these figures to be exact, find the average speed of
this car, in metres per second.
(b) In fact, the length of road was measured correct to the nearest
10 metres and the car was timed correct to the nearest second.
Find the maximum average speed of the car, correct to the
nearest metre per second.   [M]

Examination exercise 3: Algebra (non-calculator)

1. (a) Solve the equations:
(i) $7 - 2x = 9$

(ii) $\dfrac{9}{y} = 5$

(b) Factorise the expression $6z^2 - 9z$.
(c) Multiply out the brackets giving your answer in its simplest
form: $(2x + 5)(3x - 7)$ [S]

2. Solve the equation:

$$\frac{(x + 1)}{2} - \frac{(2x + 1)}{3} = 1 \qquad\qquad [M]$$

3. The cost, £C, of making n articles is given by the formula
$C = a + bn$ where a and b are constants.
The cost of making 4 articles is £20 and the cost of making 7 articles is £29.
Write down two equations in a and b.
Solve these equations to find the values of a and b.   [L]

4. A craftsman can be paid in one of two ways:

Method A   A down payment of £300, then £4 an hour.
Method B   £9 an hour.

(a) He works for n hours. Write down, in terms of n, an expression for his earnings using:
    (i) *Method A*   (ii) *Method B*.
(b) Write down an inequality that will be true if his earnings by *Method B* are greater than his earnings by *Method A*.
(c) Solve your inequality for n.   [M]

5. (a) Calculate
    (i) $1^3 + 2^3 + 3^3 - (1 + 2 + 3)^2$
    (ii) $1^3 + 2^3 + 3^3 + 4^3 - (1 + 2 + 3 + 4)^2$.
(b) Comment on your results for part (a).   [M]

6. Sian is making a pattern of tiles by surrounding black tiles with white tiles.

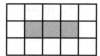

(a) Give a formula that Sian could use to find the number of white tiles required to surround n black tiles.
(b) If Sian has 81 white tiles, what is the greatest number of black tiles that can be surrounded?   [N]

7. PULL-UPS

At the end of a one minute 'pull-ups' contest the following facts were published.
The top five contestants completed a total of 163 pull-ups between them.
The winning contestant had completed only one more pull-up than the contestant who was second.
Three contestants tied for third place, each having completed only one fewer pull-up than the contestant who was second.
How many pull-ups were completed by the winning contestant?   [S]

8.

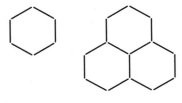 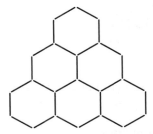

Bridget is doing an investigation which involves making patterns of hexagons with sticks. The first three of these patterns are shown above.
(a) Copy and complete this table.

number of rows, n	1	2	3	4	5
number of hexagons, h	1	3			
number of sticks, s	6				

(b) Write down a formula for h in terms of n.
(c) The formula for s in terms of n is of the form

$$s = an^2 + bn,$$

where a and b are constants.
Find the value of a and the value of b.   [M]

9. Some pupils were making up number puzzles.
One pupil's number puzzle was written like this:

$$2(3x - 1) < 18$$

What is the largest whole number that this pupil could have thought of?   [N]

10.

Pattern 1   Pattern 2   Pattern 3

and so on.

Each pattern in this sequence consists of circles and stars.
(a) Copy and complete the table:

Pattern	Number of Circles	Number of Stars
1	2	1
2	3	4
3	4	9
4		
5		

(a) (i) How many circles are in Pattern 20?
(ii) How many stars are in Pattern 20?

(c) One of these patterns has 9 circles.
How many stars are in this pattern?
(d) For Pattern n, write down in terms of n,
 (i) the number of circles,
 (ii) the number of stars.
The total number of circles and stars in Pattern n is 241.
(e) (i) Write down an equation in n.
 (ii) Solve this equation.
 (iii) **Hence**, state the pattern number.   [N.I.]

11 (a) Solve the simultaneous equations:

$$3x - 4y = 10$$
$$5x + 7y = 3$$

(b) In a physics experiment three variables, u, v and f are connected
by the formula

$$f = \frac{uv}{u + v}$$

Make v the subject of the formula.   [L]

12. The voltage, V volts, available from a 12 volt battery of internal
resistance r ohms when connected to apparatus of resistance R
ohms, is given by

$$V = \frac{12R}{(r + R)}$$

(a) Find V when $r = 1.5$ and $R = 6$.
(b) Express R in terms of V and r.   [M]

13. The current, l amps, in a circuit containing a resistor of resistance
R ohms, is given by

$$l = \frac{12}{R + 2}$$

Express R in terms of l.   [M]

14. (a) Simplify:

 (i) $(2x^5)^3$ (ii) $\dfrac{x^4 y^3}{xy^5}$

(b) Factorise completely:

$$4xy^2 - 8x^2 y$$

(c) Simplify:

 (i) $\dfrac{x^2 - 3x - xy + 3y}{x^2 - 9}$ (ii) $\dfrac{\sqrt[3]{a^2} \times \sqrt{a}}{a \times \sqrt[6]{a}}$

(d) Find the value of x for which

$$x^{-3} = 27$$ [N.I.]

15. A stone is dropped down a water well.
The time taken in seconds, for the stone to drop from ground level
to the water level is measured. The depth of a well is proportional
to the square of the time taken for a stone to drop down it. One
day, when the water level in the well was 100 m below ground level,
the time taken was 4 seconds. After heavy rain another stone took 3
seconds to drop from ground level to the water level.
What was the depth from ground level to water level after the heavy
rain?   [L]

16. (a) To what power must x^2 be raised to give

    (i) x^6?   (ii) $\dfrac{1}{x^2}$?   (iii) x^{100}?

(b) Write down the square root of $121x^{16}$.   [N]

17. The rectangle ABCD
has a perimeter of 12 cm
and side AD $= x$ cm.

(a) Write down, in terms of x, an expression for the area of the
rectangle.
(b) The rectangle ABCD has an area of 6·44 cm². Show that
$x^2 - 6x + 6\cdot44 = 0$.
(c) Complete the table of values for $y = x^2 - 6x + 6\cdot44$.
Draw the graph of y against x. Take x from 0 to 6 and y from
-4 to 8.

x	0	1	2	3	4	5	6
y		1·44	$-1\cdot56$	$-2\cdot56$	$-1\cdot56$		

(d) Read off from your graph an approximate value for the width
of the rectangle.   [N]

18. The diagram is the speed-time graph of a
bus on a short journey between two bus stops.
Calculate:
(a) the acceleration, in m/s², of the bus during
the first 50 seconds,
(b) the total distance, in m, travelled in the
journey,
(c) the average speed, in m/s, of the bus for
the whole journey.   [L]

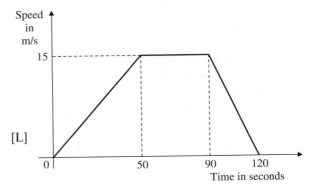

19. A car accelerates from rest at $4\,\text{m/s}^2$ for 5 seconds, then moves with constant speed for 10 seconds, then decelerates at $2\,\text{m/s}^2$ back to rest.
(a) Draw a speed-time graph for the journey of the car.
(b) Calculate the total distance travelled by the car on this journey. [L]

20. At time t seconds, the velocity, v metres per second, of a particle moving along a straight line is given by $v = 2t^2 - 9t + 8$.
Some values of v and t are given in the table

t	0	1	2	3	4	5	6
v	8		-2	-1	4		26

(a) Calculate the value of v when
    (i) $t = 1$ (ii) $t = 5$.
(b) Taking 2 cm to represent 1 unit on the t axis and 1 cm to represent 2 units on the v axis, draw the graph of $v = 2t^2 - 9t + 8$ for values of t from 0 to 6.
(c) Use your graph to estimate:
    (i) the values of t for which the velocity is zero,
    (ii) the value of t for which the acceleration is zero.
(d) Use the trapezium rule, with four equal intervals, to calculate the approximate distance travelled from $t = 4$ to $t = 6$. [L]

21. (a) Express $x^2 + 5x - 7$ in the form $(x + a)^2 + b$.
(b) Hence, solve $x^2 + 5x - 7 = 0$. [N]

22. Write as a single fraction in its simplest form
$$\frac{2x - 8}{x^2 - 16} + \frac{10}{x^2 + 3x - 4}.$$ [L]

23. This is a sketch of the curve with equation $y = f(x)$.
The vertex of the curve is A(2, 12).
Write down the coordinates of the vertex for each of the curves having the following equations.
(a) $y = f(x + 6)$
(b) $y = f(x + 3)$
(c) $y = f(-x)$
(d) $y = f(4x)$

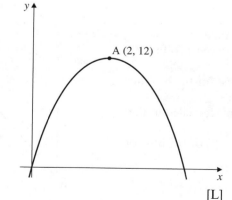

[L]

24. Two graphs are drawn below.

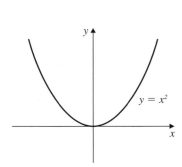
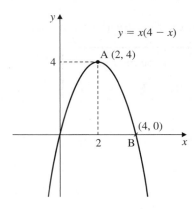

(a) Sketch the graph of $y = x^2 - 2$.
(b) Sketch the graph of $y = (x - 2)^2$.
(c) Sketch the graph of $y = 2x(4 - 2x)$, showing the new coordinates of points A and B.

Examination exercise 4: Algebra (calculators allowed)

1. The positive number n satisfies the equation

$$n^2 + n = 80$$

(a) Write down two consecutive whole numbers between which n must lie.
(b) Use trial and improvement and a calculator to find the value of n correct to 3 significant figures.

2.

D	E	E	D
E	D	D	E
E	D	D	E
D	E	E	D

C			C
	M	M	
	M	M	
C			C

A 'chessboard' consists of four different types of square. These are denoted by C, corner squares, of which there are four,
E, edge squares, which lie along the edge of the board, but are not corner squares,
M, middle squares, which are neither corner nor edge squares,
D, diagonal squares, which lie along both diagonals of the square.
(Some squares may be more than one type of square simultaneously.)
The diagrams above show the same 4 × 4 chessboard labelled in two different ways.

The table below shows the number of each type of square for various size boards.

(a) Complete the table on the right for the 6 × 6 board.

	Type of square			
Size of board	C	E	M	D
3 × 3	4	4	1	5
4 × 4	4	8	4	8
5 × 5	4	12	9	9
6 × 6				

(b) How many middle squares will an 8 × 8 board have?

(c) A board has 48 edge squares. How many diagonal squares does it have?

(d) A board has 225 middle squares. How many edge squares does it have?

(e) Find the number of each type of square for a 100 × 100 board.

[N]

3. Using the method of trial and improvement, solve the equation

$$x^3 - 4x - 1 = 30$$

correct to one decimal place. You must show your working.   [L]

4. A scientist has a theory that the time taken for a comet to disintegrate in the earth's atmosphere can be calculated from the equation

$100t = \sqrt{m}$ where t = time taken (seconds),
                    m = mass of comet (tonnes).

(a) A comet took 4·35 seconds to disintegrate.
    Use the equation to calculate its mass.

(b) In November a comet of estimated mass $8·5 \times 10^6$ tonnes was to enter the Earth's atmosphere.
    How long would it take to disintegrate?   [N]

5. (a) On a grid with values of x and y from 0 to 8, draw the lines

$$x = 2, \quad y = 4, \quad y = 8 - x.$$

(b) Shade and label, with the letter R, the region for which the points (x, y) satisfy the three inequalities

$$x \geqslant 2, \quad y \leqslant 4, \quad y \leqslant 8 - x.$$ [M]

6. The figure shows a large rectangle PQRS divided into four smaller rectangles.
   (a) Find, in terms of a and b, the perimeter of the rectangle PQRS. Give your answer in its simplest form.
   (b) (i) On the figure (right), write inside each of the smaller rectangles its area in terms of a and b.
       (ii) Find, in terms of a and b, the area of the rectangle PQRS. Give your answer in its simplest form.
   (c) If PQRS is a square, find b in terms of a.

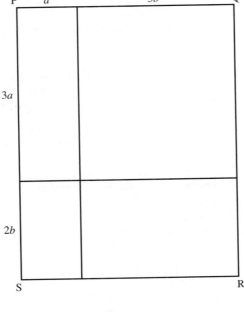

[N.I.]

7. This sketch shows part of the graph with equation

$$y = pq^x,$$

where p and q are constants.
The points with coordinates $(0, 8)$, $(1, 18)$ and $(1.5, k)$ lie on the graph.
Calculate the value of p, q and k.

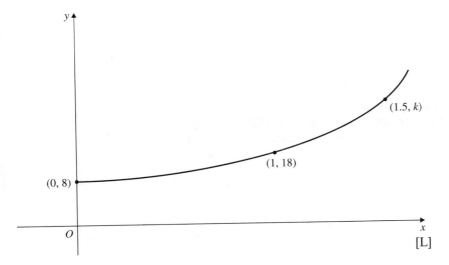

[L]

8. x, $2x + y$, $3x + 2y$, $4x + 3y$ are the first four terms of a sequence.
   (a) Write down the fifth term of the sequence.
   The sum of the first four terms is 5 and the fifth term has the value 1.
   (b) Find the values of x and y.
   (c) (i) In terms of n, x, y, write down the nth term of the sequence.
       (ii) Using the values of x, y found in (b), find the value of n for which the nth term is 0.   [N.I.]

9. The sum of the squares of three consecutive whole numbers is 6914. The middle number is n.
   (a) Write down expressions for the other two numbers.
   (b) By squaring the algebraic expressions for the three numbers and adding them together, create an equation in n.
   (c) (i) Solve your equation to find n.
       (ii) Write down the three consecutive numbers.   [L]

10.

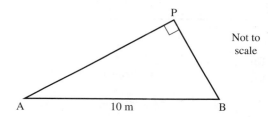

P

Not to scale

A   10 m   B

Derrick has to mark out a triangular flower bed ABP, as shown in the diagram above. The distance AB must be 10 m and the angle APB must be 90°. The lengths of the other two sides, AP and BP, must total 13 m.
   (a) Taking the length of AP as x metres, form an equation in x, and show that it simplifies to $2x^2 - 26x + 69 = 0$.
   (b) Solve the equation to find the two possible values of x, correct to 2 decimal places.   [M]

11. The formula for the surface area, A, of a closed circular cylinder of radius r and height h is

$$A = 2\pi r(r + h).$$

   (a) Make h the subject of the formula.
   (b) Find the radius, in cm to one decimal place, of a cylinder whose surface area is $120\pi \, \text{cm}^2$ and whose height is 12 cm.   [L]

12. (a) If $216^{\frac{1}{3}} = 6^x$ find x.

   (b) If $\dfrac{5^{-2}}{5^{-4}} = 5^y$ find y.

   (c) Evaluate $\sqrt{\dfrac{17\cdot85 - 12\cdot52}{27\cdot84 - 1\cdot84^5}}$.   [S]

13. The mass, m grams, of a radioactive chemical after t years is given
by $m = 80 \times 0 \cdot 5^t$.
(a) Find the mass after 3 years.
(b) Find the mass after 6 months.
(c) What was the initial mass? [M]

14. The perimeter of
the rectangle ABCD
is 42 cm. The width
BC is x cm.

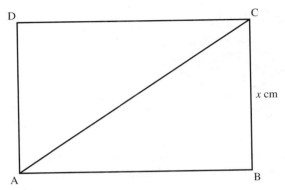

(a) Write, in terms of x, an expression for the length of AB.
(b) The diagonal AC is 15 cm. Calculate the length and width of the
rectangle. [S]

15. $y = x^3 - 4x - 1$
(a) Complete the table of values

x	-2	-1	0	1	2	3
y		2	-1			14

(b) On the grid, draw the graph
of $y = x^3 - 4x - 1$
(c) By drawing a suitable straight
line on the grid, estimate a
solution to the equation

$$x^3 - 4x - 3 = -4 \qquad [L]$$

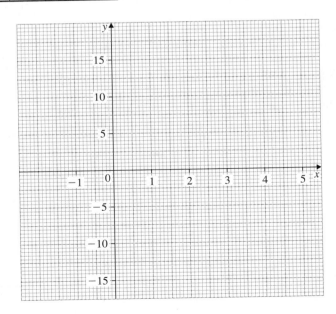

Examination exercise 5: Shape, space and measures (non calculator)

1.

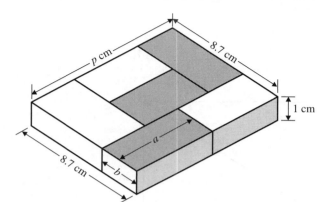

The drawing shows six identically shaped rectangular blocks of modelling clay — it is not drawn to scale. The length of each block is *a* and its width is *b*. Find the width *b* and hence the values of *a* and *p*.   [N]

2. A rectangular match box measures 12 cm by 5 cm by 3 cm. Each match is a cuboid, 5 cm by 2 mm by 2 mm. What is the greatest number of matches which can be fitted into the box?   [M]

3. Winston Jones makes a path across his rectangular front garden, as shown in the diagram.

$AB = 10\,m$, $AF = 2\,m$, $FE = 3\,m$

The edges of AC and FD of the path are parallel.
(a) Find the length of ED.
(b) Calculate the area of the path.   [M]

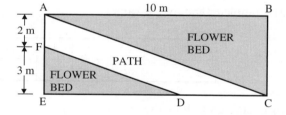

4. Squares ABFG and BCDE are drawn on the sides of an obtuse-angled triangle ABC. Are any of the following statements true? Give your reasons.
   (i) Triangle ABC is congruent to triangle FBE.
   (ii) Triangle ABC is similar to triangle FBE.
   (iii) Triangle ABC is equal in area to triangle FBE.
        [N]

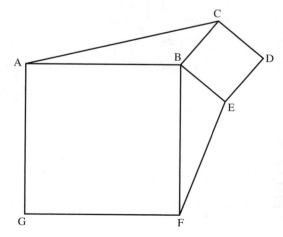

5. The list below contains an expression for the volume and an expression for the surface area of the object shown.

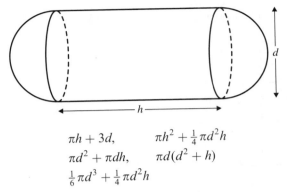

$$\pi h + 3d, \qquad \pi h^2 + \tfrac{1}{4}\pi d^2 h$$
$$\pi d^2 + \pi dh, \qquad \pi d(d^2 + h)$$
$$\tfrac{1}{6}\pi d^3 + \tfrac{1}{4}\pi d^2 h$$

(a) Write down the expression for the volume.
(b) Write down the expression for the surface area.

6. In the diagram, ABCD is a cyclic quadrilateral. The line SBT is the tangent at B to the circle. The lines BA produced and CD produced meet at E and EA = AC. Angle CBT = 42° and angle EDA = 53°. (To obtain full marks in this question you must give reasons for your answers.)
Calculate the sizes of the angles.
(a) angle CAB
(b) angle ABC
(c) angle ECA
(d) angle EAD

[L]

7. Calculate the size of:
(a) angle ABC
(b) angle AOD
(c) angle ACD
(d) angle OAD.

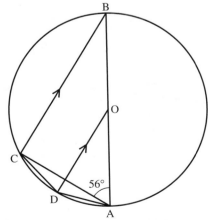

O is the centre and AB is a diameter of the circle.
OD is parallel to BC, and angle BAC = 56°.

The tangents to the circle at B and C meet at T.
Calculate the size of
(e) angle BOT.

[N.I.]

8. The diagram shows the net of a 3-D shape.

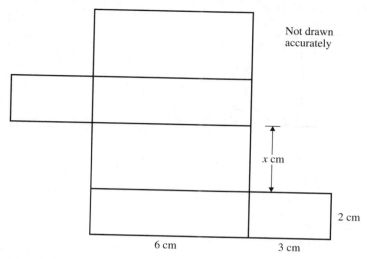

Not drawn
accurately

x cm

2 cm

6 cm   3 cm

(a) (i) What is the value of x?
(ii) Calculate the perimeter of this net.
(iii) Calculate the area of this net.
(b) What is the name of the 3-D shape?
(c) Calculate the total length of all the edges of this 3-D shape. [N.I.]

9. A is the point $(0, 4)$.

$$\overrightarrow{AB} = \begin{pmatrix} 3 \\ 2 \end{pmatrix}$$

(a) Find the coordinates of B.

C is the point $(3, 4)$. BD is a diagonal of the parallelogram $ABCD$.

(b) Express $\overrightarrow{BD}$ as a column vector.

$$\overrightarrow{CE} = \begin{pmatrix} 1 \\ -3 \end{pmatrix}$$

(c) Calculate the length of AE. [L]

10.

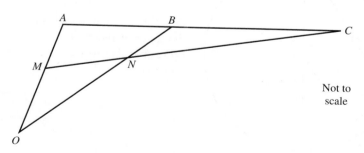

Not to
scale

In the diagram, O is the origin, ABC is a straight line and M is the mid-point of OA.

$$\overrightarrow{OA} = \mathbf{a}, \quad \overrightarrow{OB} = \mathbf{b} \quad \text{and} \quad \overrightarrow{AC} = 3\overrightarrow{AB}.$$

(a) Find, in terms of **a** and/or **b**, in their simplest forms,

   (i) $\overrightarrow{MA}$ (ii) $\overrightarrow{AB}$ (iii) $\overrightarrow{AC}$ (iv) $\overrightarrow{MC}$

   (v) the position vector of C.   [IGCSE]

11.

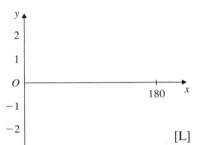

$$\sin 30° = 0\cdot5$$

(a) Write down the value of:
   (i) $\sin 150°$
   (ii) $\sin 210°$
   (iii) $\sin 330°$.
(b) (i) On the axes, sketch the
       graph of $y = \sin 2x°$
       for $0 \leqslant x \leqslant 180$.
    (ii) Solve $\sin 2x° = -0\cdot5$ for
       $0 \leqslant x \leqslant 180$.

[L]

12.

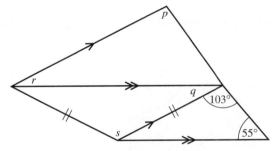

(a) Calculate the angles:
   (i) p (ii) q (iii) r (iv) s
(b) In the quadrilateral $ABCD$, M is the mid-point of BC, AM is
    parallel to DC and DM is parallel to AB.

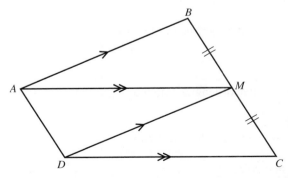

Which of the conditions for congruent triangles would prove
triangles ABM and DMC congruent?   [N.I.]

13. In Figure 1, OAB is a quadrant of a circle, radius r, and OPA is a
semicircle drawn on OA as diameter.

   (a) Calculate:

      (i) the area of the semicircle OPA, in terms of r and π,

      (ii) the area of the shaded part of Figure 1, in terms of r and π.

   (b) What is the size of angle OPA? Give your reason.

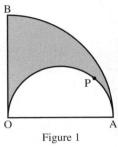

Figure 1

   (c) In Figure 2, P has been moved to P′, so that BP′A is a straight
line. Explain why a circle with OB as diameter will pass
through P′.   [N]

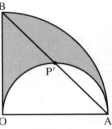

Figure 2

Examination exercise 6: Shape, space and measures (calculators allowed)

1. The diagram shows a square
garden of side 14 m. E, F, G and
H are the mid-points of AB, BC,
CD and DA respectively and EF,
FG, GH and HE are arcs of a
circle of radius 7 m.
The middle section of the garden
is to be grassed as a lawn and the
four side sections (shown shaded)
are to be flower beds.

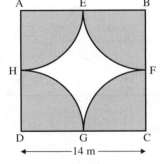

Calculate:

   (a) the total area of the flower beds [take π as $\frac{22}{7}$ or 3·14],

   (b) the area of the lawn,

   (c) the amount of grass seed needed given that 80 grams of seed are
needed per square metre.   [W]

2. A metal ingot is in the form of a solid cylinder of length 7 cm and
radius 3 cm.

   (a) Calculate the volume, in cm³, of the ingot.

   The ingot is to be melted down and used to make cylindrical coins
of thickness 3 mm and radius 12 mm.

   (b) Calculate the volume, in mm³, of each coin.

   (c) Calculate the number of coins which can be made from the
ingot, assuming that there is no wastage of metal.   [M]

3. The diagram shows a box in the form of a
cuboid, of internal dimensions 24 cm by 7 cm
by 60 cm. Emma wishes to place the longest
possible thin straight rod inside the box.
(a) Show a possible position of the rod in the
diagram.
(b) Calculate the length of the rod.   [M]

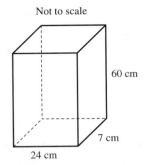

Not to scale

60 cm

7 cm

24 cm

4. The diagram shows two sightings made from
a point O of a helicopter flying at a height of
1600 metres. At the first sighting, the
helicopter was due East of O and the angle of
elevation was 58°. One minute later it was
still due East of O but the angle of elevation
was 45°. Calculate the speed of the helicopter
in kilometres per hour.   [N]

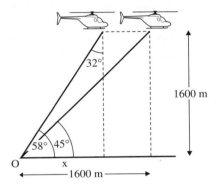

32°

1600 m

58° 45°

O   x

1600 m

5. The diagram represents a car bonnet, AB, hinged at A and propped
open by a stay, CD.

AC = 23 cm, AB = 47 cm, AD = 18 cm and CD = 20 cm.
(a) Calculate the size of angle CAD correct to the nearest degree.
(b) Find the height of B above the level of AC, giving your answer
correct to the nearest centimetre.   [M]

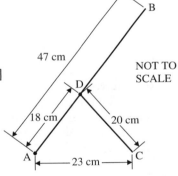

B

47 cm

NOT TO
SCALE

D

18 cm

20 cm

A   23 cm   C

6. A parallelogram *ABCD* is comprised of two right-angled
triangles *ACB* and *ACD*.
AB is 90 cm and *BC* is 54 cm.
Calculate:
(a) the length of *AC*
(b) the length of BD.   [N.I.]

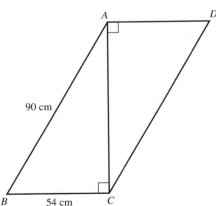

A   D

90 cm

B   54 cm   C

7.

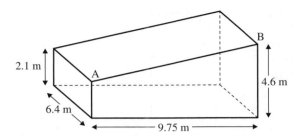

The diagram represents a squash court. The end walls and the floor
are rectangular. The four walls are vertical and made of glass. The
floor is horizontal and made of wood. The top is open.
(a) Find the total area, in m², of glass needed to build the court.
Give your answer correct to three significant figures.
(b) Find the angle of elevation of B from A. Give your answer
correct to the nearest degree. [M]

8. (a) Sketch the graph of $y = \tan x°$ on the given axes, for
$-90 \leqslant x \leqslant 450$.

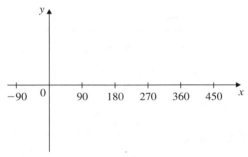

(b) Use your calculator to find the value of x between 0 and 90 for
which $\tan x° = 2$.
(c) Use your graph and the answer to part (b) to solve the equations:
(i) $\tan x° = 2$ for $180 \leqslant x \leqslant 360$
(ii) $\tan x° = -2$ for $180 \leqslant x \leqslant 360$. [M]

9. The horizontal base of the solid pyramid VABC is an isosceles
triangle ABC. The edge VC is vertical. AC = BC = 13 cm,
AB = 10 cm and VC = 8 cm.
(a) Calculate the size of the angle between the
edge VB and the horizontal base.
(b) Calculate the size of angle ABC.
(c) Calculate the length of the line drawn from C
to meet AB at 90°.
(d) Calculate the size of the angle between the
face AVB and the horizontal base.

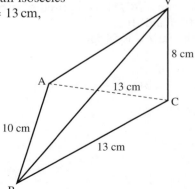

10.

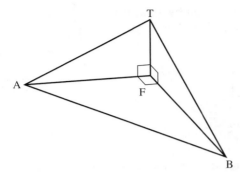

Alan is standing at the point A which is 600 m due West of the point F, the foot of a vertical television mast FT. His friend Brett is standing at the point B, 400 m due South of F.
The points A, B and F are in a horizontal plane. The angle of elevation of T from A is 13°.
(a) Calculate the height of the mast.
(b) Calculate the angle of elevation of T from B.
Given that C is the point on the line AB which is nearest to F,
(c) calculate:
  (i) the length of CF
  (ii) the angle of elevation of T from C.   [L]

11. There is an athletics competition at Suki's school. Suki is told to mark out an area of the sports field that is to be used for the shot putt competition. The area consists of two parts:
A circle of radius 1 metre from where the shot is thrown.
A sector of angle 30° and radius 26 metres which overlaps the throwing circle as shown.

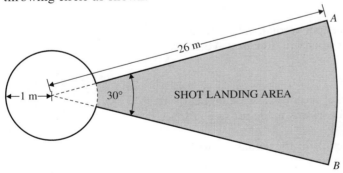

(a) Find the area of the circle from which the shot is thrown.
    Take π to be 3·14 or use the π key on your calculator.
The throwing circle is made of concrete to a depth of 5 cm.
(b) Calculate the volume of concrete needed in cubic metres.
So that no spectators are within range of the shot when it is thrown,
Suki paints the arc *AB* of the shot landing area.
(c) What is the length of the arc she paints?
(d) Find the shaded area into which the shot can land.   [S]

12. A rectangular block of steel measures 1 m by 20 cm by 4 cm. It is melted down to make ball bearings as spheres of radius 4 mm.

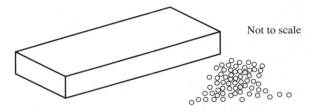

Not to scale

(a) Find the volume of the block of steel in cubic centimetres.
(b) Find the volume of a ball bearing in cubic millimetres.
(c) How many ball bearings can be made from the block of steel?

[S]

13. The diagram shows a circle, centre O, of radius 12 metres. The chord AB is 10 metres long.
(a) Calculate:
  (i) the size, correct to the nearest degree, of the angle AOB
  (ii) the area of the shaded segment.
  (You may take π to be 3·142.)
(b) The cross-section of a river may be taken to be the same as the shaded segment.

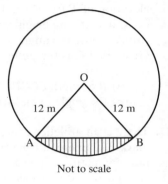

Not to scale

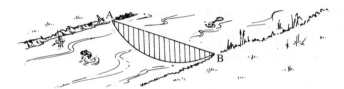

If the river is flowing at 0·8 metres per second, calculate the volume of water, in cubic metres, that will flow past AB in one hour.   [M]

14 When a new car came onto the market recently, a scale model of it, similar in every respect to the actual car, was made for sale in toy shops.
The table below compares certain details of the actual car with the scale model. Find the values of x, y and z.

	Car	Model
Length	420 cm	12 cm
Width	x cm	5 cm
Area of windscreen	8330 cm^2	y cm^2
Capacity of boot	z cm^3	19·2 cm^3

[L]

15. An ice-cream firm is to introduce a new ice cream cone called the *Concerto*. The ice cream completely fills the cone which has a base diameter of 7 cm and a height of 14 cm.
  (a) Calculate the volume of the ice cream needed to fill the *Concerto*. (Give your answer in cm^3 to the nearest whole number.)
  The ice cream used is manufactured in 10 litre containers.
  (b) Calculate the number of *Concertos* which will be filled from each container assuming that there is no wastage.
  The firm is also to produce a smaller version of the *Concerto* called the *Concertino*. The *Concertino* is geometrically similar to the *Concerto* but only $\frac{4}{5}$ of its height.
  (c) Calculate the volume of ice cream in one *Concertino*.   [L]

16. If $\cos x = 0.5$, find three possible values of x.

17. A wedge of cheese is in the shape of a prism as shown. It is cut into two equal volumes by a single cut, parallel to one face. Calculate the value of x.   [N]

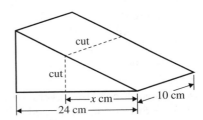

18.

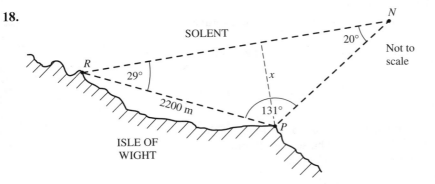

The diagram shows part of the coast of the Isle of Wight and the Solent. The point R represents the position of Ryde, point P the position of Puckpool Point and point N the position of No Man's Land Fort.

RP = 2200 metres, angle NRP = 29° and angle RPN = 131°.

  (a) Calculate the distance RN, correct to the nearest 10 metres.
  (b) A boat sails directly from Ryde to No Man's Land Fort. Calculate, correct to the nearest 10 metres, its closest distance to Puckpool Point.   [M]

19. (a) Part of the road system of a modern town was laid
    out in the form of a semicircle, radius a, as shown.
    B is due East of A.
    Imran and Johnny cycled in opposite directions
    around the semicircle, starting together at A and
    arriving at P at the same time.
    The arc length BP = x.

    Find, in terms of a and x:
    (i) the distance cycled by Imran
    (ii) the distance cycled by Johnny.

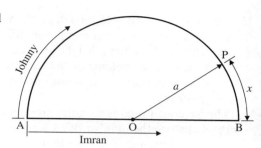

(b) If they both rode at the same speed, show that $x = \dfrac{a(\pi - 2)}{2}$.

(c) Calculate the size of the angle POB in degrees.

Examination exercise 7: Handling data (non calculator)

1. The annual salaries of the workers in a factory were as follows:

    20 apprentices.......................... £ 4 500 each
    15 semi-skilled workers................ £ 9 000 each
    10 skilled workers..................... £13 500 each
     2 foremen............................ £15 999 each
     1 manager £25 000

    At discussions on pay rises, average salaries for everyone in the
    factory were discussed and quoted.
    (a) Jack, the union representative, quoted the modal salary as the
        average. What was the modal salary?
    (b) Wilma, one of the foremen, quoted the median as the average.
        What was the median salary?
    (c) Rashid, the manager, quoted the mean as the average. What
        was the mean salary?   [N]

2. The median of $x - 4$, x, $2x$ and $2x + 12$ is 9, where x is a positive
   integer. Find the value of x.   [IGCSE]

3. A teacher thinks that the nearer her pupils sit to the front of the
   class, the higher their test results will be. The scatter diagrams show
   the results for her second and fourth year classes.

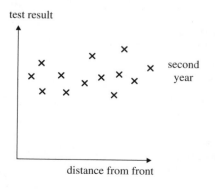

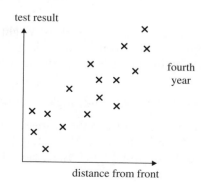

For each class, say whether the teacher's ideas are right or wrong.
If she is wrong, describe what conclusions you could come to
about that class.   [M]

4. Isobel is doing a survey on the use made of the school library.
   Write three suitable questions for a questionnaire. Each question
   must allow for at least three possible responses.   [S]

5. (a) The shoe sizes of 28 boys are recorded in the table below.

Shoe size	6	7	8	9	10
Frequency	9	6	6	5	2

      (i) Write down the median shoe size.
     (ii) Find the range of the shoe sizes.
    (iii) The shoe sizes of seven girls have a median of 5 and a range
          of 6. One of these girls takes size 11 shoes.
          Give a possible list of the shoe sizes of the seven girls.   [M]

6. The table shows the results of a short test.

Mark	0	1	2	3	4	5
Number of students	1	3	10	x	6	3

   The mode of the marks is 2 and the median mark is 3.
   Find the possible values of x.   [IGCSE]

7. The ages of all the 100 employees in a company A are shown.

Age (years)	16–	21–	31–	41–	51–	61–	81–
Frequency	8	12	25	30	20	5	0
Frequency Density	1·6	1·2					0

   (a) (i) Complete the table of frequency densities.
       (ii) On graph paper draw a histogram to illustrate the age
            distribution of the employees in Company A.

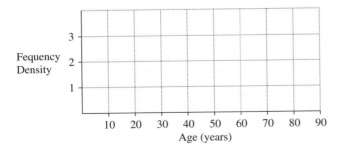

Company B has the following age distribution of its employees.

Age (years)	16–30	31–40	41–50	51–80
% of employees	8	85	7	0

(b) Write down two comparisons about the age distributions of the employees of the two companies. [S]

8. Your school is conducting a survey to investigate pupil attitudes to 'compulsory school uniform' and you have been asked to choose a sample of pupils for the survey. State, with reasons, **two** factors which you consider to be important in your choice of sample so that the survey represents fairly the opinions of all the pupils in your school. [W]

9. It is required to investigate the television viewing preferences of pupils in a mixed comprehensive school with 1000 pupils.
   (a) (i) Describe how you would obtain a simple representative sample of 100 of the pupils in order to find out their preferences.
       (ii) Describe how you would obtain a stratified representative sample of 100 of the pupils in order to find out their preferences.
       (iii) State, giving a reason, which of the above samples would give a more reliable picture of the television viewing preferences of the pupils in the school.
   (b) Give a reason why it would be necessary to take a sample when conducting a survey of the television viewing preferences of people in Wales. [W]

10. The table shows the 1961 figures for adult literacy and life expectancy in 8 countries.

Country	Adult literacy %	Life expectancy (years)
Chile	88	63
Ethiopia	10	38
India	29	49
Iran	49	51
Italy	90	71
Malaysia	58	58
Paraguay	75	61
Saudi Arabia	15	45

(a) On graph paper draw a scattergraph of these figures.
(b) From the scattergraph, state what correlation, if any, there is between the 'Adult literacy' figures and the 'Life expectancy' figures.

(c) Draw a 'line of best fit'.
Using this, estimate the life expectancy of an adult from a country with 42% adult literacy.
(d) Do you think that it is a good idea to draw conclusions from this set of figures?
Give a reason for your answer.   [N.I.]

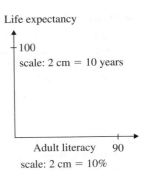

Life expectancy

100
scale: 2 cm = 10 years

Adult literacy   90
scale: 2 cm = 10%

11. (a) Calculate the standard deviation of 2, 8, 8, 10.
The standard deviation of 2, 3, 5, 7, 9 is 2·56, correct to 3 significant figures.
(b) Calculate the standard deviation of
    (i) 7, 8, 10, 12, 14.
    (ii) 6, 9, 15, 21, 27.   [L]

12. Three students A, B, C were each asked to toss a coin 30 times and record whether it came down heads (H) or tails (T) each time.
They said their results were as follows:

Student A: T T T H T H T T T H T T T H T T H T T T T H H T
               T T T H T T
Student B: H T H T H T H T H T H T H T H T H T H T H T H T H
               T H T H T H T
Student C: H T H T H H H H T H T T H H H H T T H T H T H T T
               T H H T H T T

One of the students used an unbiased coin, one used a biased coin and the remaining student did not have a coin so just made up the results.
(a) (i) Which student used a biased coin?
    (ii) Estimate the probability that, the next time this coin is tossed, it will come down heads.
(b) Which student just made up the results? Give a reason for your answer.   [M]

13. The letters of the word 'PROCESSES' are written on individual cards and placed in a bag.

P R O C E S S E S

A letter is selected at random from the bag and then replaced.
(a) What is the probability that the letter is
    (i) E
    (ii) a vowel (the vowels are A, E, I, O, U)
    (iii) T
    (iv) not S?
The experiment is repeated 360 times.
(b) How many times would you expect the letter S to be chosen?   [N.I.]

14. Two dice, one red and one black, are thrown simultaneously. A score is found by adding together the numbers of dots shown on the two upper faces of the dice.
Find the probability, expressed as a fraction, of throwing:
(a) a score of 2   (b) a score of 9
(c) a double in which both dice show the same number.   [L]

15. In a game to select a winner from three friends Arshad, Belinda and Connie, Arshad and Belinda both roll a normal die. If Arshad scores a number greater than 2 *and* Belinda throws an odd number, then Arshad is the winner. Otherwise Arshad is eliminated and Connie then rolls the die. If the die shows an odd number Connie is the winner, otherwise Belinda is the winner.
(a) Calculate the probability that:
    (i) Arshad will be the winner
    (ii) Connie will roll the die
    (iii) Connie will be the winner.
(b) Is this a fair game? Give a reason for your answer.   [L]

16. Some college students have to take 3 examination papers, one in English, one in mathematics and one in science.
In English the probability of a pass is $\frac{7}{10}$.
In mathematics it is $\frac{5}{10}$ and in science it is $\frac{8}{10}$.
Assume that these events are independent events.
(a) Calculate the probability that a student will:
    (i) pass all three papers
    (ii) fail all three papers
    (iii) pass at least one of the papers.
A student needs to pass in at least two of the three papers to go on to the next year.
(b) Calculate the probability that a student will be able to go on to the next year.   [L]

17. A game called 'HOW'S THAT' is played between two players. One player, A, rolls a dice which has one face labelled 'HOW'S THAT' and the other five faces labelled 1, 2, 3, 4 and 6 respectively.
The numbers 1, 2, 3, 4 and 6 indicate the score obtained by Player A.
Player A continues to roll this dice, adding the scores obtained each time, until the face with 'HOW'S THAT' appears uppermost on the dice.
Player B then rolls a second dice, which has four faces which show that Player A's turn is over, and two faces which show that Player A is to continue to roll his dice.
(a) What is the probability that, on the first roll of the numbered dice, Player A will score 4?

(b) What is the probability that Player A will have scored a total of 4 after two rolls of the numbered dice, assuming 'HOW'S THAT' did not appear on either occasion?

(c) What is the probability that, after just one roll of the numbered dice by Player A, followed by one roll of the other dice by Player B, Player A's turn will be over?   [N]

18. I played a game of cards with a friend. The pack of 52 cards was shuffled and then dealt so that we both received two cards. His cards were a ten and a nine. The first card that I received was an ace.

(a) Find the probability that my second card was a picture card (that is to say a King, Queen or Jack).

(b) In fact my second card was another ace. The rules of the game allowed me to treat the two aces separately, and so have a further two cards dealt to me, one to go alongside one ace and the other to go alongside the other ace.

Calculate the probability that these two cards were both picture cards. Give your answer as a fraction in its simplest form.   [N]

19. A box of chocolates contains 9 hard centres (H) and 12 soft centres (S). One chocolate is taken at random and eaten; then a second chocolate is taken.

(a) Complete this tree diagram.

(b) Giving you answers as fractions in their lowest terms, find the probability that:

    (i) both chocolates have soft centres

    (ii) one has a hard centre and one a soft centre.   [M]

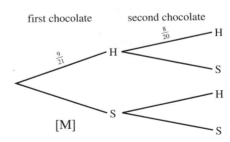

20. David and Tom each have a bag containing 5 marbles. David has 3 red and 2 green marbles, Tom has 2 red and 3 green. They play a game of 'swops' where each takes a single marble at random from his bag and places it on the floor. If the colours are the same, Tom

takes both marbles and puts them back into his bag. If the colours
are different, David wins the marbles and puts them both into his
bag.

(a) Calculate the probability:
   (i) that Tom wins the first swop
   (ii) that David wins the first swop.

(b) On the first swop, David takes out a red marble and Tom a
green marble, so that David wins and puts the marbles into his
bag. They agree to select another marble each. What is the
probability that David wins again?   [N]

Examination exercise 8: Handling data (calculators allowed)

1. The table below shows how each £1 of expenditure of Derbyshire
County Council was divided up in 1987/8.

How spent	Amount
day-to-day business	47p
repaying loans and interest	23p
investing in assets	30p

Draw a pie chart to illustrate this information, showing clearly how
you calculated the angles required.   [M]

2. Answer the whole of this question on a sheet of graph paper.
The table shows the amount of money, x, spent on books by a
group of students.

Amount spent (x)	$0 < x \leqslant 10$	$10 < x \leqslant 20$	$20 < x \leqslant 30$	$30 < x \leqslant 40$	$40 < x \leqslant 50$	$50 < x \leqslant 60$
Number of students	0	4	8	12	11	5

(a) Calculate an estimate of the mean amount of money per student
spent on books.

(b) Use the information in the table above to find the values of p, q
and r in the following cumulative frequency table.

Amount spent (x)	$x \leqslant 10$	$x \leqslant 20$	$x \leqslant 30$	$x \leqslant 40$	$x \leqslant 50$	$x \leqslant 60$
Cumulative frequency	0	4	p	q	r	40

(c) Using a scale of 2 cm to represent 10 units on each axis, draw a
cumulative frequency diagram.

(d) Use your diagram:
   (i) to estimate the median amount spent
   (ii) to find the upper and lower quartiles, and the inner-quartile
   range.   [IGCSE]

3. The histogram shows the examination results of 120 candidates.

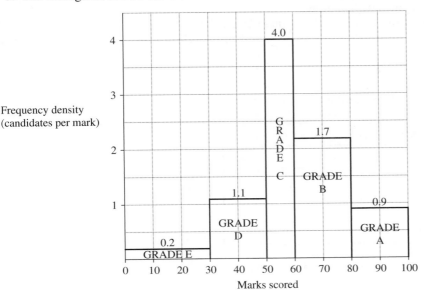

Frequency density
(candidates per mark)

Marks scored

(a) Copy and complete the table below, giving the number of
candidates who obtained each grade.

Grade	E	D	C	B	A
Number of candidates	6				

(b) If this information was presented on a pie chart, calculate the
angle which would be used for the sector representing grade E.

(c) Complete the table below, giving the cumulative total number of
candidates who obtained at the most the mark shown.

Maximum mark	30	50	60	80	100
Cumulative frequency	6				

(d) On graph paper, draw a cumulative frequency diagram to
display this information.

(e) From your graph, estimate the median mark for the
examination.   [N]

4. A poultry farmer recorded the weight (w kg) of 50 chickens at the
age of seven weeks. The results are shown in the table below.

Weight (w kg)	Number of chickens
$1 \leqslant w < 1.2$	12
$1.2 \leqslant w < 1.4$	14
$1.4 \leqslant w < 1.6$	5
$1.6 \leqslant w < 1.8$	15
$1.8 \leqslant w < 2.0$	4

(a) What is the modal class?
(b) In which class does the median weight lie?
(c) Calculate an estimate of the mean weight of these 50 chickens.

[M]

5. The following table shows a grouped frequency distribution of the waist measurements, to the nearest centimetre, of 60 people.

Waist measurement (cm)	Number of people
66–75	7
76–80	8
81–85	12
86–90	14
91–95	10
96–105	9

(a) Using the class mid-points, calculate the mean waist measurement.
(b) Complete the following cumulative frequency table.

Waist measurement less than	75·5 cm	80·5 cm	...	...	195·5 cm
Cumulative frequency	7				60

(c) On a graph draw the cumulative frequency polygon.
(d) Showing clearly how you have used your graph in (c), find estimates for:
  (i) the median waist measurement,
  (ii) the interquartile range of the waist measurements.
(e) State which of your answers in (d) is a measure of the dispersion of the waist measurements.

[W]

6. In a diving competition, each dive is marked by seven judges. To obtain the score for a dive the highest and lowest of these marks are ignored and the average (arithmetic mean) of the other five is multiplied by a figure representing the degree of difficulty of the dive.
(a) Calculate the score for a dive which the judges marked as

$$9·6, \quad 8·8, \quad 9·7, \quad 9·2, \quad 9·0, \quad 8·3, \quad 8·4$$

and for which the degree of difficulty of the dive is 2·4.
(b) Calculate the standard deviation of the five marks which were used in the calculation to obtain the score for this dive. [N]

7. The data below shows the energy levels, in kilocalories per 100 g, of 10 different snack foods (such as crisps and peanuts).

440   520   480   560   572   550   620   680   545   490

(i) Calculate the mean and standard deviation of the energy levels of these snack foods.

(ii) The energy levels, in kilocalories per 100 g, of 10 different breakfast cereals had a mean of 350 kilocalories with a standard deviation of 28 kilocalories. Which of the two types of food show great variation in energy level? Give a reason for your answer. [W]

8. The lengths, in centimetres, of 10 leaves in a sample were:

5·6,   5·8,   4·9,   6·2,   6·8,   4·8,   5·4,   5·9,   5·3,   5·2

(a) Calculate the standard deviation of these lengths.

(b) The standard deviation of another sample of 10 leaves was 0·432 cm. What difference between the two samples is shown by the two standard deviations? [S]

9. Mamoud tries to repair a broken toy. Each time he tries the probability that he succeeds is 0·8. Each time he fails he tries again.

(a) Copy and complete the tree diagram below.

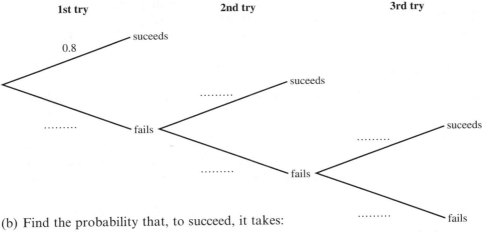

(b) Find the probability that, to succeed, it takes:
  (i) exactly two tries
  (ii) one, two or three tries
  (iii) exactly five tries.

(c) Write down a formula for the probability that he has **not** succeeded after n tries.   [IGCSE]

ANSWERS

1 Number 1

page 1 **Exercise 1**

1. (a) 1, 2, 3, 6   (b) 1, 3, 5, 15   (c) 1, 2, 3, 6, 9, 18   (d) 1, 3, 7, 21   (e) 1, 2, 4, 5, 8, 10, 20, 40
2. 2, 3, 5, 7, 11, 13, 17, 19
3. [e.g.] $2 + 3 = 5, 2 + 5 = 7\ldots$
4. 3, 11, 19, 23, 29, 31, 37, 47, 59, 61, 67, 73
5. (a) $2^3 \times 3 \times 5^2$ (b) $3^2 \times 7 \times 11$ (c) $2^5 \times 7 \times 11$
    (d) $2 \times 3^3 \times 5 \times 13$ (e) $2^5 \times 5^3$ (f) $2^2 \times 5 \times 7^2 \times 23$

6.

	Prime	Multiple of 3	Factor of 16
Number greater than 5	7	9	8
Odd number	5	3	1
Even number	2	6	4

7. (a) 7   (b) 50   (c) 1   (d) 5   **8.** (c) 18
9. (a) 30   (b) 60   (c) 30
10. 1, 4, 9, 16, 25, 36, 49, 64, 81, 100, 121, 144, 169, 196, 225
11. 1, 8, 27, 64, 125   **12.** 101, 151, 293 are prime
14. (a) $1008 = 2^4 \times 3^2 \times 7$, $840 = 2^3 \times 3 \times 5 \times 7$ **(b)** 168   **(c)** 7
15. (a) $19\,800 = 2^3 \times 3^2 \times 5^2 \times 11$, $12\,870 = 2 \times 3^2 \times 5 \times 11 \times 13$ **(b)** 990   **(c)** 22
16. 8
17. (a) (i) 1, 8, 27, 64   (ii) cube numbers   (iii) 100^3
    (b) (i) 9, 36, 100   (ii) square numbers   (iii) $55^2 (= 3025)$
18. 2, 3, 5, 7, 89, 641   **19.** Always a multiple of 4.   **20.** Not always a multiple of n.

page 4 **Exercise 2**

1. (a) 7, 22, 44, 22, $5\frac{1}{2}$ (b) 10, 25, 50, 28, 7   (c) 16, 31, 62, 40, 10   (d) $\frac{1}{2}$, $15\frac{1}{2}$, 31, 9, $2\frac{1}{4}$
2. (a) 4, 16, 48, 38, 19   (b) 5, 25, 75, 65, $32\frac{1}{2}$ (c) 6, 36, 108, 98, 49   (d) 8, 64, 192, 182, 91
3. $\times 4$, square root, -10, $\times -2$ **4.** reciprocal, $+1$, square, $\div 3$ **5.** $+3$, cube, $\div -2$, $+100$

page 5 **Exercise 3**

1. 7·91   **2.** 22·22   **3.** 7·372   **4.** 0·066   **5.** 466·2
6. 1·22   **7.** 1·67   **8.** 1·61   **9.** 16·63   **10.** 24·1
11. 26·7   **12.** 3·86   **13.** 0·001   **14.** 1·56   **15.** 0·0288
16. 2·176   **17.** 0·02   **18.** 0·0001   **19.** 355   **20.** 20
21. £12·80   **22.** 34 000   **23.** 18   **24.** 0·569   **25.** 2·335
26. 1620   **27.** 8·8   **28.** 1200   **29.** 0·001 75   **30.** 13·2
31. 200   **32.** 0·804   **33.** 0·8   **34.** 0·077   **35.** 0·0009
36. 0·01   **37.** (b) (i) 20   (ii) 184   (iii) 0·63   (iv) 540   (v) 3
38. (a) 7·4   (b) 2   (c) 0·07   (d) 0·1
39. (a) T   (b) T   (c) T   (d) F   (e) F   (f) T   (g) T   (h) F   (i) T
40. (a) >   (b) >   (c) <   (d) <   (e) >   (f) >

page 7   **Exercise 4**

1. 805   **2.** 459   **3.** 650   **4.** 1333   **5.** 2745
6. 1248   **7.** 4522   **8.** 30 368   **9.** 28 224   **10.** 8568
11. 46 800   **12.** 66 281   **13.** 57 602   **14.** 89 516   **15.** 97 525

page 7   **Exercise 5**

1. 32   **2.** 25   **3.** 18   **4.** 13   **5.** 35
6. 22 r 2   **7.** 23 r 24   **8.** 18 r 10   **9.** 27 r 18   **10.** 13 r 31
11. 35 r 6   **12.** 23 r 24   **13.** 64 r 37   **14.** 151 r 17   **15.** 2961 r 15

page 7   **Exercise 6**

1. £47·04   **2.** 46   **3.** 7592   **4.** 21, 17p change   **5.** 8
6. £80·64   **7.** £14 million **8.** £85   **9.** £21 600   **10.** 54
11. (a) 6·72   (b) 2·795   (c) 3·976   **12.** (a) 4·9   (b) 3·5   (c) 2·0
13. (b) 608

page 9   **Exercise 7**

1. (a) $\frac{3}{8}$ (b) $\frac{7}{10}$ (c) $\frac{11}{15}$
2. (a) $\frac{5}{6}$ (b) $\frac{5}{12}$ (c) $\frac{7}{12}$ (d) $\frac{9}{20}$ (e) $1\frac{11}{20}$ (f) $1\frac{1}{2}$ (g) $1\frac{7}{12}$ (h) $3\frac{7}{12}$
3. (a) $\frac{1}{6}$ (b) $\frac{5}{12}$ (c) $\frac{7}{15}$ (d) $\frac{1}{10}$ (e) $1\frac{1}{12}$ (f) $\frac{1}{6}$

4.

$+$	$\frac{1}{4}$	$\frac{3}{8}$	$\frac{1}{3}$
$\frac{1}{2}$	$\frac{3}{4}$	$\frac{7}{8}$	$\frac{5}{6}$
$\frac{1}{8}$	$\frac{3}{8}$	$\frac{1}{2}$	$\frac{11}{24}$
$\frac{1}{5}$	$\frac{9}{20}$	$\frac{23}{40}$	$\frac{8}{15}$

5. (a) $\frac{8}{15}$ (b) $\frac{5}{42}$ (c) $\frac{15}{26}$ (d) $1\frac{1}{6}$ (e) $\frac{5}{8}$ (f) $9\frac{1}{6}$ (g) $\frac{4}{15}$ (h) $\frac{3}{14}$
6. (a) $\frac{5}{12}$ (b) $4\frac{1}{2}$ (c) $1\frac{2}{3}$ (d) $\frac{5}{16}$ (e) $1\frac{7}{8}$ (f) $2\frac{5}{8}$ (g) $1\frac{9}{26}$ (h) 8
7. (From top) $\frac{1}{8}, \frac{5}{16}, \frac{3}{4}, 8$ **8.** 123 cm   **9.** $\frac{1}{5}$
10. (a) $\frac{9}{16}$ (b) $\frac{7}{20}$ (c) $1\frac{1}{5}$ (d) $1\frac{1}{6}$ (e) $\frac{1}{10}$ (f) $\frac{7}{15}$ (g) $\frac{20}{23}$ (h) $1\frac{3}{5}$
11. 5   **12.** 3   **13.** $1\frac{16}{17}$ **14.** 9
15. (a) $a = 1, b = 2$ (b) $a = 1, b = 5$ (c) $(1, 13)$ or $(3, 10)$ or $(5, 7)$ or $(7, 4)$ or $(9, 1)$

16.

$\times$	$\frac{2}{5}$	$\frac{3}{4}$	$\frac{2}{3}$
$\frac{1}{3}$	$\frac{2}{15}$	$\frac{1}{4}$	$\frac{2}{9}$
$\frac{1}{2}$	$\frac{1}{5}$	$\frac{3}{8}$	$\frac{1}{3}$
$\frac{1}{4}$	$\frac{1}{10}$	$\frac{3}{16}$	$\frac{1}{6}$

17. 144 ml   **18.** (a) $\frac{15}{16}, \frac{31}{32}$ (b) $\dfrac{2^{n-1} - 1}{2^{n-1}}$

page 12 **Exercise 8**

1. 70·56
2. 118·958
3. 451·62
4. 33678·8
5. 0·6174
6. 1068
7. 19·53
8. 18 914·4
9. 38·72
10. 0·009 79
11. 2·4
12. 11
13. 41
14. 8·9
15. 4·7
16. 56
17. 0·0201
18. 30·1
19. 1·3
20. 0·31
21. 210·21
22. 294
23. 282·131
24. 35
25. 242

page 12 **Exercise 9**

1. £8000
2. £6
3. 20 min
4. B
5. C
6. B
7. B
8. C
9. B
10. B
11. A
12. (a) 89·89
(b) 4·2
(c) 358·4
(d) 58·8
(e) 0·3
(f) 2·62
13. (a) 4·5
(b) 462
(c) 946·4
(d) 77·8
(e) 0·2
(f) 21
14. £5000
15. About 20
16. Yes
17. Yes
18. He got it wrong. Correct answer = £10·45
19. 50

page 15 **Exercise 10**

1. (a) 0·7
(b) 8·81
(c) 0·726
(d) 1·18
(e) 0·9
(f) 8·22
(g) 0·075
(h) 11·726
(i) 20·2
(j) 6·67
(k) 0·3
(l) 0·0725
2. (a) 2·7
(b) 189
(c) 3
(d) 0·36
(e) 1·7
(f) 0·0416
(g) 0·04
(h) 800
(i) 9300
(j) 8·1
(k) 6·10
(l) 8

page 17 **Exercise 11**

1. $1:3$
2. $1:6$
3. $1:50$
4. $1:1·6$
5. $1:0·75$
6. $1:0·375$
7. $2·4:1$
8. $2·5:1$
9. $0·8:1$
10. £15, £25
11. £36, £84
12. 15 kg, 75 kg, 90 kg
13. 46 min, 69 min, 69 min
14. £39
15. $5:3$
16. (a) $1:3$
(b) $2:7$
(c) $3:5$
17. £200
18. $3:7$
19. $\frac{1}{7}x$
20. 6
21. £120
22. 300 g
23. 625

page 18 **Exercise 12**

1. 12·3 km
2. 4·71 km
3. 50 cm
4. 64 cm
5. 5·25 cm
6. 40 m by 30 m; 12 cm^2
7. 1 m^2, 6 m^2
8. 0·32 km^2

page 19 **Exercise 13**

1. £1·10
2. 6 h
3. 6 days
4. $2\frac{1}{2}\,l$
5. 60 km
6. 119 g
7. 260 min
8. $2\frac{1}{4}$ weeks
9. 6 men
10. (a) 12
(b) 2100
11. 4
12. 5·6 days
13. 44 l
14. 9600
15. $\dfrac{nx}{x+1000}$
16. 5

page 20 **Exercise 14**

1. -4
2. -12
3. -11
4. -3
5. -5
6. 4
7. -5
8. -8
9. 19
10. -17
11. -4
12. -5
13. -11
14. 6
15. -4
16. 6
17. 0
18. -18
19. -3
20. -11
21. -12
22. 4
23. 4
24. 0
25. -8
26. -3
27. 3
28. -12
29. 18
30. -5
31. -66
32. 98

page 21 **Exercise 15**

1. −8	**2.** −7	**3.** 1	**4.** 1	**5.** 9

6. 11 **7.** −8
8. 42 **9.** 4 **10.** 15 **11.** −7 **12.** −9 **13.** −1 **14.** −7
15. 0 **16.** 11 **17.** −14 **18.** 0 **19.** 17 **20.** 3 **21.** −1
22. −3 **23.** 12 **24.** −9 **25.** 3 **26.** 0 **27.** 8 **28.** 2

29. (a)

+	−5	1	6	−2
3	−2	4	9	1
−2	−7	−1	4	−4
6	1	7	12	4
−10	−15	−9	−4	−12

(b)

+	−3	2	−4	7
5	2	7	1	12
−2	−5	0	−6	5
10	7	12	6	17
−6	−9	−4	−10	1

page 21 **Exercise 16**

1. −6 **2.** −4 **3.** −15 **4.** 9 **5.** −8 **6.** −15 **7.** −24 **8.** 6
9. 12 **10.** −18 **11.** −21 **12.** 25 **13.** −60 **14.** 21 **15.** 48 **16.** −16
17. −42 **18.** 20 **19.** −42 **20.** −66 **21.** −4 **22.** −3 **23.** 3 **24.** −5
25. 4 **26.** −4 **27.** −4 **28.** −1 **29.** −2 **30.** 4 **31.** −16 **32.** −2
33. −4 **34.** 5 **35.** −10 **36.** 11 **37.** 16 **38.** −2 **39.** −4 **40.** −5
41. 64 **42.** −27 **43.** −600 **44.** 40 **45.** 2 **46.** 36 **47.** −2 **48.** −8

49. (a)

×	4	−3	0	−2
−5	−20	15	0	10
2	8	−6	0	−4
10	40	−30	0	−20
−1	−4	3	0	2

(b)

×	−2	5	−1	−6
3	−6	15	−3	−18
−3	6	−15	3	18
7	−14	35	−7	−42
2	−4	10	−2	12

page 22 **Test 1**

1. −16 **2.** 64 **3.** −15 **4.** −2 **5.** 15 **6.** 18 **7.** 3
8. −6 **9.** 11 **10.** −48 **11.** −7 **12.** 9 **13.** 6 **14.** −18
15. −10 **16.** 8 **17.** −6 **18.** −30 **19.** 4 **20.** −1

page 22 **Test 2**

1. −16 **2.** 6 **3.** −13 **4.** 42 **5.** −4 **6.** −4 **7.** −12
8. −20 **9.** 6 **10.** 0 **11.** 36 **12.** −10 **13.** −7 **14.** 10
15. 6 **16.** −18 **17.** −9 **18.** 15 **19.** 1 **20.** 0

page 22 **Test 3**

1. 100 **2.** −20 **3.** −8 **4.** −7 **5.** −4 **6.** 10 **7.** 9
8. −10 **9.** 7 **10.** 35 **11.** −20 **12.** −24 **13.** −10 **14.** −7
15. −19 **16.** −1 **17.** −5 **18.** −13 **19.** 0 **20.** 8

page 23 **Test 4**

1. -5 **2.** 2 **3.** -20 **4.** 4 **5.** $-\frac{1}{2}$ **6.** -5 **7.** 1 **8.** 2
9. 6 **10.** 12 **11.** 0 **12.** -8 **13.** 1 **14.** -1000 **15.** -1

page 23 **Exercise 17**

1. 6 **2.** 17 **3.** $6^2 + 7^2 + 42^2 = 43^2$, $x^2 + (x+1)^2 + [x(x+1)]^2 = (x^2 + x + 1)^2$
4. (a) 87p (b) 240p (c) 72p (d) 15p (e) 9p
5. 200 g **6.** (a) £150 (b) £0 (c) £8 **7.** $\frac{1}{66}$ **8.** 5
9. (a) $54 \times 9 = 486$ (b) $57 \times 8 = 456$ (c) $52 \times 2 = 104$
10. (a) $\frac{1}{3}$ (b) $a = 2$, $b = 8$

page 24 **Exercise 18**

1.

$\times$	$\frac{2}{3}$	$\frac{3}{4}$	$\frac{1}{5}$
$\frac{1}{2}$	$\frac{1}{3}$	$\frac{3}{8}$	$\frac{1}{10}$
$\frac{1}{4}$	$\frac{1}{6}$	$\frac{3}{16}$	$\frac{1}{20}$
$\frac{2}{5}$	$\frac{4}{15}$	$\frac{3}{10}$	$\frac{2}{25}$

2. (a) $66667^2 = 4\,444\,488\,889$
(b) $44\,444\,448\,888\,889$
(c) $66\,666\,667$

3. (a) 100, 1 (b) 100, 2 **4.** $40 = 2^3 \times 5$, $63 = 3^2 \times 7$, $25 = 5^2$, 71 is prime
5. $5 \times 7 \times 13^2 \times 71$ **6.** 1 (127) **7.** 36, 49, 64
8. (a) 1 (b) 15 **9.** 50 **10.** 106

page 25 **Exercise 19**

1. £7055 **2.** 10
3. (a) 1, 3, 5, 7, 9, 11 (b) Step 1. Write down the number 2.
5. (a) 66 666 (b) 82 (c) 455 551

6.

8	1		9
	3	7	1
2	5	5	
7		6	8

7. (a) $\frac{3}{5}$ (b) $\frac{75}{16}$ **8.** 500 **9.** 32

page 26 **Exercise 20**

1. (a) 4 (b) 5 **2.** 21 **3.** 0·525 kg **5.** (a) $\frac{1}{994}$, $\frac{1}{949}$, $\frac{1}{499}$ (b) $\frac{91}{94}$
6. (a) £6576 (b) (i) 8 (ii) 99 **8.** 8

page 27 **Operator squares**

1. 0·49, 3·5 × 10, 40 **2.** 5·2, 4·56 × 6 → 22·8, 0·64
3. 0·7, 16 − (−19), 112, − 37·8 **4.** −12, 7 → 11, −3
5. $-\frac{1}{3}$, $-\frac{1}{20}$, $-3\frac{1}{5}$, $-1\frac{1}{4}$ **6.** 0·86, $\frac{1}{3}$ − 0·625, 0·336

2 Number 2

page 28 **Exercise 1**

1. (a) 0·25 (b) 0·4 (c) 0·375 (d) 0·41̇6̇ (e) 0·1̇6̇ (f) 0·2̇85714̇
2. (a) $\frac{1}{5}$ (b) $\frac{9}{20}$ (c) $\frac{9}{25}$ (d) $\frac{1}{8}$ (e) $1\frac{1}{20}$ (f) $\frac{7}{1000}$
3. (a) 25% (b) 10% (c) 72% (d) 7·5% (e) 2% (f) $33\frac{1}{3}$%
4. (a) 45%; $\frac{1}{2}$; 0·6 (b) 4%; $\frac{6}{16}$; 0·38 (c) 11%; 0·111; $\frac{1}{9}$ (d) 0·3; 32%; $\frac{1}{3}$
5. 0·58 6. 1·42 7. 0·65 8. 0·32 9. 0·07
10. 0·69 11. 40% 12. 25% 13. 30%, 0·33; $\frac{1}{3}$ 14. $\frac{2}{7}$; 29%; 0·3
15. $\frac{7}{11}$; 0·705; 0·71 16. 0·0005, $\frac{1}{1000}$, 0·2% 17. No 18. 96·8%
19. (a) 58·9% (b) 55% 20. Prices will still rise but not as quickly.

page 30 **Exercise 2**

1. (a) £15 (b) 900 kg (c) $2·80 (d) 125
2. £32 3. 13·2p 4. 52·8 kg
5. (a) 0·53 (b) 0·03 (c) 0·085 (d) 1·22
6. (a) £1·02 (b) £21·58 (c) £2·22 (d) £0·53
7. £248·57 8. £26 182 9. £8425·60 10. £73·03

11. (b) £6825 12. £201·40, £166 500, £2700, £199·50, £55·25
13. 325 14. £6·30
15. (a) 256 g (b) 72p (c) 14 mm (d) 30 min (e) 14·4 sec
16. (b) (i) $1·08 \times P \times 0·9$ (ii) 15%, 6% (iii) $1·25 \times Q \times 0·95$

page 32 **Exercise 3**

1. (a) 25%, profit (b) 25%, profit (c) 10%, loss (d) 20%, profit
   (e) 30%, profit (f) 7·5%, profit (g) 12%, loss (h) 54%, loss
2. 28% 3. $44\frac{4}{9}$% 4. 46·9% 5. 12% 6. $5\frac{1}{3}$% 7. 44%
8. 27·05% 9. 20% 10. 14·3% 11. 21% 12. (a) 25 000 (b) 76·3%

page 33 **Exercise 4**

1. £6200 2. £25 500 3. (a) £50 (b) £200 (c) £54 (d) £150 (e) £2400
4. 210 000 5. 85 kg 6. 8 cm × 10 cm 7. 350 g 8. £56
9. 400 kg 10. 200 11. 29 000 12. 500 cm 13. £500 14. £480 million

page 34 **Exercise 5**

1. (a) £2200 (b) £2420 (c) £2662 2. (a) £5600 (b) £7024·64
3. £13 107·96 4. (a) £36 465·19 (b) £40 202·87
5. (a) £95 400 (b) £107 191 (c) £161 176 6. (a) £14 033 (b) £734 (c) £107 946
7. £9211·88 8. No. It should be £193·48 9. 8 years 10. 11 years
11. (b) $x = 9$ (c) 9 years 12. about 12 years 13. 8% is better after more than 11 years.

page 36 **Exercise 6**

1. 3·041	**2.** 1460	**3.** 0·030 83	**4.** 47·98	**5.** 130·6
6. 0·4771	**7.** 0·3658	**8.** 37·54	**9.** 8·000	**10.** 0·6537
11. 0·037 16	**12.** 34·31	**13.** 0·7195	**14.** 3·598	**15.** 0·2445
16. 2·043	**17.** 0·3798	**18.** 0·7683	**19.** −0·5407	**20.** 0·070 40
21. 2·526	**22.** 0·094 78	**23.** 0·2110	**24.** 3·123	**25.** 2·230
26. 128·8	**27.** 4·268	**28.** 3·893	**29.** 0·6290	**30.** 0·4069
31. 9·298	**32.** 0·1010	**33.** 0·3692	**34.** 1·125	**35.** 1·677
36. 0·9767	**37.** 0·8035	**38.** 0·3528	**39.** 2·423	**40.** 1·639
41. 0·000 465 9	**42.** 0·3934	**43.** −0·7526	**44.** 2·454	**45.** 40 000
46. 0·070 49	**47.** 405 400	**48.** 471·3	**49.** 20 810	**50.** $2·218 \times 10^6$
51. $1·237 \times 10^{-24}$	**52.** 3·003	**53.** 0·035 81	**54.** 47·40	**55.** −1748
56. 0·011 38	**57.** 1757	**58.** 0·026 35	**59.** 0·1651	**60.** 5447
61. 0·006 562	**62.** 0·1330	**63.** 0·4451	**64.** 0·036 16	**65.** 19·43

66. $1·296 \times 10^{-15}$

page 40 **Exercise 9**

1. 4×10^3	**2.** 5×10^2	**3.** 7×10^4	**4.** 6×10	**5.** $2·4 \times 10^3$
6. $3·8 \times 10^2$	**7.** $4·6 \times 10^4$	**8.** $4·6 \times 10$	**9.** 9×10^5	**10.** $2·56 \times 10^3$
11. 7×10^{-3}	**12.** 4×10^{-4}	**13.** $3·5 \times 10^{-3}$	**14.** $4·21 \times 10^{-1}$	**15.** $5·5 \times 10^{-5}$
16. 1×10^{-2}	**17.** $5·64 \times 10^5$	**18.** $1·9 \times 10^7$	**19.** $1·1 \times 10^9$	**20.** $1·67 \times 10^{-24}$
21. $5·1 \times 10^8$	**22.** $2·5 \times 10^{-10}$	**23.** $6·023 \times 10^{23}$	**24.** 3×10^{10}	**25.** £$3·6 \times 10^6$

26. *c, a, b*   **27.** 13   **28.** 16

29. (b) (i) 8×10^{11} (ii) 5×10^{-7} (iii) $5·5 \times 10^{-4}$ (iv) 4×10^9 (v) 3×10^8 (vi) 4×10^{12}

30. (a) 23 000   (b) 0·03   (c) 560   (d) 800 000   (e) 0·0022

(f) 900 000 000   (g) 0·6   (h) 7000   (i) 3 140 000

31. (a) 65×10^5 (b) $2·7 \times 10^3$ (c) $3·54 \times 10^6$ (d) $2·8 \times 10^{-2}$ (e) 7×10^{-3} (f) $7·07 \times 10^{22}$

32. (a) 6×10^3 (b) $1·1 \times 10^{-1}$ (c) $4·5 \times 10^8$ (d) $8·5 \times 10^{-6}$ (e) 6×10^9 (f) 5×10^{-1}

page 42 **Exercise 10**

1. 6×10^7	**2.** 2×10^7	**3.** $1·5 \times 10^{10}$	**4.** 7×10^9	**5.** 9×10^8
6. $1·6 \times 10^9$	**7.** $4·5 \times 10$	**8.** $1·2 \times 10^{14}$	**9.** $1·86 \times 10^{11}$	**10.** 2×10^{-10}
11. $7·5 \times 10^{-12}$	**12.** $2·5 \times 10^9$	**13.** $2·7 \times 10^{16}$	**14.** $3·5 \times 10^{-6}$	**15.** $8·1 \times 10^{-21}$
16. $6·3 \times 10^{13}$	**17.** $1·5 \times 10^{16}$	**18.** 8×10^3	**19.** $2·8 \times 10^{16}$	**20.** 8×10^3
21. 2×10	**22.** 5×10^{-20}	**23.** 5×10^{-4}	**24.** 3×10^{-11}	**25.** 2×10^{-4} m

26. $5·57 \times 10^9$ **27.** Asia (0·068 people/m²)   **28.** $2·4528 \times 10^7$ km

29. 50 min   **30.** 6×10^2 **31.** $2·4 \times 10^9$ kg

32. (a) $9·46 \times 10^{12}$ km   (b) 144 million km   **33.** (a) 6×10^{101} (b) 20·5 sec   (c) $6·3 \times 10^{91}$ years

page 44 **Exercise 11**

1. Rational: $\left(\sqrt{17}\right)^2$; 3·14; $\frac{\sqrt{12}}{\sqrt{3}}$; $3^{-1} + 3^{-2}$; $\frac{22}{7}$; $\sqrt{2·25}$

3. (a) A 4, B $\sqrt{41}$ (b) A rational, B irrational   (c) Both rational   (d) B

4. (a) 6π cm, irrational   (b) 6 cm, rational   (c) 36 cm², rational

(d) 9π cm², irrational   (e) $36 - 9\pi$ cm², irrational

5. (a) $\frac{2}{9}$ (b) $\frac{29}{99}$ (c) $\frac{541}{999}$ **8.** (a) No   (b) Yes e.g. $\sqrt{8} \times \sqrt{2} = 4$

10. (a) 74   (b) irrational   **11.** rational   **12.** (e.g.) 5·29

14. e.g. (a) $\sqrt{7}$ (b) $\sqrt{3}$ (c) $\sqrt{2}$ (d) π

page 46   **Exercise 12**

1. AF, EG, BD, CH

2. (a) T	(b) T	(c) F	(d) T	(e) F	(f) T
3. (a) $3 + 2\sqrt{2}$	(b) $7 - 4\sqrt{3}$	(c) 18	(d) $14 - 6\sqrt{5}$	(e) $6 + 4\sqrt{2}$	(f) 8
4. (a) T	(b) T	(c) F	(d) T	(e) F	(f) T
5. (a) 4	(b) $4\sqrt{2}$	(c) $10\sqrt{3}$	(d) $9\sqrt{2}$	(e) $5\sqrt{5}$	(f) $5\sqrt{3}$
(g) $5\sqrt{5}$	(h) $\sqrt{3}$	(i) 2	(j) $1\frac{1}{2}$	(k) $2\frac{1}{2}$	(l) $1\frac{1}{3}$

6. (a) $2\sqrt{3}$ (b) (i) $4\sqrt{2}$ (ii) $4\sqrt{3}$ (iii) $3\sqrt{2}$ (iv) $\frac{2}{3}\sqrt{3}$

7. (a) $4\sqrt{2} - 4$ (b) (i) $2\sqrt{3} - 2$ (ii) $\sqrt{5} - 2$ (iii) $\frac{10}{3}(\sqrt{7} + 2)$

page 47   **Exercise 13**

1. 7	**2.** 13	**3.** 13	**4.** 22	**5.** 1	**6.** -1
7. 18	**8.** -4	**9.** -3	**10.** 37	**11.** 0	**12.** -4
13. -7	**14.** -2	**15.** -3	**16.** -8	**17.** -30	**18.** 16
19. -10	**20.** 0	**21.** 7	**22.** -6	**23.** -2	**24.** -7
25. -5	**26.** 3	**27.** 4	**28.** -8	**29.** -2	**30.** 2
31. 0	**32.** 4	**33.** -4	**34.** -3	**35.** -9	**36.** 4

page 48   **Exercise 14**

1. 9	**2.** 27	**3.** 4	**4.** 16	**5.** 36	**6.** 18
7. 1	**8.** 6	**9.** 2	**10.** 8	**11.** -7	**12.** 15
13. -23	**14.** 3	**15.** 32	**16.** 36	**17.** 144	**18.** -8
19. -7	**20.** 13	**21.** 5	**22.** -16	**23.** 84	**24.** 17
25. 6	**26.** 0	**27.** -25	**28.** -5		

page 48   **Exercise 15**

1. -20	**2.** 16	**3.** -42	**4.** -4	**5.** -90	**6.** -160
7. -2	**8.** -81	**9.** 4	**10.** 22	**11.** 14	**12.** 5
13. 1	**14.** $\sqrt{5}$	**15.** 4	**16.** $-6\frac{1}{2}$	**17.** 54	**18.** 25
19. 4	**20.** 312	**21.** 45	**22.** 22	**23.** 14	**24.** -36
25. -7	**26.** 1	**27.** 901	**28.** -30	**29.** -5	**30.** $7\frac{1}{2}$
31. -7	**32.** $-\frac{3}{13}$	**33.** $1\frac{1}{3}$	**34.** $-\frac{5}{36}$		

page 49   **Exercise 16**

1. 21   **2.** 1·62   **3.** 396   **4.** 650   **5.** 63·8   **6.** 9×10^{12} **7.** $10\frac{1}{2}$ **8.** 800

9. (a) $1245\,\text{km/h}$ (b) $5°\text{C}$ (c) $1008\,\text{km/h}$ **10.** (a) $ac + ab - a^2$ (b) 30

11. (a) $r - p + q$ (b) 4·1

page 50   **Exercise 17**

1. (a) $2\frac{1}{2}\,\text{h}$ (b) $3\frac{1}{8}\,\text{h}$ (c) $75\,\text{s}$ (d) $4\,\text{h}$ **2.** $156\,\text{km/h}$

3. (a) $75\,\text{km/h}$ (b) $4 \times 10^6\,\text{m/s}$ (c) $3\,\text{km/h}$

4. (a) $10\,000\,\text{m}$ (b) $56\,400\,\text{m}$ (c) $4500\,\text{m}$ (d) $50\,400\,\text{m}$

5. (a) $3·125\,\text{h}$ (b) $76·8\,\text{km/h}$ **6.** (a) $4·45\,\text{h}$ (b) $23·6\,\text{km/h}$

7. (a) $8\,\text{m/s}$ (b) $7·6\,\text{m/s}$ (c) $102·63\,\text{s}$

8. $1230\,\text{km/h}$ **9.** $3\,\text{h}$ **10.** $100\,\text{s}$ **11.** $1\frac{1}{2}$ minutes   **12.** $600\,\text{m}$

13. $53\frac{1}{3}\,\text{s}$ **14.** $5\,\text{cm/s}$ **15.** $60\,\text{s}$ **16.** $120\,\text{mph}$

page 52 **Exercise 18**

1. $10 \, \text{g/cm}^3$ **2.** 240 g   **3.** $35 \, \text{m}^3$ **4.** 0·6 kg   **5.** $250 \, \text{people/km}^2$
6. (a) $6000 \, \text{people/km}^2$ (b) $5 \cdot 58 \times 10^{10}$
7. 1000 kg   **8.** $3000 \, \text{kg/m}^3$ **9.** $0 \cdot 8 \, \text{g/cm}^3$
10. (a) 20 m/s   (b) 108 km/h   (c) 1·2 cm/s   (d) 90 m/s

page 54 **Exercise 19**

1. 195·5 cm   **2.** 36·5 kg   **3.** 3·25 kg   **4.** 95·55 m   **5.** 28·65 s
6. (a) 1·5, 2·5   (b) 2·25, 2·35   (c) 63·5, 64·5   (d) 13·55, 13·65
7. B   **8.** C   **9.** (a) Not necessarily   (b) 1 cm
10. (a) 16·5, 17·5   (b) 255·5, 256·5   (c) 2·35, 2·45   (d) 0·335, 0·345   (e) 2·035, 2·045
     (f) 11·95, 12·05   (g) 81·35, 81·45   (h) 0·25, 0·35   (i) 0·65, 0·75   (j) 51500, 52500
11. No, max. card length 11·55 cm min. envelope length 11·5 cm

page 57 **Exercise 20**

1. (a) 7·5, 8·5, 10·5 cm   (b) 26·5 cm   **2.** $46 \cdot 75 \, \text{cm}^2$
3. (a) 7   (b) 5   (c) 10   (d) 4   (e) 2   (f) 5   (g) 2   (h) 24
4. (a) 13   (b) 11   (c) 3   (d) 12·5
5. (i) 10·5   (ii) 4·3   **6.** (i) 11   (ii) 1   (iii) 0·6   **7.** $56 \, \text{cm}^2$
8. 55·706604   **9.** 17·198364 m/s   **10.** 3·298 8372, 2·872 2222
11. 4·101 6355   **12.** 7·163 6234
13. (a) $35\,390 \rightarrow \$38\,250$ (b) $35\,640 \rightarrow \$35\,660$ (c) Only one figure to approximate

page 59 **Exercise 21**

1. 7·15%   **2.** 4·23%   **3.** 12·9%   **4.** 0·177%   **5.** 6·15%
6. 0·0402%   **7.** 1·55%   **8.** 8·00%   **9.** 0·28%   **10.** Second

page 59 **Exercise 22**

1. (a) Fr 218   (b) $114·80   (c) 52 000 Ptas   (d) DM 4·65   (e) 6670 lire   (f) €1·386
2. (a) £45·87   (b) £1524·39   (c) £2·42   (d) £584·42   (e) £172·41   (f) £3·65
3. £3·84   **4.** £15 000, USA £15 172, France £17 800   **5.** 3·25 francs to the pound
6. 20   **7.** (a) 256   (b) 65 536   **8.** $4 \cdot 68 \times 10^5$
9. (a) $\frac{3}{19}$ (b) $\frac{1}{6}$ **10.** £10 485·76

page 60 **Exercise 23**

1. (a) 1·25 m/min   (b) 2·08 cm/s   **2.** 77·5%   **3.** 225 mm
4. (a) 35   (b) 5·14   **5.** 43·5 mm   **6.** (b) 12   **7.** 210 kg
8. (a) 10 : 45   (b) 51 min   **9.** 1105   **10.** (a) £699

page 61 **Exercise 24**

1. (a) 0·0018 km/h   (b) $1 \cdot 8 \times 10^{-3}$ **2.** 13   **3.** $£6 \cdot 3 \times 10^{10}$
4. (a) £9·79   (b) £32   (c) £44·20   (d) £45
5. (a) 323 g   (b) 23   (c) 67p   (d) 29
6. 6 m   **7.** approx 190 ml   **8.** 400   **9.** 18 : 52   **10.** £118
11. 37   **12.** 7^{77} **13.** $\frac{43}{99}$

3 Algebra 1

1. (a) $2x - 6$ (b) $(x+4)^2$ (c) $\dfrac{2(x-5)}{3}$ (d) $(w-x)^2$ (e) $\dfrac{(n+p)^3}{a}$ (f) $[3(t-1)]^2$

2. nw kg **3.** $\dfrac{x}{n}$ pence **4.** $\dfrac{y}{n}$ kg **5.** £$\dfrac{p}{5}$ **6.** $y+n+6$

7. £$w(n+r)$ **8.** $\dfrac{xn}{t}$ metres **9.** $\dfrac{100n}{x}$ **10.** $\dfrac{100n}{x+1}$ pence

1. $5x + 8$ **2.** $9x + 5$ **3.** $7x + 4$ **4.** $7x + 4$ **5.** $7x + 7$
6. $8x + 12$ **7.** $12x - 6$ **8.** $2x + 5$ **9.** $2x - 5$ **10.** $2x - 5$
11. $13a + 3b - 1$ **12.** $10m + 3n + 8$ **13.** $3p - 2q - 8$ **14.** $2s - 7t + 14$ **15.** $2a + 1$
16. $x + y + 7z$ **17.** $5x - 4y + 4z$ **18.** $5k - 4m$ **19.** $4a + 5b - 9$ **20.** $a - 4x - 5e$
21. (a) $3x^2 + 1$ (b) $2a + 5ab$ (c) $x^3 + 4x^2 - 7x$ (d) $4a^2 - 7a$
22. B and D **23.** $x^2 + 3x + 3$ **24.** $3x^2 + 6x + 5$ **25.** $2x^2 + 6x - 7$ **26.** $3x^2 + x + 12$
27. $2x^2 + x + 3$ **28.** $2x^2 - x$ **29.** $x^2 - 4x - 2$ **30.** $3x^2 - 2x - 2$ **31.** $4y^2 + 5x^2 + x$
32. $x + 12$ **33.** $5y^2 - 6y + 2$ **34.** $3ab - 3b$ **35.** $2cd - 2d^2$ **36.** $4ab - 2a^2 + 2a$
37. $2x^3 + 5x^2$ **38.** $11 + x^2 + x^3$ **39.** $3xy$ **40.** $p^2 - q^2$

1. (a) $6x$ (b) $15a$ (c) $-10t$ (d) $1000a$ (e) $2x^2$ (f) $4x^2$ (g) $2x^3$ (h) $20t^2$
2. (a) $4x + 2$ (b) $10x + 15$ (c) $a^2 - 3a$ (d) $2n^2 + 2n$ (e) $-4x - 6$ (f) $2x^2 + 2xy$
   (g) $-3a + 6$ (h) $pq - p^2$ (i) $ab + a^2$ (j) $2b^2 - 2b$ (k) $2n^2 + 4n$ (l) $6x^2 + 9x$
3. AG, BD, CE, FH
4. (a) $6ab$ (b) $3x^4$ (c) $2xy$ (d) $10pq$ (e) $15xy$ (f) $18x^3$
   (g) $24a^4$ (h) $2a^2b$ (i) $3xy^2$ (j) $5c^2d$ (k) a^2b^2 (l) $2x^2y^2$
   (m) $3d^3$ (n) $10x^2y$ (o) $-6a^2$ (p) $2a^2b$
5. $6x^2 + 12x$
6. (a) $7x + 10$ (b) $8x + 2$ (c) $5a - 3$ (d) $11a + 17$
   (e) $8a - 10$ (f) $8t + 4$ (g) $x + 4$ (h) $x + 6$
7. (a) $2x - 2$ (b) $x - 1$ (c) $x + 2$ (d) $x - 3$ (e) $x + 3$ (f) $x + 3$
8. (a) $2x^2 + 4x + 6$ (b) $2x^2 + 2x + 5$ (c) $3a^2 + 6a - 4$
   (d) $5y^2 + 4y - 3$ (e) $5x^2 + 2x$ (f) $a^2 + 2a$

1. $x^2 + 4x + 3$ **2.** $x^2 + 5x + 6$ **3.** $y^2 + 9y + 20$ **4.** $x^2 + x - 12$
5. $x^2 + 3x - 10$ **6.** $x^2 - 5x + 6$ **7.** $a^2 - 2a - 35$ **8.** $z^2 + 7z - 18$
9. $x^2 - 9$ **10.** $k^2 - 121$ **11.** $2x^2 - 5x - 3$ **12.** $3x^2 - 2x - 8$
13. $2y^2 - y - 3$ **14.** $49y^2 - 1$ **15.** $x^2 + 8x + 16$ **16.** $x^2 + 4x + 4$
17. $x^2 - 4x + 4$ **18.** $4x^2 + 4x + 1$ **19.** $2x^2 + 6x + 5$ **20.** $2x^2 + 2x + 13$
21. $5x^2 + 8x + 5$ **22.** $2y^2 - 14y + 25$ **23.** $10x - 5$ **24.** $-8x + 8$
25. (a) $4x(x+5)$ (b) $(2x+5)^2$ (c) 25 square units
26. $x^3 + 6x^2 + 11x + 6$
27. (a) $x^3 + 8x^2 + 19x + 12$ (b) $x^3 + 3x^2 + 3x + 1$ (c) $x^3 + 3x^2 - 4$
28. (d) Yes. $n = 10, 11, 21, 22$ etc.

page 68 **Exercise 5**

1. $x(x+5)$	**2.** $x(x-6)$	**3.** $x(7-x)$	**4.** $y(y+8)$
5. $y(2y+3)$	**6.** $2y(3y-2)$	**7.** $3x(x-7)$	**8.** $2a(8-a)$
9. $3c(2c-7)$	**10.** $3x(5-3x)$	**11.** $7y(8-3y)$	**12.** $x(a+b+2c)$
13. $x(x+y+3z)$	**14.** $y(x^2+y^2+z^2)$	**15.** $ab(3a+2b)$	**16.** $xy(x+y)$
17. $2a(3a+2b+c)$	**18.** $m(a+2b+m)$	**19.** $2k(x+3y+2z)$	**20.** $a(x^2+y+2b)$
21. $x(7x+1)$	**22.** $4y(y-1)$	**23.** $p(p-2)$	**24.** $2a(3a+1)$
25. $4(1-2x^2)$	**26.** $5x(1-2x^2)$	**27.** $\pi(4r+h)$	**28.** $\pi r(r+2)$
29. $\pi r(3r+h)$	**30.** $x(3y+2)$	**31.** $xk(x+k)$	**32.** $ab(a^2+2b)$
33. $bc(a-3b)$	**34.** $ae(2a-5e)$	**35.** $ab(a^2+b^2)$	**36.** $x^2y(x+y)$
37. $2xy(3y-2x)$	**38.** $3ab(b^2-a^2)$	**39.** $a^2b(2a+5b)$	**40.** $ax^2(y-2z)$
41. $2ab(x+b+a)$	**42.** $yx(a+x^2-2yx)$	**43.** (a) $2x+8y$ (b) $10a-2x$	

page 70 **Exercise 7**

1. 8	**2.** 9	**3.** 7	**4.** 10	**5.** $\frac{1}{3}$	**6.** 10	**7.** $1\frac{1}{2}$
8. -1	**9.** $-1\frac{1}{2}$	**10.** $\frac{1}{3}$	**11.** $\frac{99}{100}$	**12.** 0	**13.** 1000	**14.** $-\frac{1}{1000}$
15. 1	**16.** -7	**17.** -5	**18.** $1\frac{1}{6}$	**19.** 1	**20.** 2	**21.** -5
22. -3	**23.** $-1\frac{1}{2}$	**24.** 2	**25.** 1	**26.** $3\frac{1}{2}$	**27.** 2	**28.** -1
29. $10\frac{2}{3}$	**30.** $1\cdot1$	**31.** -1	**32.** 2			

page 70 **Exercise 8**

1. 35	**2.** 130	**3.** 14	**4.** $\frac{2}{3}$	**5.** $3\frac{1}{3}$	**6.** $-2\frac{1}{2}$	**7.** 3
8. $1\frac{1}{8}$	**9.** $\frac{3}{10}$	**10.** $-1\frac{1}{4}$	**11.** 10	**12.** 27	**13.** 20	**14.** 18
15. 28	**16.** -15	**17.** $2\frac{1}{2}$	**18.** $1\frac{1}{3}$			

page 71 **Exercise 9**

1. $-1\frac{1}{2}$	**2.** 2	**3.** $-\frac{2}{5}$	**4.** $-\frac{1}{3}$	**5.** $1\frac{2}{3}$	**6.** 6	**7.** $-\frac{2}{5}$	**8.** $-3\frac{1}{5}$	**9.** $\frac{1}{2}$	**10.** -4
11. 18	**12.** 5	**13.** 4	**14.** 3	**15.** $2\frac{3}{4}$	**16.** $-\frac{7}{22}$	**17.** $\frac{1}{4}$	**18.** 1	**19.** 4	**20.** -11

page 71 **Exercise 10**

1. $\frac{1}{3}$	**2.** $\frac{1}{5}$	**3.** $1\frac{2}{3}$	**4.** -3	**5.** $\frac{5}{11}$	**6.** -2	**7.** -7
8. $-7\frac{2}{3}$	**9.** 2	**10.** 3	**11.** 2	**12.** 3	**13.** 5	**14.** -4
15. 4	**16.** $\frac{3}{5}$	**17.** $1\frac{1}{8}$	**18.** -1	**19.** 1	**20.** 1	**21.** $1\frac{5}{7}$

page 72 **Exercise 11**

1. $3\frac{1}{3}$ cm	**2.** 12 cm	**3.** 91, 92, 93	**4.** 21, 22, 23, 24
5. 57, 59, 61	**6.** 506, 508, 510	**7.** $12\frac{1}{2}$	**8.** 20
9. $18\frac{1}{2}$, $27\frac{1}{2}$	**10.** 20°, 60°, 100°	**11.** 5, 15, 8	**12.** 5 cm
13. $59\frac{2}{3}$ kg, $64\frac{2}{3}$ kg, $72\frac{2}{3}$ kg	**14.** 7 cm	**15.** (b) $14+8x$ (c) 0·75 m	

page 74 **Exercise 12**

1. $5\frac{2}{5}$	**2.** $53\frac{1}{3}$ or $56\frac{2}{3}$	**3.** 2·9	**4.** 26, 58	**5.** 2 km	**6.** 8 km
7. 400 m	**8.** 21	**9.** 23	**10.** £3600	**11.** 15	**12.** 2 km
13. 6, 7, 8, 9	**14.** 2, 3, 4, 5	**15.** 26 cm			

page 76   **Exercise 13**

1. $\frac{1}{4}$ **2.** -3 **3.** 4   **4.** $-7\frac{2}{3}$ **5.** -43 **6.** 11   **7.** $-\frac{1}{2}$
8. 0   **9.** 1   **10.** $-1\frac{2}{3}$ **11.** 10, 8, 6   **12.** 13, 12, 5   **13.** 4 cm   **14.** 5 m

page 76   **Exercise 14**

1. 12 cm   **2.** 15×5
3. (a) 26×13 (b) 16×8 (c) 32×16 (d) 9×4.5 (e) 6.5×3.25
4. (a) 19×18 (b) 6.5×5.5 (c) 8.7×7.7

page 78   **Exercise 15**

1. (a) 9.5   (b) 7.6
2. (a) 5.1   (b) 3.8   (c) 6.7   (d) 3.4
3. (a) 3.2   (b) 2.7   (c) 3.6
4. 2.15   **5.** 8.1   **6.** 9.56
7. (a) $x + 10$ (c) $34 \times 44 \times 6$ **8.** (a) 3 cm   (b) 3.8
9. (a) $x - 1$ (b) 1.62   **10.** 6.3   **11.** 38.5
12. (a) $2x\sqrt{1 - x^2}$ (b) 0.71   **13.** 1.5

page 81   **Exercise 16**

1. $w = b + 4$ **2.** $w = 2b + 6$ **3.** $w = 2b - 12$ **4.** $m = 2t + 1$ **5.** $m = 3t + 2$
6. $s = t + 2$ **7.** (a) $p = 5n - 2$ (b) $k = 7n + 3$ (c) $w = 2n + 11$ **8.** $m = 8c + 4$
9. (a) $y = 3n + 1$ (b) $h = 4n - 3$ (c) $k = 3n + 5$
10. (a) $t = 2n + 4$ (b) $e = 3n + 11$ (c) $e = \dfrac{3t + 10}{2}$

11. (a) £7800   (b) $R = 5000 + 400N$ **12.** Number of white squares $= n^2 - n$

page 85   **Exercise 17**

1. (a) $10n$ (b) $5n$ (c) $n + 2$ (d) $2n + 1$ (e) $30n$ (f) $6n - 1$ (g) $3n - 2$
2. (a) $10 \to 23$, $n \to 2n + 3$ (b) $20 \to 61$, $n \to 3n + 1$
3. (b) 51; $5n + 1$ **4.** (b) A $3n - 2$; B $4n + 2$; C $7n - 2$
5. (a) 80   (b) 76   (c) $8n$ (d) $8n - 4$
6. (a) $6n + 2$ (b) $6n - 1$ **7.** (a) $n \to n(n + 1)$ (b) $n \to n^2 + 1$
8. $n^2 + 4$ **9.** 10^2, n^2 **10.** 10×11, $n(n + 1)$ **11.** $\frac{10}{11}$, $\dfrac{n}{n + 1}$
12. $\frac{5}{10^2}$, $\dfrac{5}{n^2}$ **13.** 10×12, $n(n + 2)$ **14.** A $3n - 4$; B $n^2 - 2$; C $n(n + 2)$
15. (a) 3, 5, 7   (b) 2, 7, 12   (c) 98, 96, 94
       (d) $\frac{1}{3}, \frac{2}{4}, \frac{3}{5}$ (e) 10, 100, 1000   (f) 2, 6, 12
16. (a) $t = n(n + 2)$ (b) $t = n^2 + 2n$
17. (a) $\dfrac{1}{n^2}$ (b) 2^n (c) 3×2^n (d) $10^n - 1$ (e) $2n(n + 1)$ (f) $\dfrac{n^2}{3^n}$

18. number of diagonals $= \dfrac{n(n - 3)}{2}$

19. (a) 5   (b) $a = -2$, $b = 5$ (c) $c = 3$, $d = 5$

page 88   **Exercise 18**

For questions **1** to **9** three points on each graph are given.
1. $(-3, -5)$, $(0, 1)$, $(1, 3)$ **2.** $(-2, -10)$, $(0, -4)$, $(3, 5)$ **3.** $(-2, 10)$, $(0, 8)$, $(4, 4)$
4. $(-2, 14)$, $(0, 10)$, $(4, 2)$ **5.** $(-3, 1)$, $(0, 2\cdot5)$, $(3, 4)$ **6.** $(-3, -15)$, $(0, -6)$, $(3, 3)$
7. $(-3, 2\cdot5)$, $(0, 4)$, $(3, 5\cdot5)$ **8.** $(-2, -7)$, $(0, -3)$, $(4, 5)$ **9.** $(-2, 18)$, $(0, 12)$, $(4, 0)$

10.

x	0	50	100	150	200	250	300
C	35	45	55	65	75	85	95

(a) 180   (b) $C = 0\cdot2x + 35$

11.

h	0	1	2	3
C	18	33	48	63

(a) $2\cdot5h$ (b) $C = 15h + 18$

12.

m	0	5000	10 000
C	200	450	700

(a) (i) £560   (ii) 2400 miles

13.

s	0	80	160
C	50	530	1010

(a) £188   (b) 158 km/h

page 90   **Exercise 19**

1. (a) $(4, 10)$ (b) $(6, 7)$ (c) $(1, 3)$ (d) $(1, 3)$ **2.** (a) 5   (b) 13   (c) $\sqrt{74}$
3. (a) $(3, 6)$ (b) $\sqrt{180}$ **4.** $(4a, a)$

page 91   **Exercise 20**

1. AB$:\frac{1}{5}$;   BC$:\frac{5}{2}$;   AC$:-\frac{4}{3}$ **2.** PQ$:\frac{4}{5}$;   PR$:-\frac{1}{6}$;   QR$:-5$
3. (a) 3   (b) $\frac{3}{2}$ (c) 4   (d) 5   **4.** $3\frac{1}{2}$
5. (a) $\dfrac{n+4}{2m-3}$ (b) $n = -4$ (c) $m = 1\frac{1}{2}$
6. (a) $1, -1; \frac{1}{2}, -2; 3, -\frac{1}{3}$ (b) -1 (c) -1 (d) the same
7. (from top) $5, -\frac{1}{2}, -\frac{2}{3}, \frac{1}{5}, -\frac{4}{3}$ **8.** $t = -1$

page 94   **Exercise 21**

1. 1, 3   **2.** 1, -2 **3.** 2, 1   **4.** 2, -5 **5.** 3, 4   **6.** $\frac{1}{2}$, 6   **7.** 3, -2
8. 2, 0   **9.** $\frac{1}{4}$, -4 **10.** -1, 3   **11.** -2, 6   **12.** -1, 2   **13.** -2, 3   **14.** -3, -4
15. $\frac{1}{2}$, 3   **16.** $-\frac{1}{3}$, 3   **17.** 4, -5 **18.** $\frac{3}{2}$, -4 **19.** 10, 0   **20.** 0, 4
21. A$: y = 3x - 4$;   B$: y = x + 2$ **22.** C$: y = \frac{2}{3}x - 2$ or $3y = 2x - 6$;   D$: y = -2x + 4$
23. (a) C and D, F and G   (b) A and E, B and H   (c) I
24. (a) A$(0, -8)$, B$(4, 0)$ (b) 2   (c) $y = 2x - 8$

page 95   **Exercise 22**

1. $y = 3x + 7$ **2.** $y = 2x - 9$ **3.** $y = -x + 5$ **4.** $y = 2x - 1$ **5.** $y = 3x + 5$
6. $y = -x + 7$ **7.** $y = \frac{1}{2}x - 3$ **8.** $y = 2x - 3$ **9.** $y = 3x - 11$ **10.** $y = -x + 5$

page 96   **Exercise 23**

1. $a = 2\cdot5$, $c = 1\cdot5$; $Z = 2\cdot5X + 1\cdot5$ **2.** $m = 2$, $c = 3$ **3.** $n = -2\cdot5$, $k = 15$

4. $m = 0\cdot2$, $c = 3$ **5.** $a = 12$, $b = 3\cdot5$; $V = \dfrac{12}{T} + 3\cdot5$ **6.** $k = 0\cdot092$

page 98   **Exercise 24**

1. (a) $45\,\text{min}$ (b) $09:15$ (c) $60\,\text{km/h}$ (d) $100\,\text{km/h}$ (e) $57\cdot1\,\text{km/h}$
2. (a) $09:00$ (b) $64\,\text{km/h}$ (c) $40\,\text{km/h}$ (d) $70\,\text{km}$ (e) $80\,\text{km/h}$
3. (b) $11:05$ **4.** (b) $12:42$ **5.** (b) $12:35$ **6.** $1\frac{1}{8}\,\text{h}$ **7.** $1\,\text{h}$

page 99   **Exercise 25**

1. B   **2.** (a) C   (b) A   (c) D   (d) B   **3.** D

4. 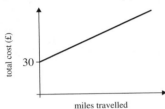   **5.**   **6.** AZ, BX, CY

7.   **8.** (a) 6 gallons   (b) 40 mpg; 30 mpg
(c) $33\frac{1}{3}$ mpg;
$5\frac{1}{2}$ gallons

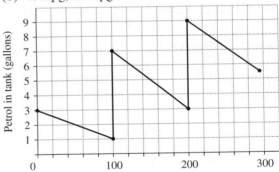

page 103   **Exercise 26**

1. (a) $(3, 7)$ (b) $(1, 3)$ (c) $(11, -1)$
2. $(2, 4)$ **3.** $(2, 3)$ **4.** $(3, 1)$ **5.** $(1, 5)$ **6.** $(5, 3)$
7. (a) $(4, 0)$ (b) $(1, 6)$ (c) $(-2, -3)$ (d) $(8, -1)$ (e) $(-0\cdot6, 1\cdot2)$

page 105   **Exercise 27**

1. $x = 2$, $y = 1$ **2.** $x = 4$, $y = 2$ **3.** $x = 3$, $y = 1$ **4.** $x = -\frac{1}{3}$, $y = -2\frac{1}{3}$
5. $x = 3$, $y = 2$ **6.** $x = 5$, $y = -2$ **7.** $x = 2$, $y = 1$ **8.** $x = 5$, $y = 3$
9. $x = 3$, $y = -1$ **10.** $a = 2$, $b = -3$ **11.** $a = 5$, $b = \frac{1}{4}$ **12.** $a = 1$, $b = 3$
13. $m = \frac{1}{2}$, $n = 4$ **14.** $w = 2$, $x = 3$ **15.** $x = 6$, $y = 3$ **16.** $x = \frac{1}{2}$, $z = -3$
17. $m = 1\frac{15}{17}$, $n = \frac{11}{17}$ **18.** $c = 1\frac{16}{23}$, $d = -2\frac{12}{23}$

page 105   **Exercise 28**

1. $x = 2$, $y = 4$ **2.** $x = 1$, $y = 4$ **3.** $x = 2$, $y = 5$ **4.** $x = 3$, $y = 7$
5. $x = 5$, $y = 2$ **6.** $a = 3$, $b = 1$ **7.** $x = -2$, $y = 3$ **8.** $x = 4$, $y = 1$

9. $x = \frac{5}{7}$, $y = 4\frac{3}{7}$　　　**10.** $x = 1$, $y = 2$　　　**11.** $x = 2$, $y = 3$　　　**12.** $x = 4$, $y = -1$

13. $x = 1$, $y = 2$　　　**14.** $a = 4$, $b = 3$　　　**15.** $x = 4$, $y = 3$　　　**16.** $x = 5$, $y = -2$

17. $x = 3$, $y = -1$　　　**18.** $x = 5$, $y = 0.2$

page 106　**Exercise 29**

1. $5\frac{1}{2}$, $9\frac{1}{2}$　　　　　　　　　**2.** 6, 3 or $2\frac{2}{5}$, $5\frac{2}{5}$　　　　　　　**3.** 4, 10

4. 10·5, 7·5　　　　　　　　**5.** $a = 2$, $c = 7$　　　　　　　**6.** $m = 4$, $c = -3$

7. $a = 30$, $b = 5$　　　　　**8.** TV £200, video £450　　　**9.** white 2 oz, brown $3\frac{1}{2}$ oz

10. 60 m, 80 m　　　　　　　**11.** 150 m, 350 m　　　　　　　**12.** 2p × 15, 5p × 25

13. 10p × 14, 50p × 7　　　**14.** 20　　　　　　　　　　　　**15.** current 4 m/s, kipper 10 m/s

16. $\frac{5}{7}$　　　　　　　　　　　　**17.** boy 10, mouse 3　　　　　　**18.** $a = 1$, $b = 2$, $c = 5$

19. $y = x^2 + 3x + 4$

page 109　**Exercise 30**

1. (a) $(-1, 3)$ and $(4, 8)$　　(b) $(3, 13)$ and $(5, 27)$　　**2.** (a) $(3, 7)$ and $(5, 17)$　　(b) $\left(\frac{1}{2}, \frac{1}{2}\right)$ and $(4, 32)$

3. $(2, 3)$ and $(-3, -2)$　　　　　　　**4.** $(4, 2)$ and $(-2, -4)$　　　　　　　**5.** $(4, 8)$ and $(1, -4)$

6. curve and line do not intersect　　　　　　**8.** $(2, 8)$

4 Shape, space and measures 1

page 111　**Exercise 1**

1. 70°　　　　　　　**2.** 70°　　　　　　　**3.** 48°　　　　　　　**4.** 40°, 140°　　　　**5.** 60°

6. 122°, 116°　　　　**7.** 135°　　　　　　**8.** 28°　　　　　　　**9.** 20°　　　　　　　**10.** 70°, 30°

11. 36°, 36°　　　　　**12.** 60°, 40°　　　　**13.** 72°　　　　　　**14.** 98°　　　　　　**15.** 80°

16. 95°, 50°　　　　　**17.** 87°, 74°　　　　**18.** 65°, 103°　　　**19.** 70°, 60°　　　　**20.** 65°

21. 46°　　　　　　　**24.** 136°　　　　　　**25.** 80°　　　　　　**26.** 66°　　　　　　**27.** $27\frac{1}{2}$°

page 113　**Exercise 2**

1. C(5, 1), D(0, 0)　　　　　　　　　　　　　　**3.** (a) 72°　　　　(b) 108°　　　　(c) 80°

4. (a) 40°　　　(b) 30°　　　(c) 110°　　　**5.** (a) 116°　　　(b) 32°　　　　(c) 58°

6. 53°　　　　　　　　　　　　　　　　　　　　**7.** (a) 26°　　　(b) 26°　　　　(c) 77°

8. 150°　　　　(b) 110°　　　　　　　　　　　**10.** (a) 54°　　　(b) 72°　　　　(c) 36°

9. 110°

11. (a) 60°　　　(b) 15°　　　(c) 75°

page 116　**Exercise 3**

1. (a) $a = 80°$, $b = 70°$, $c = 65°$, $d = 86°$, $e = 59°$　　**2.** (a) $a = 36°$　(b) 144°

3. (a) (i) 40°　(ii) 20°　(iii) 8°　(iv) 6°　　(b) (i) 140°　(ii) 160°　(iii) 172°　(iv) 174°

4. $p = 101°$, $q = 79°$, $x = 70°$, $m = 70°$, $n = 130°$　　**5.** 24　　**6.** 9　　**7.** 20

page 117　**Exercise 4**

1. (a) 1080°　　　(b) 1440°　　　　　　　　　**2.** (a) 3240°　　　(b) 162°

3. (a) 270°　　　(b) 100°　　　(c) $a = 119°$, $b = 25°$

4. (a) 150°　　　(b) 30°　　　　　　　　　　　**5.** 22　　　　　　　**6.** 20

page 119   **Exercise 5**

1. Yes, S.S.S.   **2.** Yes, S.A.S.   **3.** No   **4.** Yes, A.A.S.
5. No   **6.** Yes, A.A.S   **7.** ABD, DCA, DEB, EAC

page 127   **Exercise 8**

1. 10 cm   **2.** 4·1 cm   **3.** 10·6 cm   **4.** 5·7 cm   **5.** 4·2 cm   **6.** 9·9 cm   **7.** 4·6 cm
8. 5·2 cm   **9.** 9·8 cm   **10.** 9·8 cm   **11.** 3·5 m   **12.** 40·3 km   **13.** 12·7 cm   **14.** 5·7 cm
15. (a) 5 cm   (b) 40 cm^2 **16.** 9·5 cm   **17.** 32·6 cm   **18.** (a) $\sqrt{200}$ cm   (b) 14·3 cm^2

page 129   **Exercise 9**

1. $\sqrt{29}$ **2.** Yes   **3.** $\sqrt{44}$ **4.** $\sqrt{31}$ **5.** $\sqrt{76}$ **6.** $\sqrt{32}$ **7.** $\sqrt{44}$ **8.** $\sqrt{5}$
9. (a) 5 cm   (b) 7·8 cm   **10.** (a) 6·4 cm   (b) 13·6 cm
11. 6·3 m   **12.** 4·6 m   **13.** (a) 7·5   (b) 12·5   (c) 14·9
14. 24 cm   **15.** 10 feet   **16.** (a) 8·60   (b) 16·4
17. $x = 4$ m, 20·6 m   **18.** (a) 13, 25, 9   (d) 11, 60, 61   **19.** 29·5 cm

page 132   **Exercise 10**

1. 42 cm^2 **2.** 22 cm^2 **3.** 103 cm^2 **4.** 60·5 cm^2 **5.** 143 cm^2
6. 9 cm^2 **7.** 13 m   **8.** 15 cm   **9.** 2500
10. (cm^2) (a) 25   (b) 21   (c) 11·8   (d) 19·7   (e) 21   (f) 16·4
11. 24 cm^2 **12.** 40 cm^2 **13.** 32 cm^2 **14.** 46 cm^2 **15.** 47 cm^2 **16.** $81\frac{3}{4}$ cm^2

page 134   **Exercise 11**

1. 80 m^2 **2.** (a) 18 cm   (b) 8 cm^2
3. (a) 8·5 m   (b) 10 m   (c) 4 m   **4.** 45 cm^2
5. 2·4 cm   **6.** (a) $\frac{1}{3}$ (b) $\frac{4}{9}$ (c) 25 cm^2 **8.** 1100 m
9. 6 square units   **10.** 14 square units   **11.** 1849   **12.** 10 cm
13. (a) 60°   (b) 23·4 cm^2 **14.** 124 cm^2 **15.** 57·1 cm^2
16. 10·7 cm   **17.** 4·1 m   **18.** 4·9 m   **19.** 7·2 cm
20. (a) $\dfrac{360°}{n}$ (b) (i) $\frac{1}{2}\sin\left(\dfrac{360°}{n}\right)$ (ii) $\dfrac{n}{2}\sin\left(\dfrac{360°}{n}\right)$ (c) Area tends towards π
21. 18·7 cm

page 137   **Exercise 12**

1. 31·4 cm, 78·5 cm^2 **2.** 18·8 cm, 28·3 cm^2 **3.** 129 m, 982 m^2 **4.** 56·6 cm, 190 cm^2
5. 53·7 m, 198 m^2 **6.** 28·1 m, 54·7 m^2 **7.** 20·6 cm, 24·6 cm^2 **8.** 25·1 cm, 43·4 cm^2
9. 20·3 cm, 24·6 cm^2 **10.** 37·3 cm, 92·9 cm^2 **11.** 25·1 cm, 13·7 cm^2 **12.** 25·1 cm, 25·1 cm^2
13. 18·8 cm, 12·6 cm^2

page 138   **Exercise 13**

1. 2·19 cm   **2.** 31·8 cm   **3.** 2·65 km   **4.** 9·33 cm   **5.** 17·8 cm
6. 14·2 mm   **7.** 497 000 km^2 **8.** 21·5 cm^2 **9.** 30; (a) 1508 cm^2 (b) 508 cm^2
10. 5305   **11.** 29   **12.** (a) 40·8 m^2 (b) 6
13. (a) 80   (b) 7 **14.** 5·39 cm   **15.** 118 m^2 **16.** (a) 33·0 cm   (b) 70·9 cm^2
17. (a) 98 cm^2 (b) 14·0 cm^2 **18.** 796 m^2
19. (a) side = 4   (b) $r = 2$ (c) side = $4\sqrt{3}$ **20.** 72°   **21.** Yes   **22.** 1·72 cm

page 141 **Exercise 14**

1. 2 : 1   **2.** 112 cm^2 **3.** $\frac{7}{16}$ **4.** 0·586 m   **5.** 20%
6. 7·172 cm   **7.** 112$\frac{1}{2}$°   **8.** 60°   **9.** 55·4 cm^2 **10.** 8 cm
11. 4 : 1   **12.** (a) 2·41   (b) 1·85   (c) $1 + \sqrt{2}$

page 144 **Exercise 15**

1. (a) 2·1 cm   (b) 7·9 cm   (c) 8·2 cm   **2.** 15·4 cm
3. (a) 8·7 cm^2 (b) 40·6 cm^2 (c) 18·8 cm^2
4. (a) 7·1 cm^2 (b) 19·5 cm^2 **5.** 17·8 cm   **6.** 74·2 cm^3
7. 5·9 cm^2 **8.** (a) 4·01 cm   (b) 75°
9. (a) 12 cm   (b) 30°   **10.** (a) 30°   (b) 10·5 cm
11. (a) 85·9°   (b) 57·3°   (c) 6·3 cm   **12.** 30·6 cm^2 **13.** 57·3°
14. 36°   **15.** (a) 6·1 cm   (b) 27·6 m   (c) 28·6 cm^2 **16.** 1850 cm
17. (a) 18 cm   (b) 38·2°   **18.** (a) 10 cm   (b) 43·0°   **19.** (b) 66·8°   **20.** $x = 38·1°$

page 149 **Exercise 16**

1. (a) 14·5 cm   (b) 72·6 cm^2 (c) 24·5 cm^2 (d) 48·1 cm^2
2. (a) 5·08 cm^2 (b) 82·8 m^2 (c) 5·14 cm^2
3. (a) 60°, 9·06 cm^2 (b) 106·3°, 11·2 cm^2 **4.** 3 cm   **5.** 3·97 cm
6. (a) 13·5 cm^2 (b) 405 cm^3 **7.** (a) 130 cm^2 (b) 184 cm^2 **8.** 19·6 cm^2
9. 0·313 r^2 **10.** (a) 8·37 cm   (b) 54·5 cm   (c) 10·4 cm   **11.** 81·2 cm^2

page 151 **Exercise 17**

1. (a) 30 cm^3 (b) 168 cm^3 (c) 110 cm^3 (d) 94·5 cm^3 (e) 754 cm^3 (f) 283 cm^3
2. (a) 503 cm^3 (b) 760 m^3 **3.** 10 litres   **4.** 7·1 kg
5. 7·5 g/cm^3 **6.** 358 m^3 **7.** 3·98 cm
8. 6·37 cm   **9.** 1·89 cm   **10.** 9·77 cm   **11.** 7·38 cm
12. 1273 cm   **13.** 4·24 litres   **14.** 106 cm/s   **15.** 1570 cm^3, 12·57 kg
16. No   **17.** 1·19 cm   **18.** 53 times   **19.** 191 cm

page 154 **Exercise 18**

1. 40 cm^3 **2.** 33·5 cm^3 **3.** 66·0 cm^3 **4.** 89·8 cm^3 **5.** 339 cm^3
6. 144 cm^3 **7.** 4·71 kg   **8.** 262 cm^3 **9.** 359 cm^3 **10.** 235 cm^3
11. 5 m   **12.** 0·0036 mm   **13.** 10 balls   **14.** 415 cm^3 **15.** 106 s
16. 488 cm^3 **17.** 37$\frac{1}{2}$ million   **18.** 1·05 cm^3 **19.** 10$\frac{2}{3}$ cm^3 **20.** 1930 g

page 157 **Exercise 19**

1. (a) 3·91 cm   (b) 2·43 cm   (c) 7·16 cm   **2.** 6·45 cm   **3.** 23·9 cm
4. (a) 125   (b) 2·7 × 10^7 **5.** (a) 0·36 cm   (b) 0·427 cm
6. (a) 6·69 cm   (b) 39·1 cm   **7.** 4·19 cm^3 **8.** 53·6 cm^3
9. 74·4 cm^3 **10.** 123 cm^3 **11.** 54 500 cm^3 **12.** (a) 2·9 m^3 (b) 1·7 m
13. (a) 16π cm   (b) 8 cm   (c) 6 cm   **14.** 2720 cm^3

page 159 **Exercise 20**

1. (All cm^2) (a) 36π (b) 40π (c) 60π (d) 1.4π
2. (a) 40π (b) 28π (c) 48π (d) 24π
3. £3870   4. £331   5. 303 cm^2 6. 675 cm^2 7. 1.62×10^8 years
8. (a) 1.59 cm   (b) 4.77 *cm*   9. 1.64 cm   10. 2.12 cm   11. 3.46 cm
12. 94.0 cm^3 13. 44.6 cm^2 14. 377 cm^2 15. $l = 20$ cm, $r = 10$ cm
16. (a) 3.72 cm   (b) 41.9 cm^2 17. 123 cm^2 18. 298 cm^3

page 162 **Exercise 21**

1. C only   2. $m = 12$ 3. $x = 9$ 4. $a = 2.5$, $e = 3$ 5. $x = 6\frac{3}{4}$
6. 3.2 cm   7. $t = 5.25$, $y = 5.6$ 8. 7.7 cm   9. No
10. (a) Yes   (b) No   (c) No   (d) Yes   (e) Yes   (f) No   (g) No   (h) Yes
11. (b) 11.2 cm   (c) 4.2 cm   12. 6   13. 6   14. $4\frac{1}{2}$

page 164 **Exercise 22**

1. 16 m   2. 3.75 m   3. 10.8 m   4. 2, 6   5. $m = 6$, $n = 6$
6. $v = 5\frac{1}{3}$, $w = 6\frac{2}{3}$ 7. $3\frac{2}{3}$, $1\frac{1}{11}$ 8. 0.618; 1.618 : 1   9. 5

page 167 **Exercise 23**

1. 16 cm^2 2. 27 cm^2 3. $11\frac{1}{4}$ cm^2 4. $14\frac{1}{2}$ cm^2 5. 128 cm^2
6. 12 cm^2 7. 8 cm   8. 18 cm   9. $4\frac{1}{2}$ cm   10. $7\frac{1}{2}$ cm
11. 9, 12, 15   12. (a) 500 000 cm^2 (b) 50 m^2 13. 270 min   14. 150
15. 360   16. $5r$ 17. (a) $16\frac{2}{3}$ cm^2 (b) $10\frac{2}{3}$ cm^2
18. (a) 25 cm^2 (b) 21 cm^2 19. 24 cm^2 20. Less (for the same mass)
21. 6.29 cm   22. $\sqrt{2}$ 23. $\frac{4}{9}$

page 170 **Exercise 24**

1. 480 cm^3 2. 540 cm^3 3. 160 cm^3 4. 4500 cm^3 5. 81 cm^3
6. 11 cm^3 7. 16 cm^3 8. $85\frac{1}{3}$ cm^3 9. 4 cm   10. 21 cm
11. 4.6 cm   12. 9 cm   13. 6.6 cm   14. $4\frac{1}{2}$ cm   15. $168\frac{3}{4}$ cm^3
16. 106.3 cm^3 17. 12 cm   18. (a) 2 : 3   (b) 8 : 27   19. 8 : 125
20. 60 cm   21. £12.80   22. 21 m, 62.5 m^3, 700 cm^2, 12, 92.5 m^2 23. 54 kg
24. 240 cm^2 25. $9\frac{3}{8}$ litres   26. $2812\frac{1}{2}$ cm^2 27. 100 g

5 Algebra 2

page 173 **Exercise 1**

1. $\dfrac{B}{A}$ 2. $\dfrac{T}{N}$ 3. $\dfrac{K}{M}$ 4. $\dfrac{4}{y}$ 5. $\dfrac{T+N}{9}$
6. $\dfrac{B-R}{A}$ 7. $\dfrac{R+T}{C}$ 8. $\dfrac{N-R^2}{L}$ 9. $\dfrac{R-S^2}{N}$ 10. 2
11. $S - B$ 12. $N - D$ 13. $T - N^2$ 14. $N + M - L$ 15. $A + R$
16. $E + A$ 17. $F + B$ 18. $F^2 + B^2$ 19. $L + B$ 20. $N + T$

21. 2

22. $4\frac{1}{2}$

23. $\dfrac{N-C}{A}$

24. $\dfrac{L-D}{B}$

25. $\dfrac{F-E}{D}$

26. $\dfrac{H+F}{N}$

27. $\dfrac{Q-m}{V}$

28. $\dfrac{n+a+m}{t}$

29. $\dfrac{c-b}{V^2}$

30. $\dfrac{r+6}{n}$

31. $2\frac{2}{3}$

32. $\dfrac{C-AB}{A}$

33. $\dfrac{F-DE}{D}$

34. $\dfrac{a-hn}{h}$

35. $\dfrac{q+bd}{b}$

36. $\dfrac{n-rt}{r}$

page 174 **Exercise 2**

1. 12

2. 10

3. BD

4. TB

5. RN

6. bm

7. 26

8. $BT+A$

9. $AN+D$

10. B^2N-Q

11. $ge+r$

12. $4\frac{1}{2}$

13. $\dfrac{DC-B}{A}$

14. $\dfrac{pq-m}{n}$

15. $\dfrac{vS+t}{r}$

16. $\dfrac{qt+m}{z}$

17. $\dfrac{bc-m}{A}$

18. $\dfrac{AE-D}{B}$

19. $\dfrac{nh+f}{e}$

20. $\dfrac{qr-b}{g}$

21. 4

22. -2

23. 2

24. $A-B$

25. $C-E$

26. $D-H$

27. $n-m$

28. $q-t$

29. $s-b$

30. $r-v$

31. $m-t$

32. 2

33. $\dfrac{T-B}{X}$

34. $\dfrac{M-Q}{N}$

35. $\dfrac{V-T}{M}$

36. $\dfrac{N-L}{R}$

37. $\dfrac{v^2-r}{r}$

38. $\dfrac{w-t^2}{n}$

39. $\dfrac{n-2}{q}$

40. $\frac{1}{4}$

41. $-\frac{1}{7}$

42. $\dfrac{B-DE}{A}$

43. $\dfrac{D-NB}{E}$

44. $\dfrac{h-bx}{f}$

45. $\dfrac{v^2-Cd}{h}$

46. $\dfrac{NT-MB}{M}$

47. $\dfrac{mB+ef}{fN}$

48. $\dfrac{TM-EF}{T}$

49. $\dfrac{yx-zt}{y}$

50. $\dfrac{k^2m-x^2}{k^2}$

page 175 **Exercise 3**

1. $\frac{1}{2}$

2. $1\frac{2}{3}$

3. $\dfrac{B}{C}$

4. $\dfrac{T}{X}$

5. $\dfrac{v}{t}$

6. $\dfrac{n}{\sin 20°}$

7. $\dfrac{7}{\cos 30°}$

8. $\dfrac{B}{x}$

9. $\dfrac{vs}{m}$

10. $\dfrac{mb}{t}$

11. $\dfrac{B-DC}{C}$

12. $\dfrac{Q+TC}{T}$

13. $\dfrac{V+TD}{D}$

14. $\dfrac{L}{MB}$

15. $\dfrac{N}{BC}$

16. $\dfrac{m}{cd}$

17. $\dfrac{tc-b}{t}$

18. $\dfrac{xy-z}{x}$

19. 1

20. $\frac{5}{6}$

21. $\dfrac{A}{C-B}$

22. $\dfrac{V}{H-G}$

23. $\dfrac{r}{n+t}$

24. $\dfrac{b}{q-d}$

25. $\dfrac{m}{t+n}$

26. $\dfrac{b}{d-h}$

27. $\dfrac{d}{C-e}$

28. $\dfrac{m}{r-e^2}$

29. $\dfrac{n}{b-t^2}$

30. $\dfrac{d}{mn-b}$

31. $\dfrac{N-2MP}{2M}$

32. $\dfrac{B-6Ac}{6A}$

33. $\dfrac{m^2}{n-p}$

34. $\dfrac{q}{w-t}$

page 175 **Exercise 4**

1. 4

2. 11

3. $D^2 - C$

4. $\dfrac{c^2 - b}{a}$

5. $\dfrac{b^2 + t}{g}$

6. $d - t^2$

7. $n - c^2$

8. $c - g^2$

9. $\dfrac{D - B}{A}$

10. $\pm\sqrt{g}$

11. $\pm\sqrt{B}$

12. $\pm\sqrt{(M + A)}$

13. $\pm\sqrt{(C - m)}$

14. $\pm\sqrt{\dfrac{n}{m}}$

15. $\dfrac{at}{z}$

16. $\pm\sqrt{(a - n)}$

17. $\pm\sqrt{(B^2 + A)}$

18. $\pm\sqrt{(t^2 - m)}$

19. $\dfrac{M^2 - A^2 B}{A^2}$

20. $\dfrac{N}{B^2}$

21. $\pm\sqrt{(a^2 - t^2)}$

22. $\dfrac{4}{\pi^2} - t$

23. $\pm\sqrt{\left(\dfrac{C^2 + b}{a}\right)}$

24. $\pm\sqrt{(x^2 - b)}$

page 176 **Exercise 5**

1. $3\frac{2}{3}$

2. 3

3. $\dfrac{D - B}{2N}$

4. $\dfrac{E + D}{3M}$

5. $\dfrac{2b}{a - b}$

6. $\dfrac{e + c}{m + n}$

7. $\dfrac{3}{x + k}$

8. $\dfrac{C - D}{R - T}$

9. $\dfrac{z + x}{a - b}$

10. $\dfrac{nb - ma}{m - n}$

11. $\dfrac{d + xb}{x - 1}$

12. $\dfrac{a - ab}{b + 1}$

13. $\dfrac{d - c}{d + c}$

14. $\dfrac{M(b - a)}{b + a}$

15. $\dfrac{n^2 - mn}{m + n}$

16. $\dfrac{m^2 + 5}{2 - m}$

17. $\dfrac{2 + n^2}{n - 1}$

18. $\dfrac{e - b^2}{b - a}$

19. $\dfrac{3x}{a + x}$

20. $\dfrac{e - c}{a - d}$

21. $\dfrac{d}{a - b - c}$

22. $\dfrac{ab}{m + n - a}$

23. $\dfrac{s - t}{b - a}$

24. $2x$

25. $\dfrac{v}{3}$

26. $\dfrac{a(b + c)}{b - 2a}$

27. $\dfrac{5x}{3}$

28. $-\dfrac{4z}{5}$

29. $\dfrac{mn}{p^2 - m}$

30. $\dfrac{mn + n}{4 + m}$

page 176 **Exercise 6**

1. $-\left(\dfrac{by + c}{a}\right)$

2. $\pm\sqrt{\left(\dfrac{e^2 + ab}{a}\right)}$

3. $\dfrac{n^2}{m^2} + m$

4. $\dfrac{a - b}{1 + b}$

5. $3y$

6. $\dfrac{a}{e^2 + c}$

7. $-\left(\dfrac{a + lm}{m}\right)$

8. $\dfrac{t^2 g}{4\pi^2}$

9. $\dfrac{4\pi^2 d}{t^2}$

10. $\pm\sqrt{\dfrac{a}{3}}$

11. $\pm\sqrt{\left(\dfrac{t^2 e - ba}{b}\right)}$

12. $\dfrac{1}{a^2 - 1}$

13. $\dfrac{a + b}{x}$

14. $\pm\sqrt{(x^4 - b^2)}$

15. $\dfrac{c - a}{b}$

16. $\dfrac{a^2 - b}{a + 1}$

17. $\pm\sqrt{\left(\dfrac{G^2}{16\pi^2} - T^2\right)}$

18. $-\left(\dfrac{ax + c}{b}\right)$

19. $\dfrac{1 + x^2}{1 - x^2}$

20. $\pm\sqrt{\left(\dfrac{a^2 m}{b^2} + n\right)}$

21. $\dfrac{P - M}{E}$

22. $\dfrac{RP - Q}{R}$

23. $\dfrac{z - t^2}{x}$

24. $(g - e)^2 - f$

page 178 **Exercise 7**

1. (a) $3 < 7$ (b) $0 > -2$ (c) $3 \cdot 1 > 3 \cdot 01$ (d) $-3 > -5$ (e) $100\,\text{mm} < 1\,\text{m}$ (f) $1\,\text{kg} > 1\,\text{lb}$

2. (a) $x > 2$ (b) $x \leqslant 5$ (c) $x < 100$ (d) $-2 \leqslant x \leqslant 2$ (e) $x > -6$ (f) $3 < x \leqslant 8$

3. (a)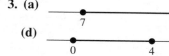

(b)

(c)

(d)

(e)

4. (a) $A \geqslant 16$ (b) $3 < A \leqslant 70$ (c) $150 \leqslant T \leqslant 175$ (d) $h \geqslant 1.75\,\text{m}$
5. (a) True   (b) True   (c) True
7. $3 \leqslant x < 5$

page 179 *Exercise 8*

1. $x > 13$ **2.** $x < -1$ **3.** $x < 12$ **4.** $x \leqslant 2\frac{1}{2}$

5. $x > 3$ **6.** $x \geqslant 8$ **7.** $x < \frac{1}{4}$ **8.** $x \geqslant -3$

9. $x < -8$ **10.** $x < 4$ **11.** $x > -9$ **12.** $x < 8$

13. $x > 3$ **14.** $x \geqslant 1$ **15.** $x < 1$ **16.** $x > 2\frac{1}{3}$

17. $x < -3$ **18.** $x > 7\frac{1}{2}$ **19.** $x > 0$ **20.** $x < 0$

21. $x > 5$ **22.** $5 \leqslant x \leqslant 9$ **23.** $-1 < x < 4$ **24.** $\frac{11}{2} \leqslant x \leqslant 6$

25. $\frac{4}{3} < x < 8$ **26.** $-8 < x < 2$ **27.** $\frac{1}{4} < x < \frac{6}{5}$ **28.** 24

page 180 *Exercise 9*

1. $x > 8$ **2.** 1, 2, 3, 4, 5, 6   **3.** 7, 11, 13, 17, 19
4. 4, 9, 16, 25, 36, 49   **5.** $-4, -3, -2, -1$ **6.** 2, 3, 4, ... 12
7. 2, 3, 5, 7, 11   **8.** 2, 4, 6, ... 18   **9.** 1, 2, 3, 4
10. 5   **11.** 16, -16, 20, -5 **12.** $>$
13. $\frac{1}{2}$ (or others)   **14.** 19   **15.** 17
16. $x > 3\frac{2}{3}$ **17.** (b) (i) $-10 < x < 10$ (ii) $-9 < x < 9$ (iii) $x > 6$ or $x < -6$
18. $-5 < x < 5$ **19.** $-4 \leqslant x \leqslant 4$ **20.** $x > 1, x < -1$
21. $x \geqslant 6, x \leqslant -6$ **22.** all values except zero   **23.** $-2 < x < 2$
24. (a) $2 \leqslant pq \leqslant 40$ (b) $\frac{1}{2} \leqslant \dfrac{p}{q} \leqslant 10$ (c) $-2 \leqslant p - q \leqslant 9$ (d) $3 \leqslant p + q \leqslant 14$
25. 7   **26.** 5   **27.** 6
28. 3, 4, 5   **29.** $30° < x < 90°$ **30.** $0 < x < 75.5°$

page 182 *Exercise 10*

1. $x \geqslant 3$ **2.** $y \leqslant 2\frac{1}{2}$ **3.** $1 \leqslant x \leqslant 6$
4. $x < 7, y < 5$ **5.** $y \geqslant x$ **6.** $x + y \leqslant 10$
7. $2x - y \leqslant 3$ **8.** $y \leqslant x, x \leqslant 8, y \geqslant -2$
9. (a) $x + y \leqslant 7, x \geqslant 0, y \geqslant x$ (b) $x + y \leqslant 6, y \geqslant 0, y \leqslant x + 2$
10. 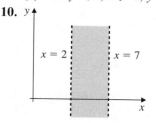   **11.** 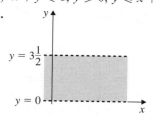   **12.**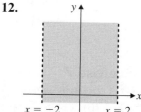

13.

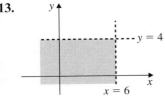

14.

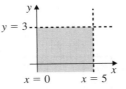

15.

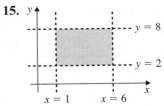

16.

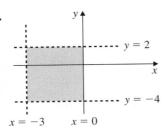

17.

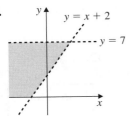

18.

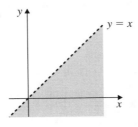

19.

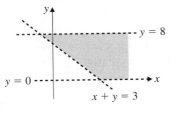

20.

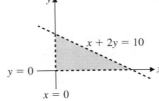

21.

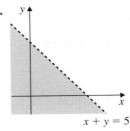

22.

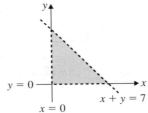

23.

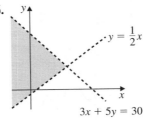

24.

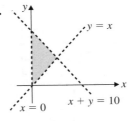

25.

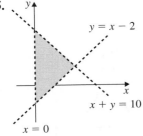

26.

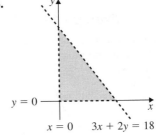

27.

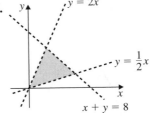

28. A: $x + y \leqslant 5$, $y \geqslant x + 1$ B: $x + y \leqslant 5$, $y \leqslant x + 1$
      C: $x + y \geqslant 5$, $y \leqslant x + 1$ D: $x + y \geqslant 5$, $y \geqslant x + 1$

29. (2, 6), (3, 5), (3, 4), (4, 4), (4, 3), (5, 3), (6, 2)

page 186 **Exercise 11**

1. (a) $S = ke$ (b) $v = kt$ (c) $x = kz^2$ (d) $y = k\sqrt{x}$ (e) $T = k\sqrt{L}$
2. (a) 9   (b) $2\frac{2}{3}$ 3. (a) 35   (b) 11   4. (a) 75   (b) 4

5.
x	1	3	4	$5\frac{1}{2}$
z	4	12	16	22

6.
r	1	2	4	$1\frac{1}{2}$
V	4	32	256	$13\frac{1}{2}$

7. 333 N/cm^2 8. 180 m; 2 s   9. 675 J; $\sqrt{\frac{4}{3}}$ cm   10. 9000 N; 25 m/s
11. $p \propto w^3$ 12. $15^4 : 1$ (50 625 : 1)

page 188 **Exercise 12**

1. (a) $x = \dfrac{k}{y}$ (b) $s = \dfrac{k}{t^2}$ (c) $t = \dfrac{k}{\sqrt{q}}$ (d) $m = \dfrac{k}{w}$ (e) $z = \dfrac{k}{t^2}$
2. (a) 6   (b) $\frac{1}{2}$ 3. (a) 6   (b) $1\frac{1}{2}$ 4. (a) 1   (b) 4
5. (a) 36   (b) ± 4 6. (a) 6   (b) 16

7.
y	2	4	1	$\frac{1}{4}$
z	8	4	16	64

8.
t	2	5	20	10
v	25	4	$\frac{1}{4}$	1

9. (a) 6 (b) 50   10. 2·5 m^3; 200 N/m^2 11. 3 h; 48 men   12. 6 cm
13. 80   14. 2 days; 200 days   15. (a) 20 min   (b) 2 min
16. $k = 100,$ $n = 3$

x	1	2	4	10
z	100	$12\frac{1}{2}$	1·5625	$\frac{1}{10}$

17. $k = 12$ $n = 2$

v	1	4	36	10000
y	12	6	2	$\frac{3}{25}$

page 191 **Exercise 13**

1. (a) quadratic, negative x^2 (b) cubic, positive x^3 (c) reciprocal
   (d) cubic, negative x^3 (e) quadratic, positive x^2 (f) exponential

2. (a) 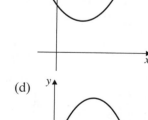   (b) 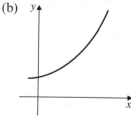   (c)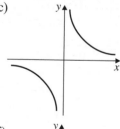

   (d) 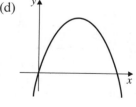   (e) 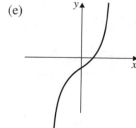   (f)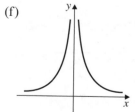

3. (i) c   (ii) b   (iii) a   (iv) e   (v) f   (vi) d
4. (a) 1   (b) 2   (c) 1   (d) 3

page 193 **Exercise 14**

Tables for graphs are given.

1.

x	−3	−2	−1	0	1	2	3
y	3	0	−1	0	3	8	15

2.

x	−3	−2	−1	0	1	2	3
y	−3	−4	−3	0	5	12	21

3.

x	−3	−2	−1	0	1	2	3
y	18	10	4	0	−2	−2	0

4.

x	−3	−2	−1	0	1	2	3
y	11	6	3	2	3	6	11

5.

x	−3	−2	−1	0	1	2	3
y	2	−3	−6	−7	−6	−3	2

6.

x	−3	−2	−1	0	1	2	3
y	4	0	−2	−2	0	4	10

7.

x	−4	−3	−2	−1	0	1	2	3
y	−5	−9	−11	−11	−9	−5	1	9

8.

x	−2	−1	0	1	2	3	4
y	6	0	−4	−6	−6	−4	0

9.

x	0	1	2	3	4	5	6
y	7	3	1	1	3	7	13

10.

x	−1	0	1	2	3	4	5
y	8	0	−4	−4	0	8	20

11.

x	−4	−3	−2	−1	0	1	2
y	14	3	−4	−7	−6	−1	8

12.

x	−1	0	1	2	3
y	14	5	2	5	14

13.

x	−3	−2	−1	0	1	2	3
y	−10	−4	0	2	2	0	−4

14.

x	−5	−4	−3	−2	−1	0	1	2
y	−9	−3	1	3	3	1	−3	−9

15.

x	−2	−1	0	1	2	3	4	5
y	−7	−1	3	5	5	3	−1	−7

16.

x	−3	−2	−1	0	1	2	3
y	−2	5	8	7	2	−7	−20

17.

x	−3	−2	−1	0	1	2	3
y	−15	−4	3	6	5	0	−9

18.

x	−2	−1	0	1	2	3
y	−8	3	8	7	0	−13

19.

x	−1	0	1	2	3	4	5	6
y	5	0	−3	−4	−3	0	5	12

20.

x	−3	−2	−1	0	1	2	3
y	22	7	0	−5	−6	−3	4

page 194 **Exercise 15**

For questions 1–9 some points on the curves are given.

1. (1, 12), (2, 6), (3, 4), (4, 3), (6, 2), (10, 1·2) **2.** (1, 9), (2, $4\frac{1}{2}$), (3, 3), (6, $1\frac{1}{2}$), (9, 1)

3. (0, 12), (1, 6), (2, 4), (3, 3), (5, 2), (8, $1\frac{1}{2}$)

4. (−4, −1), (−2, −$1\frac{1}{3}$), (0, −2), (2, −4), (3, −8), ($3\frac{1}{2}$, −16)

5. (−$3\frac{1}{2}$, −7), (−3, −3), (−2, −1), (0, 0), (4, $\frac{1}{2}$) **6.** (0, 8), (1, $4\frac{1}{2}$), (6, 2), (8, $1\frac{7}{9}$)

7. (1, 11), (2, 7), (4, $6\frac{1}{2}$), (5, 7), (6, $7\frac{2}{3}$)

8. (−3, $\frac{1}{27}$), (−2, $\frac{1}{9}$), (−1, $\frac{1}{3}$), (0, 1), (1, 3), (2, 8), (3, 27)

9. (−4, 16), (−3, 8), (−2, 4), (−1, 1), (0, 1), (1, $\frac{1}{2}$), (2, $\frac{1}{4}$), (3, $\frac{1}{8}$)

10. (a) $7\frac{1}{4}$ (b) 3·8 and 0·8 **11.** (a) 0·75 (b) 1·2

12. (a) 3·1 (b) 3·3 **13.** (a) −2·5 (b) 1·85

14. points on curve:

  (−6, −0·16), (−4, −0·24), (−3, −0·03), (−1, −$\frac{1}{2}$), (0, 0), (1, $\frac{1}{2}$), (2, 0·4), (4, 0·24), (6, 0·16)

15. (a) 245 (b) 40·8 (c) $25 < x < 67$

page 196 **Exercise 16**

1. (a) 10·7 cm^2 (b) 1·7 cm × 5·3 cm (c) 12·25 cm^2 (d) 3·5 cm × 3·5 cm (e) square

2. 15 m × 30 m **3.** (a) 2·5 s (b) 31·3 m (c) $2 < t < 3$

4. (a) 108 m/s (b) 1·4 s (c) $2·3 < t < 3·6$

5.

x	−4	−3	−2	−1	−$\frac{1}{2}$	−$\frac{1}{4}$	$\frac{1}{4}$	$\frac{1}{2}$	1	2	3	4
y	−$\frac{1}{4}$	−$\frac{1}{3}$	−$\frac{1}{2}$	−1	−2	−4	4	2	1	$\frac{1}{2}$	$\frac{1}{3}$	$\frac{1}{4}$

6.

x	−4	−3	−2	−1	−$\frac{1}{2}$	−$\frac{1}{4}$	$\frac{1}{4}$	$\frac{1}{2}$	1	2	3	4
y	−$4\frac{1}{4}$	−$3\frac{1}{3}$	−$2\frac{1}{2}$	−2	−$2\frac{1}{2}$	−$4\frac{1}{4}$	$4\frac{1}{4}$	$2\frac{1}{2}$	2	$2\frac{1}{2}$	$3\frac{1}{3}$	$4\frac{1}{4}$

7.

x	−4	−3	−2	−1	−$\frac{1}{2}$	$\frac{1}{2}$	1	2	3	4	5	6	7
y	−$\frac{1}{64}$	−$\frac{1}{24}$	−$\frac{1}{8}$	−$\frac{1}{2}$	−1·4	2·8	2	2	$2\frac{2}{3}$	4	6·4	$10\frac{2}{3}$	18·3

8. (a) 2000 (b) 270 (c) $1·6 \leqslant x \leqslant 2·4$

9. 96 min **10.** (b) After 76/77 years i.e. 2056/57 **11.** (a) 0, 2·9 (b) −0·65, 1·35, 5·3

page 200 **Exercise 17**

1. (a) −0·4, 2·4 (b) −0·8, 3·8 (c) −1, 3 (d) −0·4, 2·4

2. (a) −2·6, 1·6 (b) 0·75, 2·75 (c) −2, 2 (d) 0·7

3. −0·3, 3·3 **4.** 0·6, 3·4 **5.** 0·3, 3·7 **6.** (c) (i) 1, 4 (ii) 0·3, 3·7

7. (a) $y = 3$ (b) $y = -2$ (c) $y = x + 4$ (d) $y = x$ (e) $y = 6$

8. (a) $y = 6$ (b) $y = 0$ (c) $y = 4$ (d) $y = 2x$ (e) $y = 2x + 4$

9. (a) $y = -4$ (b) $y = 2x$ (c) $y = x - 2$ (d) $y = -3$ (e) $y = 2$

10. (b) (i) 0·3, −3·3 (ii) 1·5, −4·5 (iii) −3, 1

11. (a) −1·65, 3·65 (b) −1·3, 2·3 (c) −1·45, 3·45 **12.** (a) 1·7, 5·3 (b) 0·2, 4·8

13. (a) −2·35, 0·85 (b) −2·7, 1·7 **14.** (a) 3·35 (b) 2·4, 7·6 (c) 4·25

15. (a) ±3·74 (c) ±2·83 **16.** (a) 1 (b) 2 (c) 2 (d) 1 (e) 2 (f) 3

17. $0 < k < 4$ **18.** (a) 1·75 (b) 0, ±1·4

19. (a) 2·6 (b) 0·45, 3·3 (c) 0·64; $2^{2·5} = 5·66$

page 204 **Exercise 18**

1. 3^4
2. $4^2 \times 5^3$
3. 3×7^3
4. $2^3 \times 7$
5. 10^{-3}
6. $2^{-2} \times 3^{-3}$
7. $15^{\frac{1}{2}}$
8. $3^{\frac{1}{3}}$
9. a^8
9. $10^{\frac{1}{3}}$
11. 11^{-1}
12. $5^{\frac{3}{2}}$
13. T
14. F
15. T
16. T
17. F
18. F
19. F
20. T
21. T
22. T
23. F
24. F
25. F
26. T
27. T
28. T
29. T
30. T
31. T
32. F
33. 5^6
34. 6^5
35. 2^{13}
36. 3^4
37. 6^5
38. 8
39. 6^4
40. 2^5
41. 2^{19}
42. 10^{100}
43. 5^{-5}
44. 3^{-10}
45. 3^7
46. 2^7
47. 7^2
48. 5^{-1}

page 205 **Exercise 19**

1. 3^6
2. 5^{12}
3. 7^{10}
4. x^6
5. 2^{-2}
6. 1
7. 7^2
8. y^2
9. 2^{21}
10. 10^{99}
11. x^7
12. y^{13}
13. z^4
14. z^{100}
15. m
16. e^{-5}
17. y^2
18. w^6
19. y
20. x^{10}
21. 1
22. w^{-5}
23. w^{-5}
24. x^7
25. a^8
26. k^3
27. 1
28. x^{29}
29. y^2
30. x^6
31. z^4
32. t^{-4}
33. $4x^6$
34. $16y^{10}$
35. $6x^4$
36. $10y^5$
37. $15a^4$
38. $8a^3$
39. 3
40. $4y^2$
41. $\frac{5}{2}y$
42. $32a^4$
43. $108x^5$
44. $4z^{-3}$
45. $2x^{-4}$
46. $\frac{5}{2}y^5$
47. 1
48. $21w^{-3}$
49. $2n^4$
50. $2x$

page 206 **Exercise 20**

1. 27
2. 1
3. $\frac{1}{9}$
4. 25
5. 2
6. 4
7. 9
8. 2
9. 27
10. 3
11. $\frac{1}{3}$
12. $\frac{1}{2}$
13. 1
14. $\frac{1}{5}$
15. 10
16. 8
17. 32
18. 4
20. (a) $\frac{1}{2}$
   (b) $-\frac{1}{2}$
   (c) $\frac{3}{2}$
21. (a) $\frac{1}{9}$
   (b) $\frac{1}{8}$
   (c) 100^6
22. 10
23. 1000
24. $\frac{1}{1000}$
25. $\frac{1}{9}$
26. 1
27. $1\frac{1}{2}$
28. $\frac{1}{25}$
29. $\frac{1}{10}$
30. $\frac{1}{4}$
31. $\frac{1}{4}$
32. 100 000
33. 1
34. $\frac{1}{32}$
35. 0·1
36. 0·2
37. 1·5
38. 1
39. 9
40. $1\frac{1}{2}$
41. $\frac{3}{10}$
42. 64
43. $\frac{1}{100}$
44. $1\frac{2}{3}$
45. $\frac{1}{100}$
46. (a) $\dfrac{1}{n}$

page 207 **Exercise 21**

1. 3
2. 4
3. -1
4. -2
5. 3
6. 3
7. 1
8. $\frac{1}{5}$
9. 0
10. -4
11. 2
12. -5
13. 1
14. $\frac{1}{18}$
15. (a) $(0, 1)$
   (b) $x = 0$
16. (a) -3
   (b) 0   (c) 5
17. (a) $2\frac{1}{2}$
   (b) 1000
   (c) $\frac{7}{4}$
18. (a) 4
   (b) -1
   (c) 0
19. 2, 4
20. (a) 3·60
   (b) 5·44
21. (b) 1, 7, 9
22. 21
23. (a) 512
   (b) 6 h
   (c) 2^{21}
24. (a) 976   (b) 20
25. (a) £2519·42
   (b) 9 years
26. $x^6 = 1$; 2, -2

6 Shape, space and measures 2

page 212 **Exercise 2**

1. (a), (b) and (d)
5. (a) prism, triangle cross-section; 6 vertices, 5 faces
  (b) square-based pyramid; 5 vertices, 5 faces
7. (a) 6 (b) 4 (c) $a = 10, b = 6, c = 10, d = 10$ (d) $64 \, \text{cm}^3$
8. $a = \sqrt{2}, b = \sqrt{2}, c = \sqrt{3}, x = \sqrt{2}, y = \sqrt{3}$

page 214 **Exercise 3**

1. (a) 1 (b) none **2.** (a) 1 (b) none **3.** (a) 4 (b) 4
4. (a) 2 (b) 2 **5.** (a) 0 (b) 6 **6.** (a) 0 (b) 2
7. (a) 0 (b) 2 **8.** (a) 4 (b) 4 **9.** (a) 0 (b) 4
10. (a) 4 (b) 4 **11.** (a) 6 (b) 6 **12.** (a) infinite (b) infinite

page 215 **Exercise 4**

1. 3 **2.** (a) 1 (b) 1 (c) 2 **3.** 9 **5.** 4

page 217 **Exercise 5**

1. 4·54 **2.** 3·50 **3.** 3·71 **4.** 6·62 **5.** 8·01 **6.** 31·9
7. 45·4 **8.** 4·34 **9.** 17·1 **10.** 13·2 **11.** 38·1 **12.** 3·15
13. 516 **14.** 79·1 **15.** 5·84 **16.** 2·56 **17.** 18·3 **18.** 8·65
19. 11·9 **20.** 10·6 **21.** 5, 5·55 **22.** 13·1, 27·8 **23.** 41·3 **24.** 8·82
25. 20·4 **26.** 26·2

page 219 **Exercise 6**

1. 36·9° **2.** 44·4° **3.** 48·2° **4.** 60° **5.** 36·9°
6. 50·2° **7.** 29·0° **8.** 56·4° **9.** 38·9° **10.** 43·9°
11. 41·8° **12.** 39·3° **13.** 60·3° **14.** 50·5° **15.** 13·6°
16. 34·8° **17.** 60·0° **18.** 42·0° **19.** 36·9° **20.** 51·3°
21. 19·6° **22.** 17·9° **23.** 32·5° **24.** 59·6° **25.** 17·8°

page 221 **Exercise 7**

1. 19·5° **2.** 4·12 m **3.** 7° **4.** 91·2° **5.** 1·29 m → 2·11 m
6. 36·4° **7.** 10·3 cm **8.** 2·6 m **9.** (a) 26·0 km (b) 23·4 km
10. (a) 88·6 km (b) 179·3 km **11.** (a) 484 km (b) 858 km (c) 986 km, 060·6°
12. 954 km, 133° **13.** 71° **14.** 67·1 m **15.** 76·5 m/s
16. 83·2 km **17.** 60° **18.** No **19.** 72°, 8·2 cm

page 223 **Exercise 8**

1. (a) 3·9 m (b) 8·6 m **2.** 56·3° **3.** 35·5° **4.** 71·6° **5.** 91·8° **6.** 180 m
7. 11·1 m, 11·1 s, 222 m **8.** 2·9 m **9.** 4·4 m **10.** $(\sin A)^2 + (\cos A)^2 = 1$

11. (a) $\dfrac{360°}{n}$ (b) $\dfrac{180°}{n}$ (c) $AB = 2 \sin\left(\dfrac{180}{n}\right)$ (d) $2n \sin\left(\dfrac{180}{n}\right)$ (e) perimeter → 2π

12. 313 m **13.** 25·6 cm **14.** 27·1 cm

page 226 **Exercise 9**

1. (a) 13 cm   (b) 13·6 cm   (c) 17·1°
2. (a) 4·04 m   (b) 38·9°   (c) 11·2 m   (d) 19·9°
3. (a) 8·49 cm   (b) 8·49 cm   (c) 10·4 cm   (d) 35·3°
4. (a) 14·1 cm   (b) 18·7 cm   (c) 69·3°
5. (a) 11·3 cm, 12·4 cm   (b) 23·8°   **6.** (a) 10 cm, 9·43 cm   (b) 26·6°   (c) 32·5°
7. (a) 4·47 m   (b) 7·48 m   (c) 15·5°   **8.** (a) 58·0°
9. (a) 57·5°   (b) 61·0°   **10.** (a) $h \tan 65°$ (b) $h \tan 57°$ (c) 22·7 m
11 22·6 m   **12.** 55·0 m

page 230 **Exercise 10**

1. (a) L   (b) V   (c) A   (d) L   (e) V   (f) A
2. (a) 2   (b) 2   (c) 3   (d) 1   (e) 2   (f) 3
3. (a) 2   (b) 3   (c) 2   (d) 1   (e) 3   (f) 1   (g) 0   (h) 2
4. (a) $wh + \dfrac{\pi}{6}wh$ (b) $2h + w + \frac{5}{4}w$
5. (a) A   (b) L   (c) V   (d) A   (e) F   (f) A   (g) V   (h) A
6. (a) L   (b) I   (c) V   (d) A   (e) V   (f) A
   (g) I   (h) V   (i) L   (j) L   (k) a number   (l) I
7. 2   **8.** (a) 2   (b) 3   (c) 1   (d) 2, 1   (e) 3, 3   (f) 2
9. $\dfrac{\pi}{2}a$ is a length, not an area.

page 232 **Exercise 11**

3. (c) (8, 8) (8, −6) (−8, 6)   **4.** (f) (3, 5) (7, 5) (7, 7)
5. (f) (8, −2) (6, −6) (8, −6)   **6.** (c) (−2, 1) (−2, −1) (1, −2)
7. (e) (−5, 2) (−5, 6) (−3, 5)
8. (b)   (i) 90° ACW (0, 0)   (ii) 180° (2, 1)   (iii) 90° CW (2, 0)
        (iv) 180° $(3\frac{1}{2}, 2\frac{1}{2})$ (v) 90° ACW (6, 1)   (vi) 90° CW (1, 3)

page 234 **Exercise 12**

1. (a) $\begin{pmatrix} 7 \\ 3 \end{pmatrix}$ (b) $\begin{pmatrix} 0 \\ -9 \end{pmatrix}$ (c) $\begin{pmatrix} 9 \\ 10 \end{pmatrix}$ (d) $\begin{pmatrix} -10 \\ 3 \end{pmatrix}$ (e) $\begin{pmatrix} -1 \\ 13 \end{pmatrix}$

(f) $\begin{pmatrix} 10 \\ 0 \end{pmatrix}$ (g) $\begin{pmatrix} -9 \\ -4 \end{pmatrix}$ (h) $\begin{pmatrix} -10 \\ 0 \end{pmatrix}$

12. (b) $\frac{3}{2}$
13. (a) (6, 1), sf 2   (b) (7, 1), sf 3   (c) (4, 1), sf 4   (d) (0, 1), sf 2   (e) (10, 1), sf $\frac{2}{3}$ (f) (4, 1), sf $\frac{1}{4}$
14. (a) Rotation 90° clockwise, centre (0, −2)
    (b) Reflection in $y = x$
    (c) Translation $\begin{pmatrix} 3 \\ 7 \end{pmatrix}$
    (d) Enlargement, scale factor 2, centre (−5, 5)
    (e) Translation $\begin{pmatrix} -7 \\ -3 \end{pmatrix}$
    (f) Reflection in $y = x$

15. (a) Rotation 90° clockwise, centre (4, −2)

(b) Translation $\begin{pmatrix} 8 \\ 2 \end{pmatrix}$

(c) Reflection in $y = x$

(d) Enlargement, scale factor $\frac{1}{2}$, centre (7, −7)

(e) Rotation 90° anticlockwise, centre (−8, 0)

(f) Enlargement, scale factor 2, centre (−1, −9)

(g) Rotation 90° anticlockwise, centre (7, 3)

16. (a) Enlargement, scale factor $1\frac{1}{2}$, centre (1, −4)

(b) Rotation 90° clockwise, centre (0, −4)

(c) Reflection in $y = -x$

(d) Translation $\begin{pmatrix} 11 \\ 10 \end{pmatrix}$

(e) Enlargement, scale factor $\frac{1}{2}$, centre (−3, 8)

(f) Rotation 90° anticlockwise, centre $(\frac{1}{2}, 6\frac{1}{2})$

(g) Enlargement, scale factor 3, centre (−2, 5)

page 236   **Exercise 13**

1. (c) Reflection in $x = 4$

2. (c) Rotation 90° clockwise, centre (0, 0)

3. (a) (−4, 4)   (b) (2, −2)   (c) (0, 0)   (d) (0, 4)

4. (a) (−2, 5)   (b) (−4, 0)   (c) (2, −2)   (d) (1, −1)

5. (a) reflection in y-axis

(b) rotation 180°, centre (−2, 2)

(c) rotation 90° clockwise, centre (2, 2)

6. (a) rotation 90° clockwise, centre (0, 0)

(b) translation $\begin{pmatrix} -2 \\ 5 \end{pmatrix}$

(c) rotation 90° anticlockwise, centre (2, −4)

(d) rotation 90° anticlockwise, centre $(-\frac{1}{2}, 3\frac{1}{2})$

7. (a) rotation 90° anticlockwise, centre (2, 2)

(b) enlargement, scale factor $\frac{1}{2}$, centre (8, 6)

(c) rotation 90° clockwise, centre $(-\frac{1}{2}, -3\frac{1}{2})$

8. $\mathbf{A}^{-1}$: reflection in $x = 2$

$\mathbf{B}^{-1}$: B

$\mathbf{C}^{-1}$: translation $\begin{pmatrix} 6 \\ -2 \end{pmatrix}$

$\mathbf{D}^{-1}$: D

$\mathbf{E}^{-1}$: E

$\mathbf{F}^{-1}$: translation $\begin{pmatrix} -4 \\ -3 \end{pmatrix}$

$\mathbf{G}^{-1}$: 90° rotation anticlockwise, centre (0, 0)

$\mathbf{H}^{-1}$: enlargement, scale factor 2, centre (0, 0)

page 239   **Exercise 14**

1. d   **2. 2c**   **3. 3c**   **4. 3d**   **5. 5d**   **6. 3c**   **7.** −2**d**
8. −2**c**   **9.** −3**c**   **10.** −**c**   **11. c** + **d**   **12. c** + 2**d**   **13.** 2**c** + **d**   **14.** 3**c** + **d**
15. 2**c** + 2**d**   **16.** $\overrightarrow{QI}$ **17.** $\overrightarrow{QU}$ **18.** $\overrightarrow{QH}$ **19.** $\overrightarrow{QB}$ **20.** $\overrightarrow{QF}$ **21.** $\overrightarrow{QJ}$
22. (a) 2**a** + **b**   (b) 2**a** + 2**b**   (c) −**a** − **b**   (d) 4**a** + 2**b**   (e) 2**a** − 2**b**   (f) 2**a** + **b**
23. (a) $\overrightarrow{CO}$ (b) $\overrightarrow{TN}$ (c) $\overrightarrow{FT}$ (d) $\overrightarrow{KC}$
24. (a) −**a**   (b) **a** + **b**   (c) 2**a** − **b**   (d) −**a** + **b**
25. (a) **a** + **b**   (b) **a** − 2**b**   (c) −**a** + **b**   (d) −**a** − **b**
26. (a) −**a** − **b**   (b) 3**a** − **b**   (c) 2**a** − **b**   (d) −2**a** + **b**
27. (a) **a** − 2**b**   (b) **a** − **b**   (c) 2**a**   (d) −2**a** + 3**b**
28. (a) and (d)   **29.** AD, BE, CF

page 241   **Exercise 15**

1. (a) **a**   (b) −**a** + **b**   (c) 2**b**   (d) −2**a**   (e) −2**a** + 2**b**(f) −**a** + **b**
  (g) **a** + **b**   (h) **b**   (i) −**b** + 2**a**   (j) −2**b** + **a**
2. (a) **a**   (b) −**a** + **b**   (c) 3**b**   (d) −2**a**   (e) −2**a** + 3**b**(f) −**a** + $\frac{3}{2}$**b**
  (g) **a** + $\frac{3}{2}$**b**   (h) $\frac{3}{2}$**b**   (i) −**b** + 2**a**   (j) −3**b** + **a**
3. (a) 2**a**   (b) −**a** + **b**   (c) 2**b**   (d) −3**a**   (e) −3**a** + 2**b**(f) −$\frac{3}{2}$**a** + **b**
  (g) $\frac{3}{2}$**a** + **b**   (h) $\frac{1}{2}$**a** + **b**   (i) −**b** + 3**a**   (j) −2**b** + **a**
4. (a) $\frac{1}{2}$**a**   (b) −**a** + **b**   (c) 4**b**   (d) −$\frac{3}{2}$**a**   (e) −$\frac{3}{2}$**a** + 4**b**   (f) −**a** + $\frac{8}{3}$**b**
  (g) $\frac{1}{2}$**a** + $\frac{8}{3}$**b**   (h) −$\frac{1}{2}$**a** + $\frac{8}{3}$**b**   (i) $\frac{3}{2}$**a** − **b**   (j) **a** − 4**b**
5. $\frac{1}{2}$**s** − $\frac{1}{2}$**t**   **6.** $\frac{1}{3}$**a** + $\frac{2}{3}$**b**   **7. a** + **c** − **b**   **8.** 2**m** + 2**n**
9. (a) **b** − **a**   (b) **b** − **a**   (c) 2**b** − 2**a**   (d) **b** − 2**a**   (e) **b** − 2**a**   (f) 2**b** − 3**a**
10. (a) **y** − **z**   (b) $\frac{1}{2}$**y** − $\frac{1}{2}$**z**   (c) $\frac{1}{2}$**y** + $\frac{1}{2}$**z**   (d) −**x** + $\frac{1}{2}$**y** + $\frac{1}{2}$**z**
  (e) −$\frac{2}{3}$**x** + $\frac{1}{3}$**y** + $\frac{1}{3}$**z**   (f) $\frac{1}{3}$**x** + $\frac{1}{3}$**y** + $\frac{1}{3}$**z**
11. (a) (i) 2**b** − 2**a**   (ii) 2**c** − 2**b**   (iii) **b**   (iv) **c** − **a**   (v) **c** − **a**
  (b) parallel and equal in length   (c) parallelogram
12. (a) (i) **a** − **b**   (ii) $\frac{1}{3}$**a** − $\frac{1}{3}$**b**   (iii) $\frac{2}{3}$**b** − $\frac{1}{6}$**a** $\left[=\frac{1}{6}(4\mathbf{b} - \mathbf{a})\right]$ (iv) 2**b** − $\frac{1}{2}$**a** $\left[=\frac{1}{2}(4\mathbf{b} - \mathbf{a})\right]$
  (b) CD is parallel to DE and both lines pass through point D

page 245   **Exercise 16**

4. sin 70° = sin 110°, cos 60° = cos 300°, sin 50° = sin 130°
5. tan 45° = tan 225°, sin 180° = cos 270°, cos 30° = cos 330°, tan 60° = tan 240°
6. 162°   **7.** 153°   **8.** (a) 140°   (b) 110°   (c) 50°
9. 290°   **10.** 315°   **11.** (a) 350°   (b) 304°   (c) 60°
12. 220°   **13.** 160°   **14.** 82°   **15.** 315°   **16.** 240°   **17.** 250°
18. sin 270° = cos 180°, tan 200° = tan 20°, sin 90° = cos 360°, tan 150° = tan 330°

page 246   **Exercise 17**

1. 58·0°, 122·0°   **2.** 20·5°, 159·5°   **3.** 53·1°, 306·9°   **4.** 45°, 225°
5. (a) 19·8°, 160·2°   (b) 72°, 108°   (c) 30°, 150°   (d) 60°, 120°
6. (a) 46·1°, 133·9°   (b) 72·5°, 287·5°   (c) 78·7°, 258·7°   (d) 220·5°, 319·5°
7. 30°, 150°, 210°, 330°   **8.** 45°, 135°, 225°, 315°
10. (a) 0 < x < 180°   (b) 90° < x < 270°
12. (a) 90°, 450°, 810° etc.   (b) 90°, 270°, 450°, 630° etc.

13. (a) 0   (b) 1
16. (a) 45°, 225°   (b) 30°, 60°, 210°, 240°   (c) 15°, 75°, 135°, 195°, 255°, 315°
17. (a) 40°, 140°   (b) 30°, 150°
18. (a) (i) 16°, 111°   (ii) 153°   (b) 2·24   (c) 63°
19. (a) 48°, 205°   (b) 37°, 217°

page 249 **Exercise 18**

1. 6·38 m   **2.** 12·5 m   **3.** 5·17 cm   **4.** 40·4 cm   **5.** 7·81 m
6. 6·68 m   **7.** 8·61 cm   **8.** 9·97 cm   **9.** 8·52 cm   **10.** 15·2 cm
11. 35·8°   **12.** 42·9°   **13.** 32·3°   **14.** 37·8°   **15.** 35·5°, 48·5°
16. 68·8°   **17.** 64·6°   **18.** 34·2°   **19.** 50·6°   **20.** 39·1°
21. 39·5°   **22.** 21·6°   **23.** 72·5°, 107·5°

page 252 **Exercise 19**

1. 6·24   **2.** 6·05   **3.** 5·47   **4.** 9·27   **5.** 10·1
6. 8·99   **7.** 5·87   **8.** 4·24   **9.** 11·9   **10.** 154
11. 25·2°   **12.** 71·4°   **13.** 115·0°   **14.** 111·1°   **15.** 24·0°
16. 92·5°   **17.** 99·9°   **18.** 38·2°   **19.** 137·8°   **20.** 34·0°

page 253 **Exercise 20**

1. (a) 50·2 km   (b) 054·7°   **2.** 35·6 km   **3.** 25·2 m
4. (a) 10·3 cm   (b) 11·6 cm   (c) 4·86 cm   (d) 37·2°   (e) 94·1°   (f) 42·6°
5. 92·9°   **6.** 40·4 m   **7.** 54·7 m   **8.** 14·5 cm
9. (a) 9·8 km   (b) 085·7°   **10.** (a) 29·6 km   (b) 050·5°   **11.** 141 km
12. (a) 10·8 m   (b) 72·6°   (c) 32·6°   **13.** 378 km, 048·4°
14. (a) 62·2°   (b) 2·33 km   **15.** 101·5°   **16.** 9 m
17. (b) 4, 2   **19.** 9·64 m   **20.** 131·8°
21. (a) 5·66 cm   (b) 4·47 cm   (c) 3·66 cm   **22.** 70·2°

page 257 **Exercise 21**

1. $a = 27°, b = 30°$ **2.** $c = 20°, d = 45°$ **3.** $c = 58°, d = 41°, e = 30°$
4. $f = 40°, g = 55°, h = 55°$ **5.** $a = 32°, b = 80°, c = 43°$ **6.** $c = 34°, y = 34°$
7. 43°   **8.** 92°   **9.** 42°
10. $c = 46°, d = 44°$ **11.** $e = 49°, f = 41°$ **12.** $g = 76°, h = 52°$
13. 48°   **14.** 32°   **15.** 22°
16. $a = 36°, x = 36°$

page 259 **Exercise 22**

1. $a = 94°, b = 75°$ **2.** $c = 101°, d = 84°$ **3.** $x = 92°, y = 116°$
4. $c = 60°, d = 45°$ **5.** 37°   **6.** 118°
7. $e = 36°, f = 72°$ **8.** 35°   **9.** 18°
10. 90°   **11.** 30°   **12.** $22\frac{1}{2}°$
13. $n = 58°, t = 64°, w = 45°$ **14.** $a = 32°, b = 40°, c = 40°$ **15.** $a = 18°, c = 72°$
16. 55°   **17.** $e = 41°, f = 41°, g = 41°$ **18.** 8°
19. $x = 30°, y = 115°$ **20.** $x = 80°, z = 10°$

page 261 **Exercise 23**

1. $a = 18°$
2. $53°$
3. $77°$
4. $x = 40°$, $y = 65°$, $z = 25°$
5. $c = 30°$, $e = 15°$
6. $f = 50°$, $g = 40°$
7. $h = 70°$, $k = 40°$, $i = 40°$
8. $m = 108°$, $n = 36°$
9. $x = 50°$, $y = 68°$
10. $a = 74°$, $b = 32°$
11. $e = 36°$
12. $k = 63°$, $m = 54°$
13. $k = 50°$, $m = 50°$, $n = 80°$, $p = 80°$
14. (a) p (b) $2p$ (c) $90 - 2p$
15. $x = 70°$, $y = 20°$, $z = 55°$

page 262 **Exercise 24**

1. $137°$
2. $120°$
3. $c = d = 30°$, $e = 27°$
4. $49°$
5. $18°$
6. $110°$
7. $i = 52°$, $j = 128°$
8. $k = 45°$, $l = 135°$
9. $m = 35°$, $n = 50°$
10. $69°$
11. $78°$
12. $r = 200°$, $s = 100°$
13. $45°$
14. $b = 30°$, $c = 60°$
15. $72°$
16. $e = 19°$
17. $50°$
18. $g = 58°$, $h = 90°$
19. $i = 45°$
20. $j = 96°$, $k = 42°$
21. $m = 28°$

7 Algebra 3

page 265 **Exercise 1**

1. (a) 2 (b) -4
2. (a) 3 (accept $2·4 \to 3·6$) (b) -5 (accept $-3·5 \to -6·5$) (c) $1·5$
3. $3·3$ (accept $2·8 \to 3·8$)
4. (a) $-2\frac{1}{2}$ (b) $-\frac{2}{5}$ (c) gradient $\to 0$
5. (a) (i) 4 (ii) 8

page 267 **Exercise 2**

1. 22 square units
2. (a) about 360/370 (b) about 130/140
3. $7·47$ **4.** $13·4$
5. $23·5$ **6.** (b) $38·5$ (c) less
7. (a) 58 (b) greater
8. (a) $x = 1$, $x = 4$ (b) about $7\frac{1}{2}$ square units

page 269 **Exercise 3**

1. (a) $1\frac{1}{2}$ m/s^2 (b) 675 m
2. (a) 600 m (b) 225 m (c) -2 m/s^2
3. (a) 600 m (b) $387\frac{1}{2}$ m (c) 0 m/s^2
4. (a) 20 m/s (b) 750 m
5. (a) 8 s (b) 496 m (c) $12·4$ m/s
6. A5, B3, C1, D2, E4, F6
7. (a) $0·75$ m/s^2 (b) 680 m (both approximate)
8. (a) $0·35$ m/s^2 (b) 260 m (both approximate)
9. $1·0 <$ speed $< 1·1$ m/s
10. (a) 50 m/s (b) 20 s
11. (a) 20 m/s (b) 20 s

page 273 **Exercise 4**

1. $(x + 2)(x + 5)$
2. $(x + 3)(x + 4)$
3. $(x + 3)(x + 5)$
4. $(x + 3)(x + 7)$
5. $(x + 2)(x + 6)$
6. $(y + 5)(y + 7)$
7. $(y + 3)(y + 8)$
8. $(y + 5)(y + 5)$
9. $(y + 3)(y + 12)$
10. $(a + 2)(a - 5)$
11. $(a + 3)(a - 4)$
12. $(z + 3)(z - 2)$
13. $(x + 5)(x - 7)$
14. $(x + 3)(x - 8)$
15. $(x - 2)(x - 4)$
16. $(y - 2)(y - 3)$
17. $(x - 3)(x - 5)$
18. $(a + 2)(a - 3)$
19. $(a + 5)(a + 9)$
20. $(b + 3)(b - 7)$

21. $(x-4)(x-4)$

22. $(y+1)(y+1)$

23. $(y-7)(y+4)$

24. $(x-5)(x+4)$

25. $(x-20)(x+12)$

26. $(x-15)(x-11)$

27. $(y+12)(y-9)$

28. $(x-7)(x+7)$

29. $(x-3)(x+3)$

30. $(x-4)(x+4)$

31. $2(x+2)(x+4)$

32. (a) $2(x-3)(x+5)$ (b) $3(x+2)(x+5)$ (c) $3(x+3)(x+5)$

      (d) $2(n-5)(n+2)$ (e) $5(a-2)(a+3)$ (f) $4(x-4)(x+4)$

page 273   **Exercise 5**

1. $(2x+3)(x+1)$

2. $(2x+1)(x+3)$

3. $(3x+1)(x+2)$

4. $(2x+3)(x+4)$

5. $(3x+2)(x+2)$

6. $(2x+5)(x+1)$

7. $(3x+1)(x-2)$

8. $(2x+5)(x-3)$

9. $(2x+7)(x-3)$

10. $(3x+4)(x-7)$

11. $(2x+1)(3x+2)$

12. $(3x-2)(x-3)$

13. $(y-2)(3y-5)$

14. $(2y+3)(3y-1)$

15. $(5x+2)(2x+1)$

16. $(6x-1)(x-3)$

17. $(4x+1)(2x-3)$

18. $(3x+2)(4x+5)$

19. $(4y-3)(y-5)$

20. $(2x-5)(3x-6)$

page 274   **Exercise 6**

1. $(y-a)(y+a)$

2. $(m-n)(m+n)$

3. $(x-t)(x+t)$

4. $(y-1)(y+1)$

5. $(x-3)(x+3)$

6. $(a-5)(a+5)$

7. $(x-\frac{1}{2})(x+\frac{1}{2})$

8. $(x-\frac{1}{3})(x+\frac{1}{3})$

9. $(2x-y)(2x+y)$

10. $(a-2b)(a+2b)$

11. $(5x-2y)(5x+2y)$

12. $(3x-4y)(3x+4y)$

13. $\left(2x-\dfrac{z}{10}\right)\left(2x+\dfrac{z}{10}\right)$ **14.** $x(x-1)(x+1)$

15. $a(a-b)(a+b)$

16. $x(2x-1)(2x+1)$

17. $2x(2x-y)(2x+y)$ **18.** $y(y-3)(y+3)$

19. $1\,200\,000$

20. $12\,000\,000$

21. 103×97

page 274   **Exercise 7**

1. $-3, -4$

2. $-2, -5$

3. $3, -5$

4. $2, -3$

5. $2, 6$

6. $-3, -7$

7. $2, 3$

8. $5, -1$

9. $-7, 2$

10. $-\frac{1}{2}, 2$

11. $\frac{2}{3}, -4$

12. $1\frac{1}{2}, -5$

13. $\frac{2}{3}, 1\frac{1}{2}$

14. $\frac{1}{4}, 7$

15. $\frac{3}{5}, -\frac{1}{2}$

16. $7, 8$

17. $\frac{5}{6}, \frac{1}{2}$

18. $7, -9$

19. $-1, -1$

20. $3, 3$

21. $-5, -5$

22. $7, 7$

23. $-\frac{1}{3}, \frac{1}{2}$

24. $-1\frac{1}{4}, 2$

25. $13, -5$

26. $-3, \frac{1}{6}$

27. $\frac{1}{10}, -2$

28. $1, 1$

29. $\frac{2}{9}, -\frac{1}{4}$

30. $-\frac{1}{4}, \frac{3}{5}$

31. $\pm 1, \pm 2$

32. $\pm 2, \pm 3$

33. $\pm \frac{1}{2}, \pm 2$

34. $1, 2$

page 275   **Exercise 8**

1. $0, 3$

2. $0, -7$

3. $0, 1$

4. $0, \frac{1}{3}$

5. $4, -4$

6. $7, -7$

7. $\frac{1}{2}, -\frac{1}{2}$

8. $\frac{2}{3}, -\frac{2}{3}$

9. $0, -1\frac{1}{2}$

10. $0, 1\frac{1}{2}$

11. $0, 5\frac{1}{2}$

12. $\frac{1}{4}, -\frac{1}{4}$

13. $\frac{1}{2}, -\frac{1}{2}$

14. $0, \frac{5}{8}$

15. $0, \frac{1}{12}$

16. $0, 6$

17. $0, 11$

18. $0, 1\frac{1}{2}$

19. $0, 1$

20. $0, 4$

page 276   **Exercise 9**

1. $-\frac{1}{2}, -5$

2. $-\frac{2}{3}, -3$

3. $-\frac{1}{2}, -\frac{2}{3}$

4. $\frac{1}{3}, 3$

5. $\frac{2}{5}, 1$

6. $\frac{1}{3}, 1\frac{1}{2}$

7. $-0.63, -2.37$

8. $-0.27, -3.73$

9. $0.72, 0.28$

10. $6.70, 0.30$

11. $0.19, -2.69$

12. $0.85, -1.18$

13. $0.61, -3.28$

14. $-1\frac{2}{3}, 4$

15. $-1\frac{1}{2}, 5$

16. $3.56, -0.56$

17. $0.16, -3.16$

18. $-\frac{1}{2}, 2\frac{1}{3}$

19. $-\frac{1}{3}, -8$

20. $1\frac{2}{3}, -1$

21. $2.28, 0.22$

22. $-0.35, -5.65$

23. $-\frac{2}{3}, \frac{1}{2}$

24. $-0.58, 2.58$

25. (a) $0.2, 4.8$

26. If $b^2 - 4ac > 0$

Answers 469

page 276 *Exercise 10*

1. $-3, 2$ **2.** $-3, -7$ **3.** $-\frac{1}{2}, 2$ **4.** $1, 4$ **5.** $-1\frac{2}{3}, \frac{1}{2}$ **6.** $-0\cdot39, -4\cdot28$

7. $-0\cdot16, 6\cdot16$ **8.** 3 **9.** $2, -1\frac{1}{3}$ **10.** $-3, -1$ **11.** $0\cdot66, -22\cdot66$

12. $-7, 2$ **13.** $\frac{1}{4}, 7$ **14.** $-\frac{1}{2}, \frac{3}{5}$ **15.** $0, 3\frac{1}{2}$ **16.** $-\frac{1}{4}, \frac{1}{4}$ **17.** $-2\cdot77, 1\cdot27$

18. $-\frac{2}{3}, 1$ **19.** $-\frac{1}{2}, 2$ **20.** $0, 3$ **21.** Yes

22. (a) $3, -\frac{1}{2}$ (b) $6, -4$ **23.** (a) -1 (b) $0\cdot6258$ (c) $0\cdot5961$ (d) $0\cdot2210$

page 278 *Exercise 11*

1. $(x+4)^2 - 16$ **2.** $(x-6)^2 - 36$ **3.** $(x+\frac{1}{2})^2 - \frac{1}{4}$ **4.** $(x+2)^2 - 3$ **5.** $(x-3)^2$

6. $(x+1)^2 - 16$ **7.** $(x+4)^2 - 11$ **8.** $(x-5)^2 - 25$ **9.** $(x+\frac{3}{2})^2 - \frac{9}{4}$

10. (a) $2, 6$ (b) $-3, -7$ (c) $5, -1$

11. (a) $0\cdot65, -4\cdot65$ (b) $3\cdot56, -0\cdot56$ (c) $0\cdot08, -12\cdot08$

15. (a) 3 (b) -2 (c) $\frac{1}{3}$ **16.** (a) $\frac{3}{4}$ (b) $\frac{1}{2}$ (c) $\frac{4}{3}$ **17.** $(x+2)^2$

page 280 *Exercise 12*

1. $8, 11$ **2.** $11, 13$ **3.** 12 cm   **4.** 6 cm   **5.** $x = 11$

6. $8, 9, 10$ **7.** (d) $3, 4, 5$ **8.** 157 km   **9.** 4 **10.** 10 cm $\times 24$ cm

11. $4\cdot3$ s   **12.** $2\cdot5$ cm   **13.** 1 **14.** 9 cm or 13 cm

15. $n(n+2)$; 15th term   **16.** 2 **17.** 3 **18.** 6 cm

19. $\dfrac{40}{x}$ h, $\dfrac{40}{x-2}$ h, 10 km/h   **20.** 4 km/h   **21.** 20 mph   **22.** 4

23. $2, 5$ **24.** $\frac{3}{4}$ **25.** (c) $x = 595$ mm, $y = 841$ mm   (d) $297\cdot5$ mm

page 283 *Exercise 13*

1. (a) $\frac{5}{7}$ (b) $5y$ (c) $\frac{1}{2}$ (d) 4 (e) $\dfrac{x}{2y}$ (f) 2 (g) $\dfrac{a}{2}$ (h) $\dfrac{2b}{3}$

2. AG, BE, CH, DF

3. (a) a (b) $\dfrac{10m}{3}$ (c) $\dfrac{2xy}{9}$ (d) $\dfrac{a}{q}$ (e) $\frac{8}{3}$ (f) $\dfrac{y^2}{5x}$ (g) $2a$ (h) 1

4. (a) $\dfrac{a}{5b}$ (b) a (c) $\frac{7}{8}$ (d) $\dfrac{3}{4-x}$ (e) $\dfrac{5+2x}{3}$ (f) $\dfrac{3x+1}{x}$ (g) $\dfrac{4+5a}{5}$ (h) $\dfrac{b}{3+2a}$

5. (a) 6 (b) 4 (c) $2a$ **6.** AF, BH, CE, DG

7. (a) $\dfrac{x+2y}{3xy}$ (b) $\dfrac{6-b}{2a}$ (c) $\dfrac{2b+4a}{b}$ (d) $x-2$

8. (a) $\dfrac{x+2}{x-3}$ (b) $\dfrac{x}{x+1}$ (c) $\dfrac{x+4}{2(x-5)}$ (d) $\dfrac{x+5}{x-2}$ (e) $\dfrac{x+3}{x+2}$ (f) $\dfrac{x+5}{x-2}$

page 284 *Exercise 14*

1. (a) $\dfrac{3x}{5}$ (b) $\dfrac{3}{x}$ (c) $\dfrac{4x}{7}$ (d) $\dfrac{4}{7x}$ (e) $\dfrac{7x}{8}$ (f) $\dfrac{7}{8x}$ (g) $\dfrac{5x}{6}$ (h) $\dfrac{5}{6x}$

2. AF, BH, CD, EG

3. (a) $\dfrac{23x}{20}$ (b) $\dfrac{23}{20x}$ (c) $\dfrac{x}{12}$ (d) $\dfrac{1}{12x}$ (e) $\dfrac{5x+2}{6}$ (f) $\dfrac{7x+2}{12}$

4. (a) $\dfrac{1-2x}{12}$ (b) $\dfrac{2x-9}{15}$ (c) $\dfrac{3x+12}{14}$ **5.** (a) $\dfrac{x}{10}$ (b) $3x$ (c) 11

6. (a) $\dfrac{3x+1}{x(x+1)}$ (b) $\dfrac{7x-8}{x(x-2)}$ (c) $\dfrac{8x+9}{(x-2)(x+3)}$

    (d) $\dfrac{4x+11}{(x+1)(x+2)}$ (e) $\dfrac{-3x-17}{(x+3)(x-1)}$ (f) $\dfrac{11-x}{(x+1)(x-2)}$

7. $\frac{1}{4}$ **8.** $\frac{1}{5}$

page 288 **Exercise 15**

1.

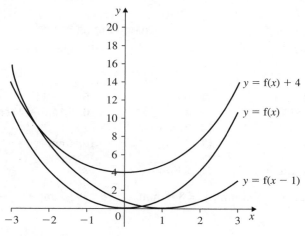

2.

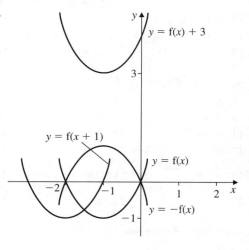

3. $(0, -2)$, $(7, 0)$

4. (a) $A'(-2, -1)$, $B'(0, -3)$, $C'(2, 0)$
    (b) $A'(0, 1)$, $B'(2, 3)$, $C'(4, 0)$
    (c) $A'(-1, 1)$, $B'(0, 3)$, $C'(1, 0)$

5.

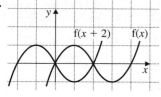

6.

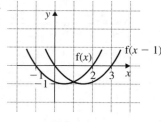

7.

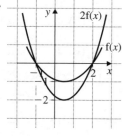

8.

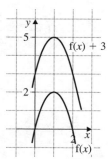

9.

10.

11.

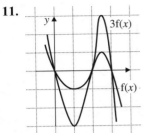

12.

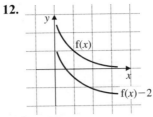

13.

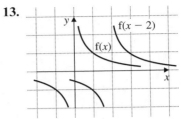

14.

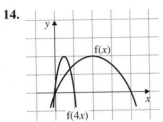

15.

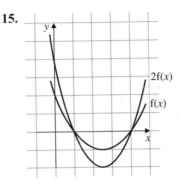

16.

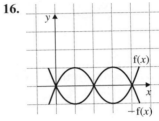

page 289   ***Exercise 16***

1.

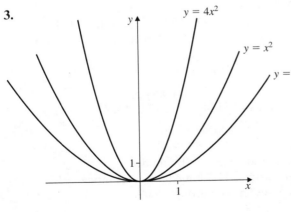

2.

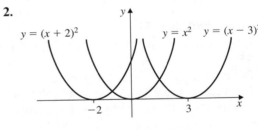

3.

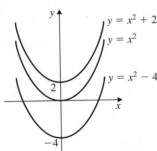

4.

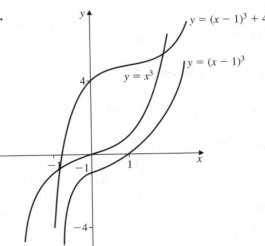

5.

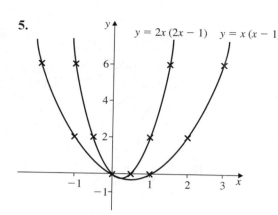

$y = 2x(2x - 1)$ $y = x(x - 1)$

6.

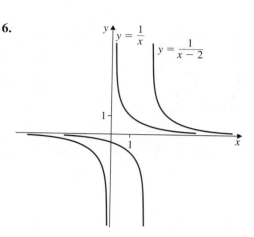

$y = \dfrac{1}{x}$

$y = \dfrac{1}{x - 2}$

9. Stretch parallel to the x-axis, scale factor 2

10. (a) $(1, 5)$ (b) $(5, -4)$

11. (a) $y = x^2 + 3x + 5$ (b) $y = x^2 - x - 2$ (c) $y = -(x^2 + 3x)$

12. $a = 2, b = 3$

13. (a) $A'(2, 4)$ (b) $B'(0, \frac{1}{5})$, $C'(3, 1)$, $D'(-3, 1)$

8 Handling data

page 293 **Exercise 1**

1. (a) 5, 6, 4, 9   (b) 7, 9, 7, 19   (c) 8, 6·5, 9   (d) 3·5, 3·5, 4, 8

2. 45 or 2   **3.** 4

4. (a) False   (b) Possible   (c) False   (d) Possible

5. (a) mean = £33 920, median = £22 500, mode = £22 500

   (b) the mean is skewed by one large figure

6. (a) mean = 157·1 kg, median = 91 kg

   (b) mean; No : Over three-quarters of the cattle are below the mean weight

7. (a) mean = 74·5 cm, median = 91 cm   (b) Yes

8. 78 kg   **9.** 35·2 cm   **10.** (a) 2   (b) 9

11. (a) 20·4 m   (b) 12·8 m   (c) 1·66 m   **12.** 55 kg   **14.** 2, 4, 4, 4, 6

15. 12   **16.** $3\frac{2}{3}$ **17.** (a) N (b) mean = $N^2 + 2$; median = N^2 (c) 2

page 295 **Exercise 2**

1. (a) 24·3p   (b) 25p   (c) 25·6p

2. averages: 9, 8·8, 9·2, 9·2, 10, 11·2, 12·8, 13·8, 14·8, 15, 14·6, 14·8, 14·8, 13·6, 13, 13

page 297 **Exercise 3**

1. 3·38   **2.** (a) mean = 6·62; median = 8; mode = 3   (b) mode

3. (a) mean = 3·025; median = 3; mode = 3   (b) mean = 17·75; median = 17; mode = 17

4. About 11   **5.** (a) 68·25   (b) median in 55-69; modal group 55-69

6. 3·8   **7.** 180/181   **8.** (a) 9   (b) 9   (c) 15

9. (a) 5   (b) 10   (c) 10   **10.** $\dfrac{ax + by + cz}{a + b + c}$

page 302 **Exercise 4**

2. (a) 20%   (b) 20–24 and 30–34
4. (a) 62
   (b) Sport B has more heavy people. Sport A has much smaller range of weights compared to sport B.
5. (a) Plants with fertiliser are significantly taller   (b) no significant effect
8. (a) £3000   (b) £4000   (c) £6000   (d) £11 000
9. eggs 270°; milk 12°; butter 23·4°; cheese 54°; salt/pepper 0·6°   **10.** 18°, 54°, 54°, 234°
11. (a) 22·5%   (b) $x = 45°, y = 114°$ **12.** $x = 8$

page 306 **Exercise 5**

3. (a) 50   (b) 14   (c) 50 kg
5. (a) 13   (b) 78
   (c) The men generally had a lower pulse rate and their pulse rates were more
       varied than those of the women.

page 308 **Exercise 6**

2. (a) strong positive   (b) no correlation   (c) weak negative
3. (a) no correlation   (b) strong negative   (c) no correlation   (d) weak positive
4. (b) 9   **5.** 44   **6.** 6·8 km
7. (b) Bogota; High altitude   (c) About 73°C   **8.** For discussion

page 319 **Exercise 9**

1. 16, 16, 16, 6, 6   **2.** 10, 25, 10, 5, 50

page 320 **Exercise 10**

1. Pupils who do not eat school meals will not be questioned.
2. Ex-directory customers should be included.
3. His friends may not be typical.
4. People going on ferry may not be representative of the population.
5. Might not be typical of the whole week.
6. OK.
7. People are unlikely to include those at work/school etc.
8. All these people likely to have a video. Not representative of whole population.

page 322 **Exercise 11**

1. (a) 4, 9·5   (b) 5·5   **2.** (a) 1·5, 55   (b) 4   **3.** (a) 10, 21   (b) 11
4. (a) 1·5, 6   (b) 4·5   **5.** (a) 28, 6, 12   (b) 4·8, 1·5, 2·7   (c) 33, 12, 24

page 324 **Exercise 12**

1. (a) 50   (b) 30, 60   (c) 30   (d) 40
2. (a) 43   (b) 28; 57   (c) 29   (d) 34 (all approximate)
3. (a) 100   (b) 250   (c) 2250 h   (d) 750 h
4. (b) (i) 45   (ii) 17   **5.** (b) (i) about 160·5 cm   (ii) about 13 cm
6. (a) About 124   (b) IQR 52 Boris is more consistent
7. (b) (i) 7·6 h   (ii) (about) 3·3 h   (c) 90%
8. (b) France 18·5, Britain 24·5   (c) 8·5
   (d) Results from Britain bunched together more closely with a higher median.

9. (b) 10·2 cm   (c) Many failed to germinate
(d) Median is better as it discounts those which do not germinate
10. (c) 44 million   (d) A : 21, B : 59   (e) A has much younger population

page 329 **Exercise 13**

Frequency densities are given.
1. 0·25, 1, 1·4, 0·3
4. 1·8, 9·2, 7, 1·3, 0·8
7. 1·6, 3·4, 2·8, 1·1

2. 1, 0·6, 1·2, 1·7, 1·3, 0·5
5. 0·4, 1·2, 1·5, 0·9, 0·2

3. 0·5, 1·4, 1, 1, 0·2
6. 0·55, 1·8, 0·7, 0·25, 0

page 332 **Exercise 14**

1. (a) (i) 10   (ii) 20   (b) 65
6. f.d: 0·2, 0·3, 1·2, 1, 0·4

2. 55
8. (a) 12   (b) 23   (c) 29   (d) 25

3. 135   **4.** 28   **5.** 44

page 336 **Exercise 15**

1. (a) $\bar{x} = 6 \cdot 4$; s.d. $= 3 \cdot 38$ (b) $\bar{x} = 4 \cdot 9$; s.d. $= 1 \cdot 00$
    (c) $\bar{x} = 115$; s.d. $= 9 \cdot 75$ (d) $\bar{x} = 0 \cdot 16$; s.d. $= 2 \cdot 67$
2. A : $\bar{x} = 23$; s.d. $= 1 \cdot 87$ B : $\bar{x} = 23$; s.d. $= 3 \cdot 08$ Method A involves less variation in lifetimes.
3. Mizuno: $\bar{x} = 70$; s.d. $= 2 \cdot 52$. Lynx : $\bar{x} = 70$; s.d. $= 4 \cdot 20$.
    Use Mizuno for consistent results as s.d. is lower.
4. About the same mean but the marks for Birchwood were more spread out.
5. The people in town were taller and their heights were more spread out.
6. (a) $\bar{x} = 11 \cdot 08$, s.d. $= 6 \cdot 11$ (b) $\bar{x} = 5 \cdot 6$, s.d. $= 1 \cdot 99$
7. (a) mean $\approx 33 \cdot 5$; s.d. ≈ 11 (b) mean $\approx 15 \cdot 5$; s.d. $\approx 7 \cdot 6$ (c) mean $\approx 10 \cdot 65$; s.d. $\approx 5 \cdot 7$
8. (a) 8, 10, 8, 4   (b) mean $= £134$; s.d. $= £25 \cdot 8$

page 338 **Exercise 16**

1. mean $= 237 \cdot 9$ g; s.d. $= 14 \cdot 3$ g
3. mean $= 1 \cdot 825$; s.d. $= 1 \cdot 14$

2. mean $= 247 \cdot 9$ g; s.d. $= 14 \cdot 3$ g
4. mean $= 3 \cdot 65$; s.d. $= 2 \cdot 28$

9 Probability

page 341 **Exercise 1**

1. B   **2.** C   **3.** A   **4.** B or C   **5.** C or D
6. A   **7.** B   **8.** B   **9.** C   **10.** D
13. Mike with a large number of spins he should get zero with a probability of about $\frac{1}{10}$

page 343 **Exercise 2**

1. (a) $\frac{1}{13}$ (b) $\frac{1}{52}$ (c) $\frac{1}{4}$
3. (a) $\frac{5}{11}$ (b) $\frac{2}{11}$ (c) $\frac{4}{11}$
5. (a) $\frac{2}{9}$ (b) $\frac{2}{9}$ (c) $\frac{1}{9}$ (d) 0   (e) $\frac{5}{9}$
7. (a) $\frac{3}{13}$ (b) $\frac{5}{13}$ (c) $\frac{8}{13}$
9. (a) $\frac{1}{12}$ (b) $\frac{1}{40}$ (c) $\frac{1}{4}$
11. $\frac{1}{7}$

2. (a) $\frac{1}{9}$ (b) $\frac{1}{3}$ (c) $\frac{4}{9}$ (d) $\frac{2}{9}$
4. (a) $\frac{4}{17}$ (b) $\frac{3}{17}$ (c) $\frac{11}{17}$
6. (a) $\frac{1}{13}$ (b) $\frac{2}{13}$ (c) $\frac{1}{52}$ (d) $\frac{5}{52}$
8. (a) (i) $\frac{5}{13}$ (ii) $\frac{6}{13}$ (b) (i) $\frac{5}{12}$ (ii) $\frac{1}{12}$
10. (a) $\dfrac{x}{12}$ (b) 3
12. (a) (i) $\frac{1}{4}$ (ii) $\frac{1}{4}$ (iii) $\frac{1}{4}$ (b) $\frac{1}{4}$ (c) $\frac{6}{27} = \frac{2}{9}$.

page 345 **Exercise 3**

1. (a) 150   (b) 50   **2.** 25   **3.** 50   **4.** 40
5. (a) $\frac{3}{8}$ (b) 25 **6.** (a) $\frac{1}{2}$ (b) $\frac{1}{2}$ **7.** (a) 15   (b) 105
8. (a) 20   (b) 5   (c) 50   (d) 40   **9.** (a) 0·2146   (b) 1073

page 347 **Exercise 4**

1. (a) 8 ways   (b) $\frac{1}{8}$ **2.** 16 ways   **3.** (b) 4   (c) $\frac{1}{9}$
4. (a) $\frac{1}{4}$ (b) $\frac{1}{4}$ **5.** WX, WY, WZ, XY, XZ, YZ, $\frac{1}{6}$ **6.** (a) 30 outcomes   (b) $\frac{1}{2}$
7. (a) $\frac{1}{12}$ (b) $\frac{1}{36}$ (c) $\frac{5}{18}$ (d) $\frac{1}{6}$ **8.** (a) $\frac{1}{12}$ (b) $\frac{5}{36}$ (c) $\frac{2}{3}$ (d) $\frac{1}{12}$ (e) $\frac{1}{36}$
9. $\frac{1}{3}$ **10.** (a) 64   (b) $\frac{1}{64}$ **11.** Yes
12. No. Y has the best chance   **13.** lose 20p   **14.** (a) $\frac{1}{144}$ (b) $\frac{1}{18}$

page 352 **Exercise 6**

1. (a) Yes   (b) No   (c) Yes   (d) No   (e) No   **2.** AF, BC, DE
3. (a) $\frac{5}{11}$ (b) $\frac{7}{22}$ (c) $\frac{15}{22}$ (d) $\frac{17}{22}$ **4.** $\frac{1}{18}$
5. (a) 0·3   (b) 0·9   **6.** $\frac{1}{3}$ **7.** $\frac{1}{24}$ **8.** $\frac{264}{360} = \frac{11}{15}$
9. (a) 0·24, 0·89   (b) 575   **10.** (a) (i) exclusive   (ii) not exclusive   (b) $\frac{11}{15}$
11. (a) (i) exclusive   (ii) exclusive   (iii) not exclusive   (b) $\frac{3}{4}$
12. (a) 0·8   (b) 0·7   (c) not exclusive   **13.** not mutually exclusive events

page 355 **Exercise 7**

1. (a) $\frac{1}{13}, \frac{1}{6}$ (b) $\frac{1}{78}$ **2.** (a) $\frac{1}{2}$ (b) $\frac{1}{2}$ (c) $\frac{1}{4}$ **3.** $\frac{1}{10}$
4. (a) $\frac{1}{78}$ (b) $\frac{1}{104}$ (c) $\frac{1}{24}$ **5.** (a) $\frac{1}{16}$ (b) $\frac{1}{169}$ (c) $\frac{9}{169}$
6. (a) $\frac{1}{16}$ (b) $\frac{25}{144}$ **7.** (a) $\frac{1}{121}$ (b) $\frac{9}{121}$ **8.** $\frac{8}{1125}$
9. (a) $\frac{1}{288}$ (b) $\frac{1}{72}$ **10.** (a) $\frac{1}{9}$ (b) $\frac{4}{27}$
11. $\frac{1}{24}$ **12.** $\frac{1}{128}$ **13.** $\frac{1}{144}$ **14.** (a) $\left(\frac{1}{6}\right)^{20}$ (b) $\left(\frac{5}{6}\right)^{n}$ (c) $1 - \left(\frac{5}{6}\right)^{n}$

page 358 **Exercise 8**

1. (a) $\frac{49}{100}$ (b) $\frac{9}{100}$ **2.** (a) $\frac{9}{64}$ (b) $\frac{15}{64}$ **3.** (a) $\frac{7}{15}$ (b) $\frac{1}{15}$
4. (a) $\frac{2}{9}$ (b) $\frac{2}{15}$ (c) $\frac{1}{45}$ **5.** (a) $\frac{1}{12}$ (b) $\frac{1}{6}$ (c) $\frac{1}{3}$ (d) $\frac{2}{9}$
6. (a) $\frac{1}{216}$ (b) $\frac{125}{216}$ (c) $\frac{25}{72}$ (d) $\frac{91}{216}$ **7.** (a) $\frac{1}{64}$ (b) $\frac{5}{32}$ (c) $\frac{27}{64}$
8. (a) $\frac{1}{6}$ (b) $\frac{1}{30}$ (c) $\frac{1}{30}$ (d) $\frac{29}{30}$ **9.** (a) $\frac{9}{16}$ (b) $\frac{1}{16}$ **10.** $\frac{1}{3}$
11. (a) $\frac{4}{9}$ (b) $\frac{1}{24}$ **12.** (a) $\frac{3}{20}$ (b) $\frac{9}{20}$
13. (a) $\frac{3}{20} \times \frac{2}{19} \times \frac{1}{18} \left(= \frac{1}{1140} \right)$ (b) $\frac{1}{4} \times \frac{4}{19} \times \frac{1}{16} \left(= \frac{1}{114} \right)$ (c) $\frac{5}{20} \times \frac{4}{19} \times \frac{3}{18} \times \frac{2}{17}$

page 361 **Exercise 9**

1. (a) $\frac{10 \times 9}{1000 \times 999}$ (b) $\frac{990 \times 989}{1000 \times 999}$ (c) $\frac{2 \times 10 \times 990}{1000 \times 999}$ **2.** (a) $\frac{3}{20}$ (b) $\frac{7}{20}$ (c) $\frac{1}{2}$
3. (a) 5   (b) $\frac{1}{64}$ **4.** (a) $\frac{1}{220}$ (b) $\frac{1}{22}$ (c) $\frac{3}{11}$ (d) 5
5. (a) $\frac{3}{5}$ (b) $\frac{1}{3}$ (c) $\frac{2}{15}$ (d) $\frac{2}{21}$ (e) $\frac{1}{7}$ **6.** (a) 0·007 81   (b) 0·511
7. (a) $\frac{21}{506}$ (b) $\frac{455}{2024}$ (c) $\frac{945}{2024}$
8. (a) $\dfrac{x}{x+y}$ (b) $\dfrac{x(x-1)}{(x+y)(x+y-1)}$ (c) $\dfrac{2xy}{(x+y)(x+y-1)}$ (d) $\dfrac{y(y-1)}{(x+y)(x+y-1)}$

9. (a) $\frac{x}{z}$ (b) $\frac{x(x-1)}{z(z-1)}$ (c) $\frac{2x(z-x)}{z(z-1)}$ 10. (a) $\frac{1}{125}$ (b) $\frac{1}{125}$ (c) $\frac{1}{10\,000}$ (d) $\frac{3}{500}$

11. (a) $\frac{1}{49}$ (b) $\frac{1}{7}$ 12. (a) $\frac{1}{52}$ (b) 1

page 364 **Exercise 10**

1. $\frac{3}{10}$ 2. $\frac{9}{140}$ 3. (a) $\frac{1}{5}$ (b) $\frac{18}{25}$ (c) $\frac{1}{20}$ (d) $\frac{2}{25}$ (e) $\frac{77}{100}$

4. (a) 0·07   (b) 0·64   (c) 10·29   5. $\frac{4}{35}$ 6. (a) $\frac{19}{25}$ (b) 38 out of 50   7. 27

10 Using and applying mathematics

page 369 **Exercise 2**

1. (a) If today is Monday, then tomorrow is Tuesday.
   (b) If it is raining, then there are clouds in the sky.
   (c) If Abraham Lincoln was born in 1809, then Abraham Lincoln is dead.
2. (a) T   (b) T   (c) T   (d) F   (e) T   (f) F   (g) T   (h) T   (i) F   (j) T

page 375 **Investigations**

Note. It must be emphasised that the *process* of obtaining reliable results is far more important than these few results. It is not suggested that 'obtaining a formula' is the only aim of these coursework tasks. The results are given here merely as a check for teachers or students working on their own.

It is not possible to summarise the enormous number of variations which students might think of for themselves. Obviously some original thoughts will be productive while many others will soon 'dry up'.

1. With the numbers written in c columns, the difference for a $(n \times n)$ square is $(n-1)^2 \times c$.
2. From 0 to 774 miles, choose 'Hav-a-car'.
   From 775 to 1538 miles, choose 'Snowdon'.
   Over 1538 miles, choose 'Gibson'.
3. For a final score $a - b$, number of possible half-time scores $= (a+1)(b+1)$.
   e.g. For 7–5, number of half-time scores $= (7+1)(5+1) = 48$.
4. For diagram number n, number of squares $= 2n^2 - 2n + 1$.
5. For a square card, corner cut out $= \frac{1}{6}$ (size of card).

   For a rectangle $a \times 2a$, corner cut out $\cong \dfrac{a}{4\cdot732}$.

7. For n names, maximum possible number of interchanges $= \dfrac{n(n-1)}{2}$.

8. For smallest surface area, height of cylinder $= 2 \times$ radius.
9. For $n \times n \times n$: 3 green faces $= 8$
   2 green faces $= 12(n-2)$
   1 green face $= 6(n-2)^2$
   0 green face $= (n-2)^3$

10. (a) For n blacks and n whites, number of moves $= \dfrac{n(n+1)}{2}$.

    (b) For n blacks and n whites, number of moves $= \dfrac{n(n+2)}{2}$.

    (c) For n of each colour, number of moves $= \frac{3}{2}\,n(n+1)$.

11. Consider three cases of rectangles $m \times n$
   (a) m and n have no common factor. Number of squares $= m + n - 1$.
      e.g. 3×7, number $= 3 + 7 - 1 = 9$
   (b) n is a multiple of m. Number of squares $= n$ e.g. 3×12, number $= 12$
   (c) m and n share a common factor a
      so $m \times n = a(m' + n')$
      number of squares $= a(m' + n' - 1)$
      e.g. $640 \times 250 = 10(64 \times 25)$, number of squares $= 10(64 + 25 - 1) = 880$.

12. 4×4: There are 30 squares (i.e. $16 + 9 + 4 + 1$)
   8×8: There are 204 squares ($64 + 49 + 36 + 25 + 16 + 9 + 4 + 1$)
   $n \times n$: Number of squares $= 1^2 + 2^2 + 3^2 + \ldots + n^2$

$$= \frac{n}{6}(n+1)(2n+1)$$

   This could be found using method of differences or from the standard result for $\sum_1^n r^2$.

13. (a) square root twice, 'multiply by x' gives cube root of x.
   (b) square root three times, 'multiply by x' gives seventh root of x.
   (c) square root twice, 'divide by x' gives $1/(\sqrt[3]{x})$
   (d) square root twice, 'multiply by x^2' gives $(\sqrt[3]{x})^2$

The numbers in the first difference column are the squares of the terms in the original Fibonacci sequence.

14. Pick's theorem: $A = i + \frac{1}{2}p - 1$

11 Revision

page 385 **Revision exercise 1**

1. A $y = 6$, B $y = \frac{1}{2}x - 3$, C $y = 10 - x$, D $y = 3x$ **2.** (a) 5   (b) $\frac{4}{7}$ **3.** 6

4. (a) $220°$ (b) $295°$ **5.** (a) 14   (b) 18   (c) 28

6. 253   **7.** £5250   **8.** 0·375   **9.** (a) $\frac{9}{20}$ (b) $\frac{11}{24}$

10. $\dfrac{a}{b}$ **11.** (a) $1:50\,000$ (b) $1:4\,000\,000$ **12.** (a) $s = t(r + 3)$ (b) $r = \dfrac{s - 3t}{t}$

13. 10   **14.** $y \geqslant 2$, $x + y \leqslant 6$, $y \leqslant 3x$

page 385 **Revision exercise 2**

1. 5%   **2.** 21%   **3.** 6   **4.** 25

5. (a) 8   (b) 140   (c) 29   (d) 42   (e) 6   (f) -6 **6.** $\frac{1}{6}$

7. (a) $9\pi\,\text{cm}^2$ (b) $9:1$ **8.** $33\frac{1}{3}$ mph   **9.** 9

10. (a) 5·45   (b) 5   (c) 5   **11.** 6 cm   **12.** (a) $6x + 15 < 200$ (b) 29

page 386 **Revision exercise 3**

1. (a) 20·8%   (b) £240   **2.** A $4y = 3x - 16$, B $2y = x - 8$, C $2y + x = 8$, D $4y + 3x = 16$

3. (a) 0·005 m/s   (b) 1·6 s   (c) 173 km   **4.** $c = 5$, $d = -2$

5. (a) reflection in $x = \frac{1}{2}$ (b) reflection in $y = -x$ (c) rotation $180°$, centre $(1, 1)$

6. (a) A   (b) V   (c) Impossible   (d) L   **7.** $152°$ **8.** $45°$, $225°$

9. (a) $\frac{3}{5}$ (b) $w = \dfrac{k(1 - y)}{y}$ **10.** (a) 50   (b) 50

page 387   *Revision exercise 4*

1. $x < \frac{3}{4}$ **2.** $1 \cdot 5 \times 10^{11}$ **3.** (a) $(x+5)(x+3)$ (b) $(x+3)(x-2)$ (c) $5x(x-6)$

4. $20 \, \text{cm}^2$ **5.** (a) $\frac{1}{28}$ (b) $\frac{15}{28}$ (c) $\frac{3}{7}$ **6.** (a) 30   (b) 75

7. f.d: 5, 18, 26, 10, 8   **8.** $x \geqslant 0$, $y \geqslant x - 2$, $x + y \leqslant 7$

9. (a) 600   (b) 9000/10 000   (c) 3   (d) 60   **10.** (a) $40°$ (b) $100°$

page 388   *Revision exercise 5*

1. (a) $-3, 5$ (b) 2, 6   **2.** -4 **3.** (a) $30 \, \text{m/s}$ (b) $600 \, \text{m}$

4. $250 \, \text{cm}^3$ **5.** (a) $3\frac{1}{3} \, \text{cm}$ (b) $1620 \, \text{cm}^3$

6. (a) $50°$ (b) $128°$ (c) $c = 50°$, $d = 40°$ (d) $x = 10°$, $y = 40°$

7. $\frac{1}{12}$ **8.** $\frac{35}{48}$ **9.** (a) $\frac{1}{9}$ (b) $\frac{1}{12}$ (c) 0   **10.** $\dfrac{31b}{20}$

page 389   *Revision exercise 6*

1. (a) $0 \cdot 3 \, \text{m/s}^2$ (b) $1050 \, \text{m}$ (c) $40 \, \text{s}$ **2.** (a) $1 \, \text{m}^2$ (b) $1000 \, \text{cm}^3$

3. (a) $\frac{1}{2}, -\frac{1}{2}$ (b) $\frac{7}{11}$ **4.** $\frac{5}{16}$ **6.** (a) $1\frac{5}{6}$ (b) $0 \cdot 09$

7. (a) $z = x - 5y$ (b) $m = \dfrac{11}{k+3}$ (c) $z = \dfrac{T^2}{C^2}$

8. (a) A$'(0, -3)$, B$'(6, 0)$, C$'(12, -3)$ (b) A$'(3, 0)$, B$'(9, 3)$, C$'(15, 0)$ (c) A$'(0, 0)$, B$'(2, 3)$, C$'(4, 0)$

9. 13   **10.** (a) $\mathbf{a} - \mathbf{c}$ (b) $\frac{1}{2}\mathbf{a} + \mathbf{c}$ (c) $\frac{1}{2}\mathbf{a} - \frac{1}{2}\mathbf{c}$ CA is parallel to NM and CA $= 2$NM.

11. (a) 3   (b) 0, 5   (c) 0   **12.** (a) $0 \cdot 65$, $3 \cdot 85$

page 390   *Revision exercise 7*

1. 28   **2.** (a) $12 \cdot 2 \, \text{cm}$ (b) $61 \cdot 1 \, \text{cm}^2$ **3.** 23×10^9 **4.** $33 \cdot 1\%$

5. $9 \cdot 95 \, \text{cm}$ **6.** $123 \cdot 25 \, \text{cm}^2$, $101 \cdot 25 \, \text{cm}^2$ **7.** $0 \cdot 72\%$

8. (a) $0 \cdot 5601$ (b) $3 \cdot 215$ (c) $0 \cdot 6161$ (d) $0 \cdot 4743$

9. (a) 84   (b) $19 \cdot 2$ **10.** $0 \cdot 335 \, \text{m}$ **11.** $1 \cdot 24$ **12.** (a) $1 \cdot 75$ (b) $60 \cdot 3°$

page 391   *Revision exercise 8*

1. 95p for 1 lb   **2.** (a) $290°$ (b) $60°$, $300°$

3. (a) $0 \cdot 340$ (b) $4 \cdot 08 \times 10^{-6}$ (c) $64 \cdot 9$ (d) $0 \cdot 119$

4. (a) $45 \cdot 6°$ (b) $58 \cdot 0°$ (c) $3 \cdot 89 \, \text{cm}$ (d) $33 \cdot 8 \, \text{m}$ **5.** $3 \cdot 63 \, \text{cm}$

6. $3 \cdot 43 \, \text{cm}^2$, $4 \cdot 57 \, \text{cm}^2$ **7.** $273 \, \text{cm}^3$ **8.** (a) $14 \cdot 1 \, \text{cm}$ (b) $35 \cdot 3°$

9. (a) $\dfrac{x}{x+5}$ (b) $\left(\dfrac{x}{x+5}\right)^2$ **10.** 68p   **11.** $1 \cdot 552 \, \text{m}$ **12.** $4 \cdot 12 \, \text{cm}$

13. (a) 0   (b) 16   (c) -8

page 392   *Revision exercise 9*

1. $2 \cdot 1 \times 10^{24}$ **2.** $17 \, \text{kg}$ **3.** (a) $x = 14 \cdot 1$, card $= 48 \cdot 3 \, \text{cm}$ (b) $1930 \, \text{cm}^2$

4. (a) $2 \cdot 92$, $-1 \cdot 25$ (b) $5 \cdot 24$, $0 \cdot 76$ **6.** $5 \cdot 39 \, \text{cm}$ **7.** $\bar{x} = 13 \cdot 6$, s.d. $= 8 \cdot 2$

8. $45 \cdot 2 \, \text{km}$, $33 \cdot 6 \, \text{km}$ **9.** (a) $6 \cdot 63 \, \text{cm}$ (b) $41 \cdot 8°$ **10.** $5 \cdot 14 \, \text{cm}^2$

11. $41 \cdot 2 \, \text{cm}$, $9 \cdot 93 \, \text{cm}$ **12.** $8 \cdot 06 \, \text{cm}$ **13.** $89°$

page 395   **Test 1**

1. C	**2.** D	**3.** D	**4.** B	**5.** A
6. C	**7.** C	**8.** D	**9.** C	**10.** B
11. C	**12.** A	**13.** D	**14.** C	**15.** C
16. D	**17.** A	**18.** C	**19.** B	**20.** D
21. A	**22.** C	**23.** B	**24.** B	**25.** C

page 396   **Test 2**

1. B	**2.** C	**3.** A	**4.** B	**5.** D
6. C	**7.** A	**8.** D	**9.** B	**10.** C
11. B	**12.** D	**13.** A	**14.** C	**15.** C
16. D	**17.** B	**18.** A	**19.** B	**20.** B
21. C	**22.** D	**23.** A	**24.** A	**25.** B

page 398   **Test 3**

1. D	**2.** D	**3.** D	**4.** B	**5.** A
6. C	**7.** A	**8.** D	**9.** D	**10.** B
11. D	**12.** D	**13.** D	**14.** B	**15.** A
16. B	**17.** C	**18.** A	**19.** D	**20.** D
21. C	**22.** A	**23.** B	**24.** B	**25.** D

page 399   **Test 4**

1. B	**2.** B	**3.** A	**4.** C	**5.** D
6. A	**7.** B	**8.** C	**9.** D	**10.** B
11. C	**12.** A	**13.** B	**14.** B	**15.** D
16. A	**17.** C	**18.** B	**19.** B	**20.** D
21. A	**22.** C	**23.** C		

12 Examination questions

page 401   **Examination exercise 1**

1. (a) $2 \cdot 750$ (b) $57 \cdot 14$ (c) $7 \cdot 698$ (d) $107 \cdot 6$ **2.** $1 : 4000$

3. (a) $\frac{1}{9}$ (b) 24 **4.** (a) $\frac{11}{80}$ (b) $0 \cdot 1375$

5. (a) $3 \times 7 \times 13$ (b) $2 \times 7 \times 11 \times 13, 91$ **6.** 11×10^{34} **7.** $85 \, \text{s}$

8. (a) $4\frac{11}{16}$ inches   (b) $5 \, \text{s}$ **9.** (a) $2^3 \times 5^2 \times 7^2$ (b) $19\,600$

10. (a) Ruth £100, Ben £80   (b) 60% **11.** $10^4 = 10\,000, \; 11^4 = 121 \times 121 = 14\,641$

12. (b) various e.g. $\sqrt{8}, \sqrt{2}$ (c) $2^{\frac{3}{2}}$ (d) $-\frac{1}{4}$

page 402   **Examination exercise 2**

1. (a) $7 : 13$ (b) (i) $7 \cdot 3 \times 10^7$ (ii) $53\,533\,000$ **2.** £614·68   **3.** 15p

4. (a) $1 \cdot 350\,85 \times 10^8$ (b) $22 \cdot 3\%$ **5.** (a) 10% (b) £900   **6.** (a) $4 \cdot 1$

7. £2593·74   **8.** (a) Saved by 47 votes   (b) 843   **9.** (a) $480\,000$ (b) $4 \cdot 8 \times 10^5$

10. (a) $6 \cdot 5, 6 \cdot 6$ etc   (b) $\sqrt{37}, \sqrt{38}$ etc   (c) $\frac{1}{3}, \frac{1}{6}$ etc   **11.** (a) $45 \cdot 5 \times 10^{-3}$ (b) $34\,100 \, \text{cm}^3$

12. (a) (i) $a \approx 50, b \approx 30$ (ii) 75 (iii) $79 \cdot 368\,75$ (b) $4 \cdot 9 \times 10^{11}$

13. (a) $36 \cdot 6 \, \text{km/h}$ (b) $9 \cdot 2\%$ **14.** (a) (i) $146 \cdot 8 \, \text{m}$ (ii) $145 \cdot 5 \, \text{m}$ (b) $150 \, \text{m}$

15. (a) $16\frac{2}{3} \, \text{m/s}$ (b) $19 \, \text{m/s}$

page 405 **Examination exercise 3**

1. (a) $-1, 1\frac{4}{5}$ (b) $3z(2z - 3)$ (c) $6x^2 + x - 35$ **2.** -5 **3.** $a = 8, b = 3$
4. (a) $300 + 4n, 9n$ (b) $9n > 300 + 4n$ (c) $n > 60$
5. (a) $0, 0$ (b) sum of cubes = square of the sum of integers
6. (a) $2n + 6$ (b) 37 **7.** 34
8. (b) $h = \frac{1}{2}n(n + 1)$ (c) $a = 1\frac{1}{2}, b = 4\frac{1}{2}$
9. 3 **10.** (b) (i) 21 (ii) 400 (c) 64 (d) (i) $n + 1$ (ii) n^2 (e) (iii) $n = 15$

11. (a) $x = 2, y = -1$ (b) $v = \dfrac{uf}{u - f}$ **12.** (a) 9.6 (b) $\dfrac{Vr}{12 - V}$ **13.** $R = \dfrac{12}{I} - 2$

14. (a) (i) $8x^{15}$ (ii) $\dfrac{x^3}{y^2}$ (b) $4xy(y - 2x)$ (c) (i) $\dfrac{x - y}{x + 3}$ (ii) 1 (d) $\frac{1}{3}$

15. $56.25\,\text{m}$ **16.** (a) (i) 3 (ii) -1 (iii) 50 (b) $11x^8$
17. (a) $x(6 - x)$ (d) $1.4\,\text{cm}$ **18.** (a) $0.3\,\text{ms}^{-2}$ (b) $1200\,\text{m}$ (c) $10\,\text{m/s}$
19. (b) $350\,\text{m}$ **20.** (c) (i) $t = 1.2, 3.3$ (ii) $t = 2.25\,\text{s}$ (d) $27.5\,\text{m}$
21. (a) $\left(x + \frac{5}{2}\right)^2 - \frac{53}{4}$ (b) $x = 1.14, -6.14$ **22.** $\dfrac{2}{x - 1}$
23. (a) $(2, 18)$ (b) $(-1, 12)$ (c) $(-2, 12)$ (d) $\left(\frac{1}{2}, 12\right)$ **24.** (c) $A'(1, 4), B'(2, 0)$

page 411 **Examination exercise 4**

1. (a) $8, 9$ (b) 8.46 **2.** (b) 36 (c) 28 (d) 60 (e) $4, 392, 9604, 200$
3. 3.6 **4.** (a) $189\,225$ tonnes   (b) $29.2\,\text{s}$
5. (a) $8a + 10b$ (b) (ii) $3a^2 + 11ab + 6b^2$ (c) $b = 2a$ **7.** $p = 8, q = \frac{9}{4}, k = 27$
8. (a) $5x + 4y$ (b) $x = \frac{7}{5}, y = -\frac{3}{2}$ (c) (i) $nx + (n - 1)y$ (ii) 15
9. (c) $n = 48; 47, 48, 49$ **10.** (b) $9.28, 3.72$ **11.** (a) $h = \dfrac{A}{2\pi r} - r$ (b) 3.8
12. (a) 1 (b) 2 (c) 0.889 (3 s.f.)   **13.** (a) $10\,\text{g}$ (b) $56.6\,\text{g}$ (c) $80\,\text{g}$
14. (a) $21 - x$ (b) $12\,\text{cm}, 9\,\text{cm}$ **15.** (c) 0.3 or 1.8 approx.

page 416 **Examination exercise 5**

1. $b = 2.9\,\text{cm}, a = 5.8\,\text{cm}, p = 11.6\,\text{cm}$ **2.** 900
3. (a) $6\,\text{m}$ (b) $16\,\text{m}^2$ **4.** (i) No   (ii) No   (iii) Yes
5. (a) $\dfrac{\pi}{6}d^3 + \dfrac{\pi}{4}d^2h$ (b) $\pi d^2 + \pi dh$ **6.** (a) $42°$ (b) $53°$ (c) $21°$ (d) $106°$
7. (a) $34°$ (b) $34°$ (c) $17°$ (d) $73°$ (e) $56°$
8. (a) (i) 3 (ii) $44\,\text{cm}$ (iii) $72\,\text{cm}^2$ (b) cuboid   (c) $44\,\text{cm}$

9. (a) $(3, 6)$ (b) $\begin{pmatrix} -3 \\ -4 \end{pmatrix}$ (c) 5 units

10. (a) (i) $\frac{1}{2}\mathbf{a}$ (ii) $-\mathbf{a} + \mathbf{b}$ (iii) $-3\mathbf{a} + 3\mathbf{b}$ (iv) $-\frac{5}{2}\mathbf{a} + 3\mathbf{b}$ (v) $-2\mathbf{a} + 3\mathbf{b}$
11. (a) (i) 0.5 (ii) -0.5 (iii) -0.5 (b) (ii) $105°, 165°$
12. (a) (i) $103°$ (ii) $22°$ (iii) $22°$ (iv) $136°$ (b) ASA
13. (a) $\dfrac{\pi r^2}{8}$ (ii) $\dfrac{\pi r^2}{8}$ (b) $90°$

page 420 **Examination exercise 6**

1. (a) 154 m^2 (b) 42 m^2 (c) 3360 g
3. 65 cm
6. (a) 72 cm (b) 129·8 cm
8. (b) 63·4° (c) (i) 243·4° (ii) 296·6°
9. (a) 31·6° (b) 67·4° (c) 12 cm (d) 33·7°
10. (a) 139 m (b) 19·1° (c) (i) 333 m (ii) 22·7°
11. (a) 3·14 m^2 (b) 0·157 m^3 (c) 13·6 m (d) 177 m^2
12. (a) 8000 cm^3 (b) 268 mm^3 (c) 29 800
13. (a) 49°, 7·34 m^2 (b) 21 200 m^3
15. (a) 180 cm^3 (b) 55 (c) 92 cm^3
18. (a) 4850 m (b) 1070 m

2. (a) 198 cm^3 (b) 1360 mm^3 (c) 145
4. 36 km/h
5. (a) 57° (b) 39 cm
7. (a) 108 m^2 (b) 14°

14. 175, $y = 6·8$, $z = 823\,200$
16. 60°, 300°, 420° (and others) 17. 16·97
19. (a) (i) $2a + x$ (ii) $\pi a - x$ (c) 32·7°

page 426 **Examination exercise 7**

1. (a) £4500 (b) £9000 (c) £8646 2. 6
3. Second year: wrong, no correlation; fourth year: wrong, do better at back
5. (a) (i) 7 (ii) 4 (iii) various answers e.g. 5, 5, 5, 5, 6, 7, 11
6. $6 \leqslant x < 10$
7. (a) (i) 2·5, 3, 2, 0·25
   (b) (e.g.) more elderly employees in A, smaller spread of ages in B.
10. (b) Fairly strong positive correlation (c) 51 years
11. (a) 3 (b) (i) 2·56 (ii) 7·68 [i.e $3 \times 2·56$] 12. (a) (i) A (ii) $\frac{4}{15}$ (b) B
13. (a) (i) $\frac{2}{9}$ (ii) $\frac{1}{3}$ (iii) 0 (iv) $\frac{2}{3}$ (b) 120 14. (a) $\frac{1}{36}$ (b) $\frac{1}{9}$ (c) $\frac{1}{6}$
15. (a) (i) $\frac{1}{3}$ (ii) $\frac{2}{3}$ (iii) $\frac{1}{3}$ (b) Yes 16. (a) (i) $\frac{7}{25}$ (ii) $\frac{3}{100}$ (iii) $\frac{97}{100}$ (b) $\frac{3}{4}$
17. (a) $\frac{1}{6}$ (b) $\frac{1}{12}$ (c) $\frac{1}{9}$ 18. (a) $\frac{12}{49}$ (b) $\frac{11}{188}$
19. (b) (i) $\frac{11}{35}$ (ii) $\frac{18}{35}$ 20. (a) (i) $\frac{12}{25}$ (ii) $\frac{13}{25}$ (b) $\frac{1}{2}$

page 432 **Examination exercise 8**

1. Angles: 169°, 83°, 108°
2. (a) mean is about $36 [36·25] (b) $p = 12$, $q = 24$, $r = 35$
   (d) (i) about $36 (ii) 45, 27, IQR = 18
3. (a) 22, 40, 34, 18 (b) 18° (e) 58 marks
4. (a) 1·6–1·8 (b) 1·2–1·4 (c) 1·44 kg
5. (a) 86·3 cm (d) (i) 86·5 cm (ii) 12 cm (e) part (ii)
6. (a) 21·6 (b) 0·4
7. (i) mean = 545·7, s.d. = 66·1 (ii) snack foods show greater variation as s.d. is higher
8. (a) 0·579 cm (b) second sample less widely spread
9. (b) (i) 0·16 (ii) 0·992 (iii) 0·001 28 (c) $(0·2)^n$

INDEX